# Performance of Communication Systems

Springer
*Berlin*
*Heidelberg*
*New York*
*Barcelona*
*Hong Kong*
*London*
*Milano*
*Paris*
*Singapore*
*Tokyo*

Alexander Ost

# Performance of Communication Systems

## A Model-Based Approach with Matrix-Geometric Methods

Springer

Alexander Ost
Ericsson Eurolab Germany
Ericsson Allee 1
52134 Herzogenrath
Germany

*e-mail: ost@ieee.org*

D 82 (Diss. RWTH Aachen)

ISBN 3-540-41438-x Springer Verlag Berlin Heidelberg New York

Cataloging-in-Publication Data applied for

Ost, Alexander:
Performance of communication systems : a model based evaluation with matrix geometric methods / Alexander Ost. – Berlin ; Heidelberg ; New York ; Barcelona ; Hongkong ; London ; Mailand ; Paris ; Singapur ; Tokio : Springer, 2001

Zugl.: Aachen, Techn. Hochsch., Diss. 2000
ISBN 3-540-41438-X



Springer-Verlag Berlin Heidelberg New York
a member of BertelsmannSpringer Science+Business Media GmbH
http.//www.springer.de



Printed in Germany



Coverdesign: medio Technologies AG, Berlin
Typesetting: Camera-ready copy by author
SPIN: 10792251 62/3020 Printed on acid-free paper – 5 4 3 2 1 0

# Abstract

This work aims at providing approaches to guide the design of modern communication networks. During the last decades, the size and the complexity of *communication networks* increased substantially. Communication facilities have become a major infrastructural factor, and the *performance* offered by a network is of central interest. However, the evaluation of the performance of today's communication networks is very difficult due to their size and complexity. Although it is clearly possible to measure the performance of one particular configuration, this provides no guidance in choosing among alternative configurations prior to their actual deployment.

In this work, we therefore advocate a *model-based* approach, where mathematical models of actual networks are analyzed to predict the performance of a large number of different design alternatives. Clearly, these models must be able to account for the key properties of the networks under investigation. Our assessment of the modeling requirements is based on both theoretical considerations and on practical experiences made in an industrial collaboration. The computational effort for investigating models that account for all relevant network properties becomes quickly prohibitive, being the reason why most approaches in this area resort to a very high level of abstraction.

This work aims to tackle this situation by employing a special class of stochastic processes, so called quasi-birth-and-death (QBD) processes, for modeling network nodes. While QBDs are related to conventional queueing-theoretic modeling approaches, they provide the flexibility to account for a large amount of additional detail. Still, a number of very efficient, so-called *matrix-geometric* solution methods for the analysis of these models exist, and a thorough comparison of these approaches is performed in this work.

While QBDs combine a large degree of modeling expressiveness with efficient solution algorithms, their direct and manual specification is cumbersome and error-prone. In this work, we therefore adopt a high-level modeling technique which is based on the well-established stochastic Petri net formalism. Our Petri net class, called infinite-state SPNs (iSPNs), is specifically tailored towards generating QBDs; the equivalence of an iSPN-based specification to the direct specification of QBDs is formally proven, and a suitable tool environment is presented.

The applicability of the iSPN-based performance modeling framework is then illustrated by performing several case studies in the areas of World-Wide-Web and TCP/IP interaction, ATM connection management, and modeling systems subject to self-similar traffic.

The last part of this work is dedicated to the development of a framework for the analysis of a large number of interacting network nodes (each modeled by an iSPN), thereby allowing to investigate models that are directly related to the structure of modern communication networks. This is accomplished by embedding the iSPN approach in a parametric decomposition framework.

# Acknowledgements

This book is the result of my PhD studies at the Laboratory for Performance Evaluation and Distributed Systems at the Aachen University of Technology (RWTH Aachen), Germany.

In the first place I would like to thank my advisor Prof. Boudewijn Haverkort for supervising this work. Without his excellent and broad background in the performance evaluation area, this work would not have been possible. His guiding and careful advice was the important foundation of my work; with his open-minded and sound attitude, he created a research environment that can hardly be surpassed. I also thank Prof. Bernhard Walke (RWTH Aachen, Communication Networks) for taking the co-advisorship and providing valuable comments on this thesis.

Many thanks go to my colleague Henrik Bohnenkamp for being a pleasant roommate over several years and for reviewing large parts of this work. I am indebted to him and my colleagues Ramin Sadre and Rachid El Abdouni for creating a cooperative and competent atmosphere that made the Department for Performance Evaluation and Distributed Systems an enjoyable place to be. Furthermore, I would like to thank Aad van Moorsel for hosting me during my stay at Bell Labs and for providing valuable and detailed feedback on the manuscript of this book. Thanks also go to Prof. Isi Mitrani (University of Newcastle upon Tyne, UK) for reviewing parts of my work.

I gratefully acknowledge the German Research Council (DFG) for funding large parts of my research by granting a scholarship in the graduate college "Computer Science and Technology" at the RWTH Aachen; thanks also go to the college's chairman Prof. Otto Spaniol (RWTH Aachen, Communication and Distributed Systems).

Finally, I would like to thank my parents, who always encouraged me on my way. I dedicate this work to them.

# Table of Contents

## Part II. Node Analysis

# Part I

## Introduction and Motivation

# 1. Introduction

This book is about efficient communication. The role of communication, as a means for the transmission of information among people, is one of the key factors driving the development of our civilization.

There are numerous examples where new achievements in the communications area had a crucial impact on our society. The human language can be considered as one of the first steps for the efficient delivery of information between people. Similarly, the influence of other milestones like written communication, bookprinting, telegraphy and the telephone and television system was essential. Facilities for quickly accessing existing information and for the fast delivery of information have become one of the most important economical factors nowadays.

Clearly, fulfilling the ever-increasing demand for fast, cheap and reliable communications represents itself an important industry, leading to substantial technical achievements in this area. After only being used by an "information elite" in the beginning, developments like the telephone system and the Internet quickly spread and are now widely accepted, as witnessed by the ubiquitous presence of mobile phones and Internet applications like the World Wide Web.

During the last decades, one of the most important developments in the communications area can be summarized under the notion *integration and digitization*: all kinds of information are transformed to a simple digital representation which can easily be processed and conveyed by computers. In the early days of data communication, modems were employed for transforming digital data to analog tones in order to deliver data using the analog telephone system. Nowadays, the situation has completely reversed: the entire telephone system internally employs a digital representation of voice, which requires that voice is converted to a machine-readable format prior to transmission. There are basically three reasons for this development: first, the transformation of all kinds of information to a uniform digital representation enables the provision of complex services operating on the transferred information. With digitization, the technical quality of these services could be increased and the available network resources could be used more efficiently. Secondly, it is no longer necessary to establish different communication infrastructures for different information types. By using one and the same network, resources can be used more efficiently and the administration becomes much easier. Thirdly, the adoption of one uniform network architecture substantially facilitates the development of global standards that enable the provision of interconnected communication networks that go beyond national borders.

The success of ISDN (the Integrated Services Digital Network) underlines this development. For example, in 1999, the number of ISDN subscriber lines in Germany increased by 23%, while the number of non-ISDN lines decreased [58]. By the end of 1999, already 26% of all lines were ISDN channels. In view of the higher pricing of ISDN access compared to conventional telephony, this also shows that consumers are willing to pay more for better services. While ISDN is particularly successful in Germany, the proliferation of mobile telephony and especially the explosive growth of the Internet reflects this trend on a world-wide scale.

From an engineering point of view, the integration of networks by digitization has several implications. First, the network *size* increases, as there is no longer a variety of networks for different services. Secondly, networks become more "intelligent", as more bearer services are provided and complex services are offered to the end-users, leading to complex switching architectures. Thirdly, the usage pattern of the network changes, as it no longer comprises only a relatively simple service like (for example) telephony, but also bandwidth-intensive video applications or highly interactive Internet traffic. Thus, the key issues when designing and deploying telecommunication networks, namely

- Where to place switching nodes?
- How to dimension nodes and lines?

become considerably more difficult under the consideration of

- new services, not only in nature, but also in view of their (often distributed) implementation and administration,
- the increasing size of networks, and
- complex network usage patterns.

With the increasing economical importance of the communications sector, the cost-efficient deployment of communication networks has become an essential issue. Therefore, appropriate means are needed for comparing the performance of different deployment alternatives. Clearly, one approach is to simply *measure* the performance of a network configuration. Measurement-based approaches, however, represent a very expensive approach for the assessment of the network performance, as they allow the evaluation of a configuration only *after* its actual installation. Also, this approach requires the actual presence of certain traffic scenarios for determining the performance under these scenarios.

As an alternative to measurement-based approaches, the *model-based* evaluation of communication networks is very attractive. Based on a mathematical model of a communication network, performance predictions can be performed in a very cost-efficient manner for different implementation alternatives and different usage scenarios. For model-based performance evaluation, basically three approaches exist:

*Simulation approaches* are based on the development of appropriate simulation programs for mimicking the behaviour of the network components and the behaviour of the users of the network. While the development of simulation models is a straightforward approach that is widely used, it also involves several drawbacks:

- Simulation models tend to account for very much detail of the system to be evaluated. This leads to very complex models where much effort is spent in accounting for properties which often have a small effect on the overall network performance. Also, detailed information on system properties is often not available in the early design stages, leading to situations where it is tempting to make a large number of assumptions on unknown parameters instead of resorting to a suitable level of abstraction (see also the remarks in [75] in this context).
- Especially for large models, the computation time needed for obtaining performance measures can be excessive. In particular, when information related to rare events (like reliability-oriented measures) is needed, very long simulation runs are required to obtain results with reasonably tight confidence intervals.
- By simply designing programs that simulate the system behaviour, no insight is gained in how the complex set of network properties impacts the overall performance. This also implies that errors in simulation programs are hard to find, which increases the potential to obtain misleading results by simulation.

*Analytical approaches* develop a mathematical network model in order to obtain explicit relationships between system properties and performance measures. These models provide deep insights in what system parameters are crucial for obtaining good overall performance, and analytical approaches do not suffer from the computational problems that simulation incurs. However, the creation of appropriate models requires a thorough mathematical background on behalf of the modeler, and even then, it is very difficult to account for other than very high-level system properties.

*Numerical approaches* are similar to analytical models with respect to explicitly modeling relationships among network parameters and system properties. However, numerical approaches take advantage of state-of-the-art numerical algorithms in order to allow for much more complex relationships than possible in purely analytical approaches. Clearly, this also simplifies the modeling task, as it is no longer necessary to operate on explicit expressions only. An important class of approaches in this area are based on Markovian processes, which allow for an intuitive state-based system description which is amenable to efficient numerical solution techniques. Clearly, the simplification of the modeling process and the increased modeling flexibility is bought by the computational effort that is needed by the numerical solution methods, which can grow excessively large especially for very complex systems.

In this work, a numerical approach will be followed in order to design a model-based performance evaluation framework for modern telecommunication networks. One key issue that will be dealt with is to avoid the excessive computational requirements that often arise in numerical approaches. The second key concern is to provide an environment that is tailored towards the considered application area, and which can be used by system modelers without special knowledge of the underlying mathematical techniques of our analysis framework.

### Plan of the Manuscript

This work comprises three main parts. The aim of *Part I* is to clearly define the requirements that are needed for a model-based performance evaluation approach for modern communication networks. To this end, we first introduce the application area in *Chapter 2*, thereby presenting two modern network architectures and then formally deriving a set of requirements for an environment for guiding the deployment of these networks. Based on these requirements, we then develop a first modeling and evaluation approach for a case study that has been accomplished in close cooperation with an industrial partner in *Chapter 3*. The purpose of this study is two-fold: first, we verify our requirements against the requirements formulated by practitioners from industry. Secondly, the initial modeling approach helps to show which aspects need to be tackled further—also taking into account the results of existing work. One result of this investigation is that approaches are needed for describing the nodes occurring in a communication network (like switches or World Wide Web servers) in sufficient detail.

*Part II* of this work is therefore dedicated to the development of a methodology for describing the complex communication network nodes. The main reason why existing approaches do not provide sufficiently detailed node models is that the node analysis quickly becomes computationally infeasible. We thus first focus on suggesting an appropriate mathematical foundation that is also amenable to efficient analysis methods. As a suitable approach, a special class of stochastic processes, so-called quasi-birth-and-death processes is investigated in *Chapter 4*. The approach provides a large degree of modeling flexibility and analysis efficiency, however, for large practical applications, the direct development of quasi-birth-and-death models is cumbersome, if feasible at all. In *Chapter 5*, we thus develop a comfortable high-level modeling approach based on stochastic Petri nets (SPNs), which represent an established approach in the modeling area. Our class of so-called infinite-state SPNs (iSPNs) is tailored towards taking advantage of quasi-birth-and-death processes as the underlying mathematical system model, but still hides all detail associated to this special class of stochastic processes. In *Chapter 6*, we then illustrate the applicability of the iSPN framework by investigating three case studies in the areas of World-Wide-Web and TCP/IP interaction, ATM connection management, and systems subject to self-similar traffic

In *Part III* of this work, we investigate how the Petri-net-based node modeling approach can be extended towards analyzing entire *networks* of nodes. This is performed in *Chapter 7*, where we review existing approaches in this area and resolve the key issues that need to be tackled for embedding the iSPN approach in a network-oriented context, thereby taking an approach based on parametric decomposition.

We finally conclude this work with a summary of the key results and prospective areas for further research in *Chapter 8*.

# 2. Modern Telecommunication Networks

Supported by economical deregulation efforts and the increasing importance of fast information exchange in our society, the telecommunications sector has become one of the most quickly expanding industries in the past two decades. Driven by a highly competitive market situation and increasing customer demands, this field experienced a large number of technical innovations, leading to the sophisticated and complex telecommunication infrastructure as it can be found today.

Up to the mid-80s, the main objective in designing communication systems was to offer high bandwidth capacities at reasonable prices. Since this goal has been widely realized with the development of optical transmission techniques (which recently experienced a further boost with the introduction of wavelength division multiplexing [36]) and appropriate transport network architectures like ATM [34, 85] and SDH [155], another factor is becoming increasingly important in the telecommunications market: the ability to quickly develop and efficiently manage complex telecommunication services, tailored to the subscribers' needs.

In telephone networks, the service side of this development became evident with the introduction of, for example, *freephone* services (800 numbers), *caller identification* or *call forwarding* facilities, and the development of PCS (Personal Communication Systems) [134] and UPT (Universal Personal Telecommunications) [131], offering highly advanced services for supporting both personal mobility and terminal mobility. There is obviously an even larger potential for new services in broadband communication networks (usually subsumed under the buzzword "multimedia", examples being video-on-demand services and video conferencing applications).

From an operational point of view, several technical achievements accompanied this development in order to simplify the introduction of new services and to enable the inter-operability among different networks and equipment providers. Modularization and abstraction methods as well as distribution techniques have been adopted from the computer science area, leading to powerful and complex network- and service-provisioning architectures. Telecommunication networks are more and more converging to large distributed computing systems, a development which is underlined by recent suggestions for integrating data networks (in particular, the Internet) and telephone networks [11, 80, 178].

Due to the enormous complexity and the economical relevance of modern telecommunication networks, the role of model-based performance evaluation approaches for such systems has become increasingly important for network planning and optimization. As the architectural properties of such networks impact the demands to a model-based performance evaluation framework, the aim of this chapter is to summarize the most important developments in this area and to discuss their relevance from a performance-modeling point of view. Based on this, suitable modeling approaches will be developed in the subsequent chapters.

In Section 2.1, we briefly review the concept of *common channel signaling*, which enabled the development of the complex control mechanisms which can be found in today's networks. In Section 2.2 and Section 2.3, two main approaches for structuring telecommunication networks and deploying services are reviewed: the *Intelligent Network* (IN) approach and the *telecommunications infrastructure networking architecture* (TINA). While INs represent a commonly used approach for providing telephony services, TINA is a more general architecture which is currently being developed to support a broader range of services (e.g., also in the area of broadband data communications). In Section 2.4, we then summarize the main properties of these network architectures and discuss their impact on model-based performance evaluation approaches. We conclude this chapter with a summary in Section 2.5.

## 2.1 Common Channel Signaling

Common Channel Signaling Networks (CCSNs) [124] were introduced in the mid-70s as dedicated networks for conveying signaling information between the nodes of the basic communication network. Signaling information initially related to call setup and supervision, but now also comprises inter-process communication and support for distributed applications. In [169], the role of signaling has vividly been described as being

> "analogous to the central nervous system of a living organism, something to coordinate the functions while remaining completely separate from the organism's other parts and not actually performing their function."

As the key role of CCSNs has become even more emphasized with the introduction of complex service provisioning environments like Intelligent Networks (see Section 2.2), the aim of this section is to provide a brief overview on CCSN concepts. In Section 2.1.1, we first explain the motivation for deploying CCSNs. As an example, we then describe the most widespread CCSN, the Signaling System Number 7, in Section 2.1.2.

### 2.1.1 CCSN Objectives

One of the main objectives to introduce CCSNs was the support of out-of-band signaling, thus separating signaling information (for example, for connection establishment) from the actual transmission channels. In this way, the following benefits could be realized:

- Additional information transfer capabilities are provided, which are not necessarily associated to the establishment of actual subscriber connections;
- compared to in-band signaling, the trunk utilization is reduced;
- the clear separation of internal signaling information from transmission channels provided protection against fraud of signaling facilities, a major problem in the 1970s which was known as "blue boxing".

In addition, CCSNs provide connectionless communication services (thus drastically reducing for example call-setup times) and a highly reliable delivery of signaling information.

The standardization efforts for CCSNs furthermore facilitated inter-networking of telecommunication networks on an international basis, and eventually led to the development of today's most common CCSN: *Signaling System Number 7* (SS7) [108, 144], which, for example, is used for ISDN trunk signaling. In the following section, we provide a brief overview on SS7.

### 2.1.2 Signaling System Number 7

The design of the SS7 protocol stack is tightly oriented towards the OSI reference model. The protocol layers can be separated into those which deliver the basic transport services (MTP 1–3) and those supporting a specific application area (e.g., ISDN User Part). The protocol stack is shown in Figure 2.1. The functionality of the indicated layers is as follows:

- *MTP level 1 (Message Transfer Part level 1)* directly corresponds to the OSI physical layer, defining data links and terminal equipment. The recommended bit rate is 64 kb/s.
- *MTP level 2* provides for the reliable transfer of signaling messages between two directly connected signaling points.
- *MTP level 3* offers services for the transfer of signaling messages between signaling points which are not directly connected. Furthermore, it provides reconfiguration functions to cope with the failure of links and signaling points.
- The *SCCP (Signaling Connection Control Part)* extends the functions of MTP level 3 to conform to the specifications of the OSI network layer. In particular, it adds additional addressing capabilities by introducing so-called *global titles* (suitable for world-wide inter-network routing) and provides connection-oriented services.

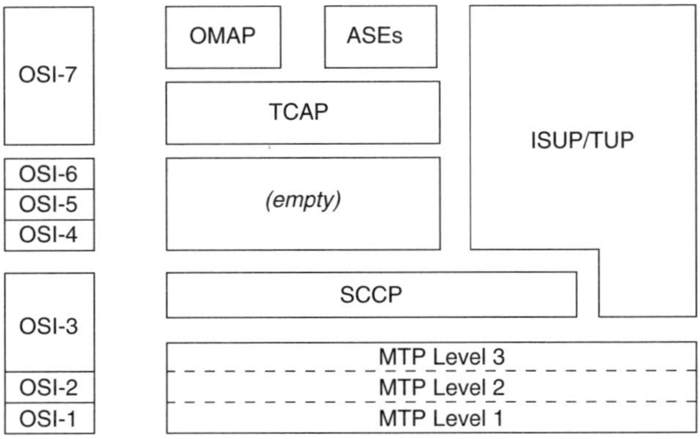


**Fig. 2.1.** The SS7 protocol architecture.

The functionality of the OSI layers 4–7 is contained in the application-oriented protocols. For example, TUP (*Telephone User Part*) provides signaling functions to support telephone networks, or ISUP supports signaling in ISDN networks. These user parts use both MTP services for inter-exchange signaling and SCCP services for end-to-end signaling.

The *Transaction Capabilities Application Part* (TCAP) supports non-circuit related information exchange on a dialogue-oriented basis. It is used for communication among distributed applications using so-called ASEs (*Application Specific Elements*) and by the *Operations, Maintenance and Administration Part* (OMAP). Example ASEs are the mobile application part (MAP), used in mobile telecommunication networks like GSM, and the Intelligent Network application protocol (INAP). Note that the TCAP layer directly accesses SCCP services; no individual layers corresponding to the OSI layers 4–6 have been specified for SS7 outside the user parts.

To summarize, SS7 provides a highly reliable and efficient architecture dedicated to the control and maintenance of telecommunication networks. With specialized ASEs, it can be tailored to specific application areas. These properties made it the *de facto* standard for controlling today's telecommunication networks, and provides the basis for complex service provisioning environments like Intelligent Networks, which will be discussed in the next section.

## 2.2 Intelligent Networks

The (IN) concept [78, 135], standardized by the International Telecommunication Union (ITU) [107], represents an approach to deal with the growing demand for complex services in telecommunication networks. Prior to IN architectures, service provision functionality was commonly implemented in those parts of a telecommunication network where it was needed—within the individual switching nodes. However, this approach proved to be more and more costly as more services were to be introduced, since the implementation of services clearly depends on the switch architecture (leading to poor code portability and a high dependence on network equipment vendors). Furthermore, changes to services had to be distributed to a large number of switches in the network. The proliferation of mobile telephony networks further aggravated the situation, as subscriber-related information could no longer be associated to fixed switches.

In Section 2.2.1, we illustrate how these problems are tackled in the IN context by describing the IN service provisioning concept. We then present the overall IN architecture in Section 2.2.2, and conclude with a brief discussion in Section 2.2.3.

### 2.2.1 Service Provision

Prior to INs, advanced services like freephone or call forwarding were realized by appropriate software in the network switching points. At first glance, this is the easiest way to implement such services, however, this approach encompasses some serious drawbacks:

- Since the service software is implemented directly in the switch, the specific hardware and operating system properties heavily influence the design process. If multi-vendor equipment is used in the network, or if new switching architectures are introduced, the software has to be ported to the corresponding architecture. This leads to a difficult and time-consuming development process, to non-reusable software, and to dependence on the manufacturers of the switches.
- Due to the provision of the service software in each switch, software maintenance becomes complicated, since changes have to be performed in all nodes of the network.
- Network-wide data (e.g., re-routing information for freephone numbers) is also kept in each node; thus, additional efforts are required to accomplish changes and to ensure data consistency.
- The replication of program and data resources in several places in the network requires the multiple installation of expensive equipment, leading to a poor overall usage of network resources.

In the IN framework, these problems are resolved by two key concepts: the detection point mechanism, and the building block approach. Both are briefly reviewed in the following paragraphs.

**Detection Points.** In the IN framework, service provision is decoupled from the actual switches. This is accomplished by the so-called *detection point* mechanism, which introduces a *basic call state model* (BCSM). The BCSM defines a set of unique states (and associated transitions) in the basic call

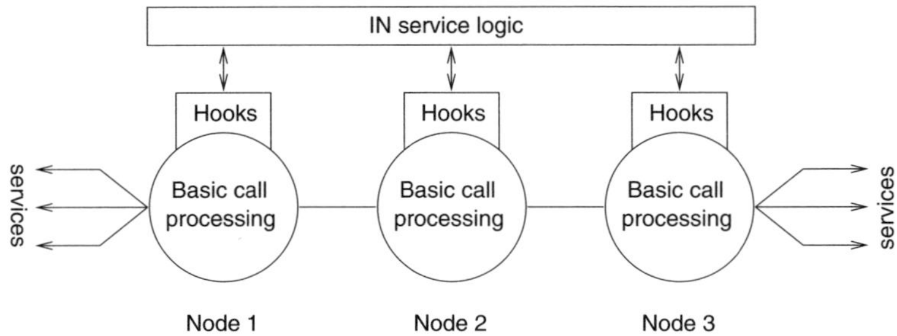


**Fig. 2.2.** The detection point mechanism.

process which characterize the current state of a call in the switching nodes (in the IN terminology, these nodes are referred to as *service switching points*, SSPs).

SSPs have the capability to detect the invocation of an IN service, for example, by checking for certain number prefixes like 0800 while a subscriber is dialing. If a request for an IN service is detected at a *point of initiation* in the BCSM, a "hook" is activated (see Figure 2.2) to invoke the IN service logic, which may physically be located on a different network node (denoted as *service control point*, SCP). In case of a remote SCP, the communication is performed via the signaling network. After IN-related service activities of the IN service logic have been performed (e.g., by mapping a freephone number to a real number), control is returned to the SSP by continuing execution at a (possibly new) *point of return* in the BCSM.

Note that in order to realize its services, the SCP may in turn access other IN nodes, for example, a *service data point* (SDP) which provides access to a database of freephone numbers. The communication among different IN nodes is accomplished by the *Intelligent Network application protocol* (INAP) which has already been mentioned in Section 2.1.2, thereby illustrating the crucial role of the signaling network in an IN.

**Building Blocks.** To overcome the dependencies of the service implementation on vendor-specific equipment, the IN framework uses the concept of *service-independent building blocks* (SIBs):

- Complex services are built from a set of small, standardized SIBs. This simplifies the implementation of new services, and leads to portable service software which is independent of a particular switching hardware.
- SIBs can reside on arbitrary nodes in the IN, thus alleviating the problem of data and software redundancies and inconsistencies.

Thus, the IN service logic is not associated to a particular switch, but implemented in a distributed fashion. This allows a more efficient usage of the available network resources, since it is no longer necessary to install identical program code and data in all network nodes. This distribution is best illustrated by the IN conceptual model, which is presented in the following section.

### 2.2.2 The IN Conceptual Model

The overall service provisioning architecture of INs is given by the IN conceptual model (INCM) [107], depicted in Figure 2.3. In this model, four different planes are defined, representing four different views on the network:

- The *service plane* shows the IN from a user-oriented point of view, illustrating the services which are offered by the network. Here, services comprise a set set of *service features*, which are realized by a set of SIBs.
- On the *global functional plane*, the IN is seen as a single monolithic structure, where service features are realized by a number of SIBs which are invoked by the detection point mechanism mentioned in Section 2.2.1. This plane is relevant to the designer of new services.
- The *distributed functional plane* determines which *functional entities* (FE) are necessary for the implementation of a particular SIB, and which information flows occur between them. This plane is of primary interest for the network designer.
- On the *physical plane*, FEs are mapped to physical nodes (*physical entities, PEs*) in the network, and information flows in the distributed functional plane correspond to physical links in this plane. This plane is important for network operators and equipment providers.

To summarize, the INCM provides a framework for realizing services from a set of standardized SIBs, and for distributing service functionality by a flexible mapping of FEs to physical network nodes. The distributed fashion in which services are implemented again underlines the crucial role of the signaling network, which is used to convey all control information among the involved network nodes.

### 2.2.3 Discussion

The IN framework greatly simplifies the design of new services, their maintenance, and the efficient use of network resources independent of vendor-specific hardware. On the one hand, this is accomplished by providing a concept for distributing service functionality among different physical nodes, on the other hand, a standardized set of SIBs allows for hardware- and vendor independence. The increased network design flexibility and the emphasized role of the signaling network in an IN evidently require appropriate methods for guiding the IN deployment process; this issue will thoroughly be discussed in Section 2.4.

Today, INs can be considered as a well-established approach, as their concepts are commonly used in both fixed and mobile telephone networks (see

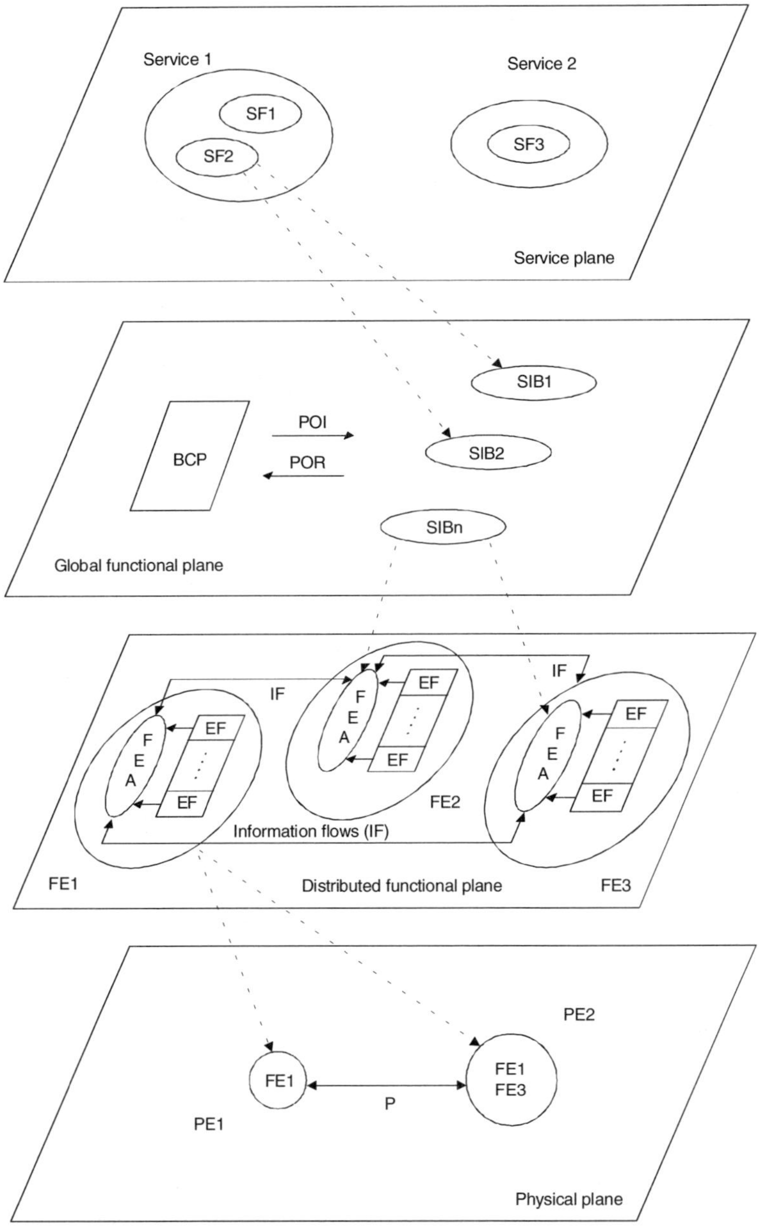


BCP Basic call process
EF Elementary function
FE Functional entity
FEA Functional entity action
IF Information flow
P Protocol

PE Physical entity
POI Point of initiation
POR Point of return
SF Service feature
SIB Service-independent building block
- - ▸ Pointer

**Fig. 2.3.** The IN conceptual model (figure after [107]).

e.g. [40, 109, 177]). However, in spite of their popularity and their strengths, it must be said that some problems still remain in practice. A main problem is due to the fact that the initial *capability set* CS-1, standardized by ITU-T in 1993 to define the first stage of the IN evolution, often proved to be insufficient for the implementation of particular services. These problems have partly been solved by the introduction of CS-2, however, the standardization of CS-2 has not been finalized until the end of 1997. Therefore, proprietary solutions have partly been introduced, contradicting the initial goal of vendor- and hardware-independent service design. Other problem areas of INs are the rather inflexible scheme of distributing network functionality in terms of predefined functional entities, and the fact that the fixed BCSM makes it difficult to implement services which go beyond simple point-to-point connections (like, for example, multiparty calls).

All of these problem areas are resolved in the TINA framework, a network architecture which was developed with similar motivations in mind like those which led to the development of INs, but aiming to provide a more flexible approach which circumvents the restrictions of INs. An overview on this approach will be presented in the next section.

## 2.3 The TINA Approach

Originating from a workshop of IN specialists in 1990, the *telecommunications infrastructure networking architecture* (TINA) consortium [186] was founded in 1992 by 40 leading companies in the information and telecommunications networking area. The main objectives of the TINA initiative are

- to define a telecommunications architecture which supports distributed processing and service provision,
- to test the effectiveness of the architecture in practice, and
- to encourage the wide distribution of the proposed architecture.

The first task has been completed by the end of 1997. Phase 2 of the TINA project, which is scheduled for completion by the end of 2000, focuses on the last two issues.

The main idea of the TINA framework is to provide an open architecture for telecommunication software applications, integrating results from both telecommunication and information technology and avoiding limitations of, for example, the IN concept. The TINA standards comprise a set of four architectures:

- A service architecture, describing the design, implementation, and management of services,
- a network architecture, focusing on the underlying transport networks,
- a management architecture, dealing with the design and implementation of software for services and network resources, and

– a computing architecture, defining the design and implementation of distributed software.

All these architectures are tied together by an overall TINA architecture. The TINA framework accommodates architectures and viewpoints which are powerful enough to account for all aspects of modern telecommunication networks, from business roles of the parties acting in such a network to the distributed deployment of service functionality; its scope is flexible enough to be integrated with and to migrate from other approaches like INs.

A summary of the entire TINA framework is beyond the scope of this work; for further information, see [18, 43, 104, 186]. In this section, we first focus on illustrating the main differences between the TINA architecture and the IN concept in Section 2.3.1. In Section 2.3.2, we then provide an overview on the TINA computing architecture, which has the largest impact on the performance of a TINA-based communication system as it deals with the deployment of service functionality in a distributed manner. In Section 2.3.3, we conclude with a brief discussion.

### 2.3.1 TINA and the IN Concept

The IN approach was designed to cope with the growing demand for complex telecommunication services in telephony environments. However, as already mentioned in Section 2.2.3, the IN approach exhibits weaknesses in the areas of service distribution and the realization of complex call scenarios. These issues can be summarized by the following two points:

*Service Software Distribution.* While the IN concept clearly separates switching aspects from service provisioning, distribution aspects of the service software itself are only considered at a very low level of abstraction, i.e., by assigning functional entities to different physical nodes. However, the service implementation should ideally be completely decoupled from network dependencies, and more fine-grained and transparent distribution methods are desired. Furthermore, the interaction of distributed service applications with network and service management parts needs to be improved.

*The Control Model.* The control of other than simple point-to-point connections is difficult in the IN framework, as the concept of a set of fixed detection points in a standard call process is often not flexible enough. Future broadband networks will support the transmission of several different types of information (e.g., audio and video) with different properties (e.g., constant and variable bit rate sources), so a more flexible channel control model is needed. Also, additional connection types like unidirectional or multipoint connections must be supported.

In the following section, we review the TINA computing architecture to provide an overview on how these areas are tackled in the TINA framework.

### 2.3.2 The TINA Computing Architecture

In order to cope with the complexity of large telecommunication systems, the TINA computing architecture adopts the five viewpoints of the reference model for open distributed processing (RM-ODP) [105], each of them dealing with different aspects of the system:

- The enterprise viewpoint defines the system purpose on a high, functional level in terms of actors and their roles and responsibilities.
- The information viewpoint deals with objects, their types, and their relation to each other.
- The computational viewpoint considers the decomposition of the system into interacting objects and their interfaces
- The engineering viewpoint focuses on the distributed deployment of applications in an abstract infrastructure.
- The technology viewpoint cares about the actual technology to support such an environment.

From a distributed processing point of view, the computational and engineering viewpoints are the most interesting ones; we therefore provide a short overview on the main TINA concepts in these areas.

**TINA Applications.** A key idea of the TINA framework is the adoption of object-oriented techniques from the computer science area and to use them for structuring complex telecommunication systems. Consequently, TINA applications are realized in terms of communicating *objects*. Their communication takes place via two kinds of interfaces, supporting either operational communications (i.e., by invoking methods and receiving responses from peer objects), or by allowing the continuous exchange of information flows by so-called *stream* interfaces, representing unidirectional bitstreams to which quality of service parameters can be associated. Therefore, at each interface, certain *services* are offered to other objects.

As the management of individual objects in a complex network is not feasible, several objects can be aggregated in so-called object *groups*, which are built, installed, maintained and managed as a unit (see also Figure 2.5). From the outside, only a subset of the services offered by objects within an object group are visible; these externally available services are referred to as *contracts*. Therefore, objects which belong to different object groups can only communicate via contracts. Note that object groups may again hold other object groups, thus allowing to structure distributed TINA applications in a hierarchical manner.

**Distribution Concepts.** TINA's core concept for the transparent implementation of applications in a distributed way is the *distributed processing environment* (DPE) (see Figure 2.4). Its purpose is to provide an abstract execution environment for running distributed applications, based on the following components:

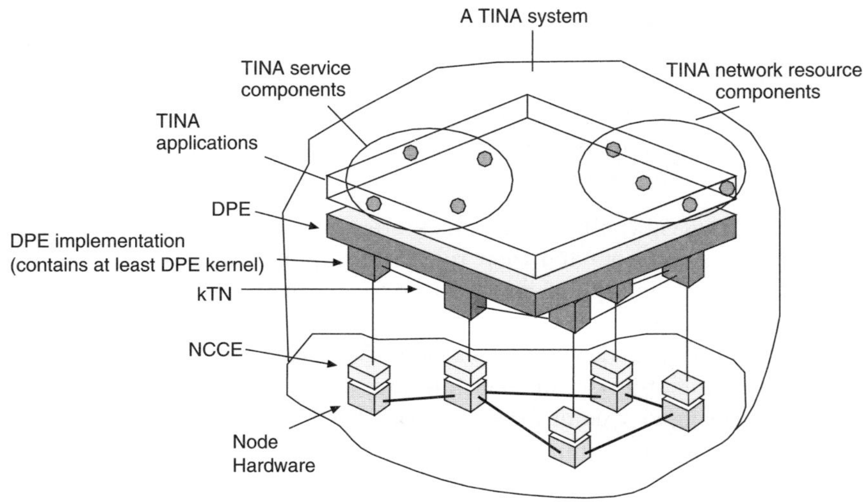


**Fig. 2.4.** The TINA DPE and the underlying infrastructure (figure after [104]).

- The DPE kernel, part of the DPE implementation on any network node, provides the basic functionality for object creation and deletion, and for communication amongst objects.
- The kernel transport network (kTN) represents a virtual network for the communication between the kernels on different network nodes. In practice, it may be mapped to a signaling network like SS7.
- Additional DPE services may be available in some nodes, and provide extended functionality to support runtime execution and communication (e.g., trading and notification services).

The DPE implementation on a network node is based on the native computing and communications environment (NCCE), representing the operating system and communication software of the actual nodes. Finally, the computing and communication services are realized by the actual network hardware resources.

For managing the distribution of application objects in the DPE, the following engineering units are introduced (see Figure 2.5):

- *DPE nodes* represent a unit of resource administration which supports the global DPE via its DPE kernel implementation. DPE nodes offer a number of computing resources, including processing, memory, and communication facilities. DPE nodes may represent any system which is capable of independent operations, ranging from single processor workstations to multiprocessor systems and clusters of workstations.
- *Capsules* are subsets of a DPE node, representing units of encapsulation and resource allocation. For example, capsules can be associated to processes in the UNIX operating system, or to individual processors in a mul-

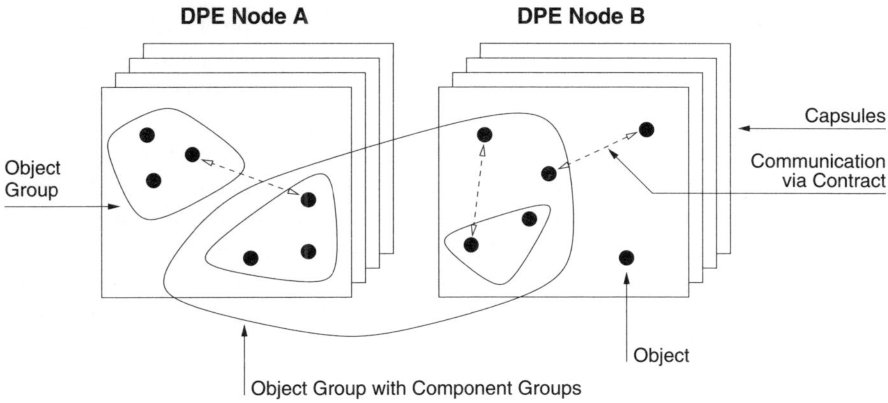


**Fig. 2.5.** TINA DPE nodes with capsules, objects, and object groups. Node that service relations between objects in identical object groups are omitted for conciseness.

tiprocessor system. Capsules provide the run-time environment for individual object instances.

Objects represent the smallest unit of distribution in the TINA framework. However, as mentioned above, object groups are managed as a unit, and therefore all objects in one object group must be located in the same capsule. Note that this is not true for component *groups*, which may be located in remote capsules. All objects associated to an object group on a particular capsule are required to be activated and migrated together, thus simplifying the management of a large number of objects.

Note that the actual transport network is not necessarily part of the DPE as depicted in Figure 2.4; it may well be realized by a set of different switches and communication links, which do not contribute to the DPE, but are controlled by applications running in the DPE.

### 2.3.3 Discussion

Compared to INs, TINA provides a much more flexible framework for deploying service functionality in a distributed fashion. TINA allows service designers to define their own units of distribution in terms of object groups, instead of restricting distribution to a set of fixed functional entities (as in the IN context). Also, the DPE provides an abstract infrastructure for the transparent distributed execution of service functionality, thereby sharply separating network- and distribution aspects from service implementation issues.

TINA is not yet a widely implemented platform, due to its complexity and the fact that the standards are still quite new. However, since TINA is supported by all major telecommunication network operators and network equipment providers, and due to the emphasis that the TINA consortium puts on spreading TINA and developing prototype implementations, it can be assumed that TINA will play a major role in future telecommunication networks.

Despite of the fact that TINA provides a more structured service provisioning framework and a more fine-grained control of distribution, the flexibility when deploying TINA networks is qualitatively similar to that provided by INs. Consequently, similar guidance is needed in the deployment process. In the following section, we therefore focus on the question how such guidance can be given, and what particular properties of IN- and TINA-like architectures must be taken into account.

## 2.4 Guiding the Network Design Process

In the previous sections, an overview on the properties of modern telecommunication architectures has been provided. Due to the increased network design flexibility, the question arises how the design of such complex networks can be guided.

Guiding methodologies are needed in the IN framework in order to balance node capacities, budget issues, estimated traffic and desired quality of service [16]. The early understanding of the impact of new services on the overall network performance needs to be known as soon as possible—ideally, already during the service specification and design phases [17, 165, 190]. The need for such guiding methods has also been noted in the context of ATM signaling network design, where, according to [203], "detailed quantitative work is needed to study network-dependent economic and performance issues". Even in the relatively new area of TINA networks, where few implementations are available yet, first studies to estimate the impact of different design options have been performed [197, 202].

Clearly, such goals can only be met with *model-based* performance evaluation approaches, since the networks to be designed are often yet to be deployed and measurement-based approaches are not possible. Also, if networks are already operational and need to be optimized, or if the impact of introducing new services is of interest, costly reconfigurations and measurements cannot be performed, since ongoing provisioning must not be disturbed.

In order to identify the specific demands to a model-based performance evaluation approach for telecommunication architectures like IN and TINA, we first summarize the differences between these architectures and conventional telecommunication networks, and describe how these differences affect the performance evaluation of such architectures in Section 2.4.1. Based on these implications, we will then derive a set of requirements for an appropriate model-based performance evaluation approach in Section 2.4.2.

### 2.4.1 Impact of Network Architectures on Performance Analysis

From an operator's point of view, the key benefit of telecommunication network architectures like IN or TINA is that new services can be designed, introduced and managed in a fast and cost-efficient manner. In essence, this task is accomplished by two main principles:

- The service provisioning *software design* adheres to strict modularization and abstraction principles in order to overcome hardware-dependencies and to simplify the service management and software maintenance.
- Service functionality is *distributed* over the network, thus simplifying its management by concentrating functionality in certain places and allowing to share expensive network equipment throughout the network.

From a performance modeling point of view, the first principle mainly addresses the design of the service provisioning software, and as such does not change the nature of the systems to be investigated. Similar design principles (though not standardized) are likely to have been applied to conventional telecommunications service software as well. Nevertheless, a structured system design may well help to develop modular performance models. The second point, however, fundamentally impacts the structure of the entire network, and largely affects its performance. The implications of distributed service provisioning can be summarized by the following three points:

*Non-Local Behaviour.* The isolated investigation of individual network nodes where services are provided locally to a fixed set of subscribers is no longer feasible. External traffic from other network nodes may impose additional load on the node under investigation; conversely, the quality of services realized by a node now also depends on other nodes, which are accessed in order to provide local services.

*Impact of the Signaling Network.* The role of the signaling network is much larger than in conventional telecommunication networks, as it provides the communication basis for the distributed service provisioning components. With the introduction of dedicated signaling facilities in conventional networks, delays due to the exchange of simple connection establishment and -release messages became negligible. However, in modern network architectures, this is no longer the case, as a large number of individual information transfers may be necessary to realize a service. The corresponding delays may sum up to substantial values which cannot be neglected.

Furthermore, the amount of signaling traffic heavily increases with the introduction of distributed architectures; according to [16], the signaling network load will increase by a factor of at least 4 in cellular networks, and it will increase another 3-4 times with advanced services like PCS. This dramatic increase will significantly impact the overall response times. Also, note that the additional amount of IN- or TINA-related signaling information also impacts the quality of non-IN and non-TINA services if they use the same signaling infrastructure.

*The Mapping Problem.* The distribution of service functionality in the network adds another level of freedom to the network design process, as decisions must be taken on which functionality is assigned to which hardware resources in the network. In the IN context, this corresponds to mapping the functional entities and information flows of the distributed functional plane to physical entities and paths in the physical place. Similarly, in the TINA framework, objects and object groups must be mapped to capsules, and the kTN must be realized by an appropriately designed signaling network.

In the following, we will refer to this issue as the *mapping problem*. Obviously, it represents an optimization problem which heavily impacts both the load of the individual network nodes and the signaling links, and therefore crucially affects the quality of the services to be provided [74].

To summarize, the distribution facilities of network architectures like IN and TINA offer a large amount of flexibility during network design. Model-based performance evaluation approaches can provide guidance in this area, as they allow to evaluate the suitability of several design alternatives and to estimate the impact of new services prior to performing any changes to the real network. In the next section, we will derive the requirements for a model-based performance evaluation environment in order to cope with the above-mentioned properties.

### 2.4.2 Requirements

Based on the three main implications of a distributed service provisioning framework mentioned in Section 2.4.1, the following three requirements for a model-based performance evaluation approach can be identified directly:

**R1 (Expressiveness and Convenience)** The approach must be capable to model the properties of individual network nodes in sufficient detail, *and* it must provide means to conveniently support this modeling flexibility. Since network nodes can be considered as service stations in both the IN and the TINA context, the modeling framework should be able to account for complex service time distributions, scheduling strategies, and overload control mechanisms. Clearly, a state-based node description (e.g., in terms of a CTMC) is the most flexible approach, however, the state space explosion problem which occurs when modeling nodes becomes even worse in a distributed environment, since the behaviour of other network nodes must be accounted for as well. In addition, the problem of specifying large state-based models conveniently has to be tackled.

**R2 (Non-Local Interactions)** As the service provisioning performance of network nodes often depends on other nodes in the same network, appropriate means must be provided for coupling individual nodes, and for modeling the additional effort needed for exchanging information with remote entities. Also, the traffic arising between different network nodes will exhibit complex distributions and correlations, due to persistent dialogues between peer entities and due to particular signaling "patterns" which are characteristic for specific services. Thus, care must be taken when describing such flows and approximating them by simple models.

**R3 (Fast Analysis)** Finally, the modeling approach must be amenable to very efficient analysis techniques. This is particularly important in view of the mapping problem, where a large number of different network configurations needs to be evaluated within the "outer" optimization procedure. Therefore, the quick analysis of individual models is crucial to deal with the mapping problem. Fast evaluation techniques are also necessary for realizing automated self-optimization and self-reconfiguration techniques in operational networks.

In addition to these requirements which directly follow from the distributed network architecture, two further requirements can be formulated. The first one accounts for the special way in which services are realized in IN and TINA-like architectures, while the second one describes what information a model-based performance evaluation approach shall provide in this particular application area:

**R4 (Service-Oriented Modeling)** The modeling approach should allow for an easy mapping of service functionality to network nodes, i.e., the approach should be able to account for the way services are designed and distributed in the network. On the one hand, this simplifies both the specification of a service (in terms of interacting building blocks or objects) and its distribution aspects in the model. On the other hand, the derivation of service-specific performance measures is facilitated during model evaluation.

**R5 (Model Evaluation)** Concerning the performance measures which are to be obtained from the analysis of network models, two different areas can be identified. First, node-specific performance measures must be available in order to identify bottlenecks in the network and to discover opportunities for local optimization within the nodes. Example measures in this area are node utilization values, moments of the queue length distribution, and buffer overflow probabilities. Generally speaking, the granularity of node-specific performance measures should be on a similar level as the model specification, e.g., for detailed state-based node descriptions individual state probabilities are likely to be of interest, while this is not the case if entire nodes are abstracted by few main properties.

Secondly, as also mentioned in [76, 82], the computation of *network-wide* end-to-end delays is of major interest. There are tight limits concerning, for example, call establishment delays in INs, which shall not exceed 1.5 seconds [106]. With complex services, where several information exchanges and database queries must be performed prior to connection setups, such values are quickly reached. Therefore, the analysis of such models must provide appropriate delay estimates.

In the remainder of this work, these five requirements will serve as the guiding principle for the development of a model-based performance evaluation approach for modern telecommunication architectures. Obviously, some of the requirements are contradictory, like the expressiveness of node models (R1) and the need for fast model evaluation (R3). Therefore, care must be taken when balancing the importance of the individual requirements. Ideally, the modeling framework is flexible enough to account for both coarse-grained models and fast solutions on the one hand, and detailed models where solution time is not of primary interest on the other hand, so that the user can choose a level of detail appropriate for the current application area and for the computational resource situation.

## 2.5 Summary and Concluding Remarks

In this chapter, a survey has been provided on the main architectural properties of modern telecommunication networks which are currently being used, and which will be used in the near future in order to satisfy the growing demand for new services in the converging areas of telephony and broadband networks.

Compared to conventional telephone networks, these architectures provide an enormous flexibility for designing services and distributing them in the network. This increased flexibility, however, leads to a wealth of design alternatives, so that guidance is needed in order to use this flexibility to provide optimal service quality at minimal costs. A model-based performance evaluation approach can provide such guidance, as it provides information on the performance of different design alternatives prior to their actual implementation.

A main property of modern network architectures is that services are no longer provided by a single piece of software running on an isolated network node, but instead are realized by a set of interacting entities, which may be distributed physically within the network. This implies that the performance analysis can no longer focus on investigating isolated nodes, since the node behaviour largely depends on other nodes as well. Based on this observation, five requirements have been derived which must be fulfilled by appropriate model-based performance evaluation approaches.

In the remainder of this work, a performance modeling framework will be developed based on these five requirements. As a first step, an intuitive initial approach will be presented in the next chapter. It will be compared to other performance-modeling approaches in the IN/TINA area in order to clarify the requirements, and to illustrate the limitations of state-of-the-art techniques in this area.

# 3. The View from Industry: First Modeling Approaches

In the previous chapter, the architectural properties of current and future telecommunication networks have been discussed in order to identify the impact of these properties on the development of appropriate modeling and performance evaluation approaches. The set of requirements which has been derived was therefore mainly motivated by technical and architectural issues.

The purpose of this chapter is two-fold. First, we aim to assess the requirements to appropriate modeling and evaluation techniques from a practitioner's point of view, and to validate these requirements against those identified in the previous chapter. To accomplish this, a cooperation has been established with an industrial partner who is actively involved in the modern telecommunications market. Philips Research Laboratories (Aachen, Germany) proved to be an ideal partner in our context due to the ongoing work in the IN field.

The second objective of this chapter is to present a prototype modeling and evaluation framework, tailored towards the specific needs of our industrial partner. In addition, care has been taken that the approach is extensible enough to account for more general requirements as well. On the basis of this prototype approach, the major problem areas in the development of appropriate modeling approaches are identified in a very practical way. In addition, the prototype serves as a baseline for a comparison with other modeling and evaluation approaches which were presented in the literature before.

The structure of this chapter is as follows: In Section 3.1, we summarize the set of requirements which have been formulated within the collaboration with Philips Research Laboratories. An initial modeling and evaluation approach is then developed in Section 3.2. The viability of the proposed approach is illustrated in a case study in Section 3.3. A comparison of this approach with related work is then performed in Section 3.4. Finally, this chapter concludes with a summary in Section 3.5.

## 3.1 Modeling and Evaluation Requirements: The Practitioner's View

The purpose of this section is to assess the requirements for a model-based performance evaluation framework for modern telecommunication networks from an application-oriented viewpoint. The importance of taking this viewpoint must not be underestimated, as it serves to validate the requirements which were derived from a more theoretical side in Section 2.4.2. In particular, this approach complements the theoretical considerations with information on the relative importance of the individual requirements, and it helps to discover limitations imposed by practical constraints (e.g., concerning the availability of detailed data for model parameterization).

For this purpose, a collaboration project with Philips Research Laboratories (Aachen, Germany) has been established. Philips has developed a sophisticated IN service creation environment (PHIDES, Philips Design Tool for Services [62]), where performance issues are currently not being considered during service creation. Interestingly, Philips also aims to employ a similar environment for the creation of services in a realtime multimedia communication system for broadcasting applications, where performance issues are of major importance. Therefore, appropriate system dimensioning is crucial for the successful deployment of these systems.

During the cooperation, we were able to approach and to understand the practical requirements of Philips [150, 91, 121]. The resulting requirements can be divided in two areas: those concerning the modeling expressiveness of the framework, and those concerning the results which need to be obtained from the model analysis. After presenting these requirements, we will briefly discuss their relation to those formulated in Section 2.4.2.

**Model Expressiveness.** Concerning the modeling flexibility, it has been found that the modeling formalism must allow the specification of a distributed service provisioning architecture, thereby

- distinguishing individual functional entities,
- describing their mapping to physical network nodes and
- the way how these functional entities communicate.

These requirements are essential to specify and compare different distribution scenarios. In addition, it has been stated that

- it must be possible to specify the load offered to the system and to specify the distribution of the overall load to individual functional entities.

Clearly, the load distribution within the system is closely related to the mapping of functional to physical entities mentioned above.

**Analysis.** Concerning the model analysis, an important point is the discovery of bottleneck components in the system. In order to identify those, it is necessary to obtain performance measures for functional entities, physical network nodes and communication links. In particular,

- utilization values are desired, as well as
- response times both on per-node level and on system level.

These results are also necessary in order to investigate the scalability of the system with respect to (for example) the number of users or the introduction of more complex services.

Once a performance bottleneck has been discovered, the modeling and analysis framework should be able to

- provide advice on how to remove a bottleneck situation, e.g., by introducing faster or more components or by re-distributing functional entities.

Another issue, which has been stressed on behalf of Philips, is that only very limited information is available on the actual design and the performance of the physical nodes and on the way services are implemented. Therefore,

- the approach must be capable to deal with very high-level numerical data on the system under investigation.

Obviously, this requirement also affects the modeling requirements, since this is where system properties are specified.

**Discussion.** The requirements determined in close cooperation with Philips are in strong accordance to those which have been derived from a more theoretical point of view in Section 2.4.2. On the modeling side, it is required that individual network nodes can be described (R1) as well as interactions among them (R2). The specification of the system load can be considered as a special case of information flows which are externally originated. The easy specification of mapping and distributing service functionality to network nodes has been formulated in R4.

On the model evaluation side, utilization and response times are desired, as already mentioned in R5. It should be emphasized that network-wide delay values are of particular interest to Philips, as these delays are those ultimately experienced by the users of the system (i.e., the customers). Guidance is also desired in the removal of bottleneck situations; for performing appropriate "what-if" studies and implementing optimization procedures, fast solution methods are therefore required, as also mentioned in R3.

In view of the good accordance of these requirements with those found in Section 2.4.2, the last requirement which has been mentioned above (concerning the availability of very high-level information only) turns out to be the most important one in the framework of our cooperation. Using the terminology of R5, it requires the modeling framework to be located on a rather coarse level of "granularity". Clearly, this also limits the amount of information which can be obtained from the analysis of these models. However, as will be illustrated by the approach developed in the next section, only little information is necessary to approximate the performance measures of interest.

## 3.2 A First Modeling Approach

In this section, a first modeling and evaluation approach is presented which is taylored towards the needs identified in collaboration with Philips (see Section 3.1). In the development of the approach, care has been taken that it provides an extensible framework which can serve as a starting point for more sophisticated solutions which will be developed in the remainder of this work.

The underlying mathematical technique of the approach is the analysis of a network of M|G|1 queues. While this is a rather simple methodology from a theoretical point of view, it turned out to be appropriate for a prototype framework, especially in view of the amount of information Philips was able to provide concerning physical node properties and service implementation details. In addition, the computational effort to solve such models is negligible, so the prototype represents a scalable solution suitable for large networks, and the danger of developing a too time-consuming initial approach has been avoided.

In Section 3.2.1, we present the model description method, followed by the model evaluation trajectory in Section 3.2.2.

### 3.2.1 Model Description

The overall model description consists of three parts. The first part describes the physical network components. The second part summarizes which kind of services exist, and how they are realized by the available physical components. The specification framework for these parts is tightly oriented towards the IN terminology and the INCM; for an overview, see Figure 3.1. Finally, the third part of the specification describes the performance measures of interest and how they can be derived from the "raw" numerical results.

**Physical System.** We assume that the whole system under consideration consists of $N$ physical nodes, referred to by values in $\{1, \cdots, N\}$. These nodes correspond, for example, to physical entities in the INCM. They provide the physical infrastructure to realize the services of one or more *virtual service devices*. In IN terminology, virtual devices correspond to functional entities on the distributed functional plane. We denote the overall number of virtual service devices by $K$.

The performance of each physical node $n \in \{1, \cdots, N\}$ is given by its processing speed $\mu_i$ (e.g., given in instructions per second or code lines per second).

Finally, a mapping is needed in order to define on which physical node a virtual service device is located, and how much physical resources are needed to satisfy a request to a virtual service device. This is accomplished by a matrix $\mathbf{L} = [L_{i,j}] \in \mathbb{R}^{K \times N}$. Its entries contain the load introduced on a physical node by a request to a virtual service device. An entry in row $i$ and

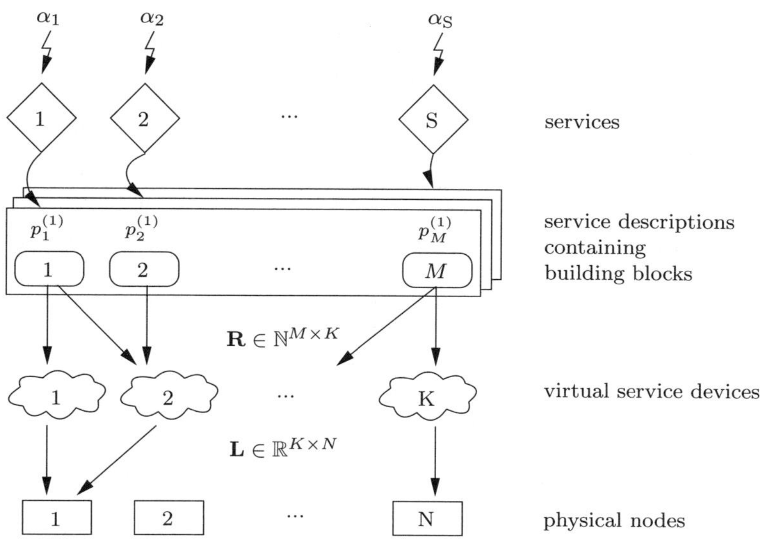


**Fig. 3.1.** Overview on the specification framework.

column $j$ represents the amount of work (e.g., in number of instructions) that a request to virtual service $i$ imposes on physical node $j$. To easily determine which virtual service devices are located on which physical node, we will also use the indicator matrix $\boldsymbol{\delta}_{\mathbf{L}} \in \{0,1\}^{K \times N}$, the elements of which are non-zero if the corresponding entry of $\mathbf{L}$ is larger than 0.

**Service Structure.** We assume that there is a total number of $S$ services $\{1, \cdots, S\}$ available in the system (like, for example, the freephone service). Each service is realized by combining service building blocks from a set of $M$ available building blocks, denoted by $\{1, \cdots, M\}$. Each such block accesses a number of virtual service devices, possibly several times. Thus, we describe the number of invocations of a virtual service device by each building block by a matrix $\mathbf{R} \in \mathbb{N}_0^{M \times K}$ (so rows in this matrix correspond to building blocks, and columns refer to service devices).

In order to model the load offered to the system under consideration, we account for the rate at which services are invoked. These rates are summarized in a vector $\boldsymbol{\alpha} = (\alpha_1, \cdots, \alpha_S\} \in \mathbb{R}^S$. In addition, we need information on how frequently each building block is visited. For service $s$, the probabilities of entering the individual building blocks are contained in the vector $\mathbf{p}^{(s)} \in \mathbb{R}^M$. Note that the entries of this vector may well be equal to zero in case a building block is not used by a service.

**Measure Specification.** To determine delays experienced by the end-user due to the execution of a service (or part of it), information is needed on which building blocks are involved in realizing this service, and to which extent delays due to user-interactions shall be accounted for. Thus, for each

| Variable | Meaning |
|---|---|
| $N \in \mathbb{N}$ | number of physical network nodes |
| $K \in \mathbb{N}$ | number of virtual service devices |
| $\mu_n \in \mathbb{R}$ | speed of physical node $n$ |
| $\mathbf{L} \in \mathbb{R}^{K \times N}$ | load matrix mapping physical to virtual components |
| $\boldsymbol{\delta}_{\mathbf{L}} \in \{0,1\}^{K \times N}$ | indicator matrix corresponding to $\mathbf{L}$ |
| $S \in \mathbb{N}$ | number of available services |
| $M \in \mathbb{N}$ | number of available building blocks |
| $\mathbf{R} \in \mathbb{N}_0^{M \times K}$ | virtual service device invocations for building blocks |
| $\boldsymbol{\alpha} \in \mathbb{R}^S$ | service invocation rates |
| $\mathbf{p}^{(s)} \in \mathbb{R}^M$ | building block probabilities for service $s$ |
| $W \in \mathbb{N}_0$ | number of performance measures |
| $\mathbf{u}^{(w)} \in \{0,1\}^M$ | building blocks covered by measure $w$ |
| $d^{(w)} \in \mathbb{R}_0^+$ | additional delay for measure $w$ |

**Table 3.1.** Variables needed for system specification.

performance measure $w \in \{1, \cdots, W\}$ the vector $\mathbf{u}^{(w)} \in \{0,1\}^M$ denotes which building blocks are involved. In addition, a variable $d^{(w)} \in \mathbb{R}$ contains fixed delays which are added in order to account for user-interactions (e.g., due to dialing).

To provide an overview on the system specification framework, a summary of all relevant variables is given in Table 3.1.

### 3.2.2 Model Evaluation

In a first step, the total rate of requests to a virtual service device has to be calculated. This is accomplished by considering the rate at which services are invoked, the probability of entering a particular building block used by the service, and the number of requests to virtual service devices in that building block. As $\mathbf{p}^{(s)}$ holds the building block probabilities for a service, the product $\mathbf{p}^{(s)}\mathbf{R}$ yields the mean number of invocations of the individual virtual service devices by service $s$. The overall virtual service device invocation rates in the system are thus obtained as a vector $\boldsymbol{\lambda}^{\mathrm{V}} = (\lambda_1^{\mathrm{V}}, \cdots, \lambda_K^{\mathrm{V}}) \in \mathbb{R}^K$, with

$$\boldsymbol{\lambda}^{\mathrm{V}} = \sum_{s=1}^{S} \alpha_s \mathbf{p}^{(s)} \mathbf{R}. \tag{3.1}$$

In a similar way, this vector can be used to derive the vector of request rates to the physical network nodes, denoted by $\boldsymbol{\lambda} = (\lambda_1, \cdots, \lambda_N) \in \mathbb{R}^N$. This is accomplished by adding the request rates to all virtual service devices accessing the same physical node. Using the indicator matrix $\boldsymbol{\delta}_{\mathbf{L}}$, this is easily derived as

$$\boldsymbol{\lambda} = \boldsymbol{\lambda}^{\mathrm{V}} \boldsymbol{\delta}_{\mathbf{L}}. \tag{3.2}$$

In order to evaluate the physical nodes not only the rate at which requests arrive must be known, but also the amount of time needed to complete a request. Since requests submitted by different virtual service devices to the same physical node may differ in size (reflected by different entries in the same column of $\mathbf{L}$), we consider the service time *distribution* of each node, described by its first two moments. This is accomplished by considering the different times needed to satisfy a request of a virtual service device and the probabilities that a request is issued by that device.

The service time $s_{k,n}$ which is needed by a physical node $n$ to satisfy a request of virtual service device $k$ is obtained by considering the amount of work as described in matrix $\mathbf{L}$ and the physical node's speed $\mu_n$:

$$s_{k,n} = \frac{L_{k,n}}{\mu_n}. \tag{3.3}$$

We denote the probability that a job arriving at physical node $n$ stems from virtual service device $k$ by $\pi_{k,n}$. It is derived by relating the arrival rate of requests originating from virtual service device $n$ to the overall arrival rate at physical node $n$:

$$\pi_{k,n} = \frac{\lambda_k^{\mathrm{V}}}{\lambda_n}. \tag{3.4}$$

Given these values, the first two moments of the service time distribution at physical node $n$ can easily be computed:

$$\mathrm{E}[S_n] = \sum_{k=1}^{K} \pi_{k,n} s_{k,n} \quad \text{and} \quad \mathrm{E}[S_n^2] = \sum_{k=1}^{K} \pi_{k,n} s_{k,n}^2. \tag{3.5}$$

The utilization of each physical node is now given by $\rho_n = \lambda_n \, \mathrm{E}[S_n]$. Using the Pollaczek-Khinchine formula for the analysis of M|G|1 queues [119], the following node-specific measures can be computed:

– Mean number of packets at each node:

$$\mathrm{E}[N_n] = \frac{\lambda_n^2 \, \mathrm{E}[S_n^2]}{2(1-\rho_n)}. \tag{3.6}$$

– Mean waiting time at node $n$:

$$\mathrm{E}[W_n] = \frac{\lambda_n \, \mathrm{E}[S_n^2]}{2(1-\rho_n)}. \tag{3.7}$$

– Average response time of node $n$:

$$\mathrm{E}[R_n] = \mathrm{E}[S_n] + \mathrm{E}[W_n]. \tag{3.8}$$

Once these mean values are available for all physical nodes, they can be used to derive the high-level delay measures specified by the vectors $\mathbf{u}^{(w)}$. This is accomplished by adding the delay value $d^{(w)}$ to the service and waiting times of all physical nodes involved in measure $\mathbf{u}^{(w)}$. The involved physical nodes are obtained from the building blocks selected by non-zero entries of $\mathbf{u}^{(w)}$ by considering the virtual service device requirements of these building blocks and the mapping of virtual service devices to physical nodes.

The number of virtual service device invocations by measure $w$, denoted by $\mathbf{r}^{\mathrm{V},(w)} = (r_1^{\mathrm{V},(w)}, \cdots, r_N^{\mathrm{V},(w)}) \in \mathbb{N}_0^K$, is readily obtained from the involved building blocks and the matrix $\mathbf{R}$ as

$$\mathbf{r}^{\mathrm{V},(w)} = \mathbf{u}^{(w)}\mathbf{R}. \tag{3.9}$$

Then, analogously to Eq. (3.2), the number of physical node invocations in measure $w$ are given by

$$\mathbf{r}^{(w)} = \mathbf{r}^{\mathrm{V},(w)}\boldsymbol{\delta}_{\mathbf{L}}. \tag{3.10}$$

Now, in order to derive the expected delay $\mathrm{E}[D_w]$ for a measure $\mathbf{u}^{(w)}$, the response times of these physical nodes need to be added, thereby accounting for their multiple usage (here, we assume that requests to different physical nodes are satisfied in a strictly sequential manner).

Note that we do not consider the overall mean response time $\mathrm{E}[R_n]$ at a physical node $n$ directly. Instead, we employ the actual service times $s_{k,n}$ for requests associated to a particular virtual service device $k$, and add the mean waiting time $\mathrm{E}[W_n]$ to this value. By employing the actual service times, much more accurate results can be obtained. Summarizing, this yields

$$\mathrm{E}[D_w] = d^{(w)} + \underbrace{\sum_{k=1}^{K}\left(r_k^{\mathrm{V},(w)}\sum_{n=1}^{N} s_{k,n}\right)}_{\substack{\text{total service time of}\\ \text{involved virtual devices}}} + \underbrace{\sum_{n=1}^{N} r_n^{(w)}\,\mathrm{E}[W_n]}_{\substack{\text{total waiting time of}\\ \text{involved physical nodes}}} \quad . \tag{3.11}$$

In addition to this measure-oriented delay value, we can similarly obtain the expected response time of a virtual service device $k$, denoted by $\mathrm{E}[R_k^{\mathrm{V}}]$. It is given as a special case of Eq. (3.11) where $r_n^{(w)}$ equals one if the physical node $n$ is used by virtual service device $k$, and zero otherwise (i.e., $\mathbf{r}^{(w)}$ equals the $k$-th row of $\boldsymbol{\delta}_{\mathbf{L}}$). Denoting these values by $r_n$, and only considering the service requirements of the virtual service device $k$, we obtain

$$\mathrm{E}[R_k^{\mathrm{V}}] = \sum_{n=1}^{N} s_{k,n} + \sum_{n=1}^{N} r_n\,\mathrm{E}[W_n]. \tag{3.12}$$

| $p_{\text{IN}}$ | $p_{\text{org}}$ | $p_{\text{SN}}$ | $p_{\text{noSN}}$ | $p_{\text{noVPN}}$ | $p_{\text{ok}}$ |
|---|---|---|---|---|---|
| 0.1 | 0.9 | 0.25 | 0.05 | 0.2 | 0.9 |

**Table 3.2.** Basic branching probabilities for the example model.

## 3.3 Application Example

In this section, the modeling and evaluation framework which was developed in Section 3.2 is applied to a small example IN network configuration. Due to the cooperation with Philips, the design of the system and its parameterization could be performed in a way that represents the real-world usage of the modeling framework very well.

For our studies, a prototype modeling and evaluation tool was designed in the C++ programming language; it computes the performance characteristics of a system which is specified in a textual manner (for an example specification, see Appendix C.1).

After describing the considered system in Section 3.3.1, numerical results from its analysis are presented in Section 3.3.2.

### 3.3.1 System Description

The example under consideration is a small IN system which provides a virtual private network (VPN) service. This service simplifies the dialing procedure for a closed group of subscribers by offering short-number access within the group. Furthermore, authorized users are also allowed to establish calls to non-VPN members.

**Service Description.** The VPN service is realized by a set of SIBs which are invoked according to the service logic program (SLP) illustrated in Figure 3.2. In order to identify individual SIBs in the SLP, we refer to them by the numbers 1 to 15. The overall visiting probabilities for each SIB in the SLP are obtained by the probabilities of entering the SLP at node 1 or 2 (denoted by $p_{\text{IN}}$ and $p_{\text{org}}$) and branching probabilities after SIBs 3 and 4. The probability of establishing a short number connection after building block 3 by entering block 5, 6 or 7 is denoted by $p_{\text{SN}}$. Alternatively, an error announcement for unknown short numbers is made with probability $p_{\text{noSN}}$, or a non-VPN call is initiated by entering node 4 with probability $p_{\text{noVPN}}$. Such a call is then authorized after node 4 with probability $p_{\text{ok}}$.

The actual branching probabilities used in our experiments are given in Table 3.2. We assume that the VPN service is invoked at rate $\alpha_{\text{VPN}}$.

**Service Realization.** In accordance with the INCM distributed functional plane, each SIB is realized by one or more functional entities. For the VPN example, four functional entities are considered:

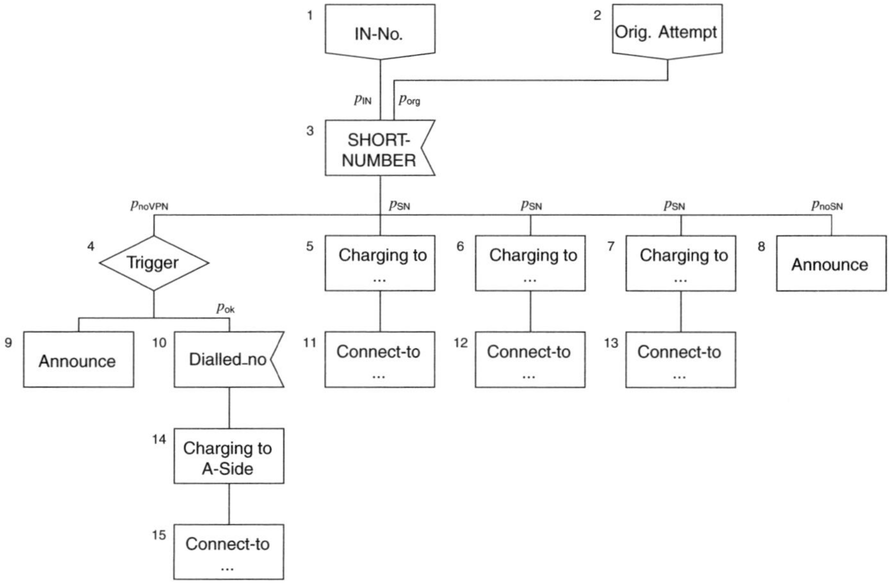


**Fig. 3.2.** Service logic program for the VPN example SLP.

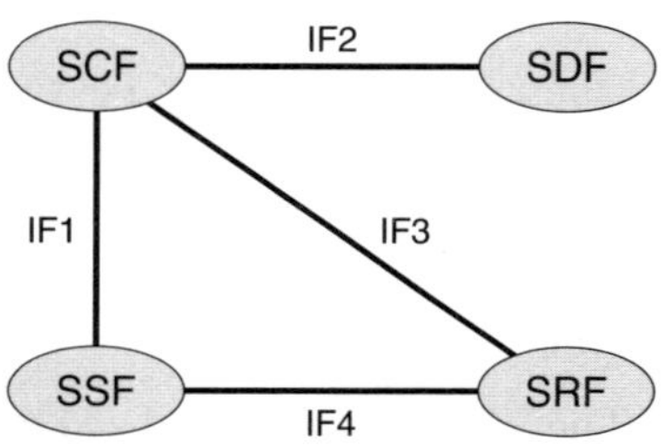


**Fig. 3.3.** Distributed functional plane view of the example system.

- A Service Control Function (SCF) realizes the VPN service by processing the SLP.
- A Service Switching Function (SSF) delivers the core switching functionality.
- A Service Data Function (SDF) provides database access, e.g., to resolve short numbers or for storing billing information.
- A Specialized Resource Function (SRF) offers access to special functionality, e.g., for playing announcements or collecting user-provided data.

As illustrated in Figure 3.3, four information flows can be distinguished between these entities. Note that the SDF is solely accessed by the SCF. Similarly, the SRF is controlled by the SCF; however, an additional information

| virtual device | SIB no. 1 | 2 | 3 | 4 | 5 | 6 | 7 | 8 | 9 | 10 | 11 | 12 | 13 | 14 | 15 |
|---|---|---|---|---|---|---|---|---|---|---|---|---|---|---|---|
| SCF | 1 | 1 | 1 | 1 | 1 | 1 | 1 | 1 | 1 | 1 | 1 | 1 | 1 | 1 | 1 |
| SSF | 1 | 1 | 1 | 0 | 0 | 0 | 0 | 1 | 1 | 1 | 1 | 1 | 1 | 0 | 1 |
| SDF | 0 | 0 | 0 | 1 | 1 | 1 | 1 | 0 | 0 | 0 | 1 | 1 | 1 | 1 | 1 |
| SRF | 0 | 0 | 0 | 0 | 0 | 0 | 0 | 1 | 1 | 0 | 0 | 0 | 0 | 0 | 0 |
| IF1 | 1 | 1 | 1 | 0 | 0 | 0 | 0 | 0 | 0 | 1 | 1 | 1 | 1 | 0 | 1 |
| IF2 | 0 | 0 | 0 | 1 | 1 | 1 | 1 | 0 | 0 | 0 | 1 | 1 | 1 | 1 | 0 |
| IF3 | 0 | 0 | 0 | 0 | 0 | 0 | 0 | 1 | 1 | 0 | 0 | 0 | 0 | 0 | 0 |
| IF4 | 0 | 0 | 0 | 0 | 0 | 0 | 0 | 1 | 1 | 0 | 0 | 0 | 0 | 0 | 0 |

**Table 3.3.** Virtual device usage of each SIB which appears in the VPN SLP; the table directly corresponds to $\mathbf{R}^{\mathrm{T}}$.

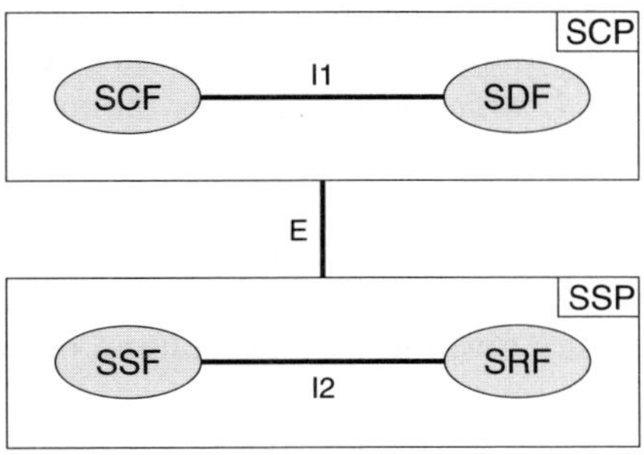


**Fig. 3.4.** Physical plane view of the example system.

flow is necessary to co-ordinate special resource functionality with the switching system.

We consider both information flows and functional entities as virtual devices in our modeling framework. The mapping of SIBs to these virtual devices which was used in our case study is summarized in Table 3.3. This table thus corresponds directly to the transpose of matrix $\mathbf{R}$.

**Physical Network Properties.** For our experiments, we assume that two physical network nodes are available for realizing virtual service devices. These nodes are interconnected via Ethernet.

In accordance with IN terminology, we refer to these two nodes as SCP (service control point, focusing on IN service provision) and SSP (service switching point, concentrating on switching aspects). As illustrated in Figure 3.4, the SCP thus realizes SCF functionality, and the SSP implements the SSF. In addition, the SDF is assigned to the SCP, and the SRF is mapped to the SSP.

Due to this partitioning, the information flows IF1 and IF3 which appear in the distributed functional plane view in Figure 3.3 are realized by the Ethernet connection between SCP and SSP, denoted by "E" in Figure 3.4. Note that in contrast to the INCM viewpoint, we also account for information

| physical node | functional entity SCF | SSF | SDF | SRF | IF1 | IF2 | IF3 | IF4 | node speed |
|---|---|---|---|---|---|---|---|---|---|
| SCP | 30 | 0 | 60 | 0 | 0 | 0 | 0 | 0 | 1000 [instr./s] |
| SSP | 0 | 40 | 0 | 40 | 0 | 0 | 0 | 0 | 1500 [instr./s] |
| E | 0 | 0 | 0 | 0 | 50 | 0 | 50 | 0 | $10^6$ [B/s] |
| I1 | 0 | 0 | 0 | 0 | 0 | 200 | 0 | 0 | $10^7$ [B/s] |
| I2 | 0 | 0 | 0 | 0 | 0 | 0 | 0 | 50 | $10^7$ [B/s] |

**Table 3.4.** Workload on physical nodes induced by functional entity invocations (corresponding to $\mathbf{L}^{\mathrm{T}}$) and physical node speeds (corresponding to $\mu_i$).

flows appearing within a node (e.g., realized by appropriate platform-specific inter-process communication facilities) and refer to them by I1 and I2.

The resulting overall mapping from virtual devices to physical nodes is summarized in Table 3.4, which also indicates the amount of work which is imposed on a physical component due to the usage of a virtual service device; this mapping therefore corresponds to the transpose of matrix $\mathbf{L}$ introduced in Section 3.2.1. The physical speed is measured in instructions per second for nodes, and in bytes per second for data transport facilities. The values used in our case study are given in the last column of Table 3.4.

### 3.3.2 Analysis

Using the methods which were introduced in Section 3.2.2, a large number of performance characteristics for the VPN example system can now me computed. In conjunction with the rather large number of parameters involved in the model specification, an overwhelming number of investigations are conceivable. Here, we focus on two related case studies only. In the first study, a bottlenecks is identified in the VPN system, and then removed by changing the mapping of virtual service devices to physical nodes. As an alternative to this approach for tackling the bottleneck situation, we increase the physical node speeds in the second case study, and compare the costs and merits of this to the reconfiguration case.

In the experiments, the following three properties are considered to describe the system performance:

1. Node response times.
   Here we focus on the average time needed until a (physical or virtual) node completes a request.
2. Response times at user level.
   As an example measure, we investigate the average time from entering a VPN short number until the corresponding connection is established. This is identified in the VPN SLP (Figure 3.2 on page 36) by (for example) SIBs 5 and 11.

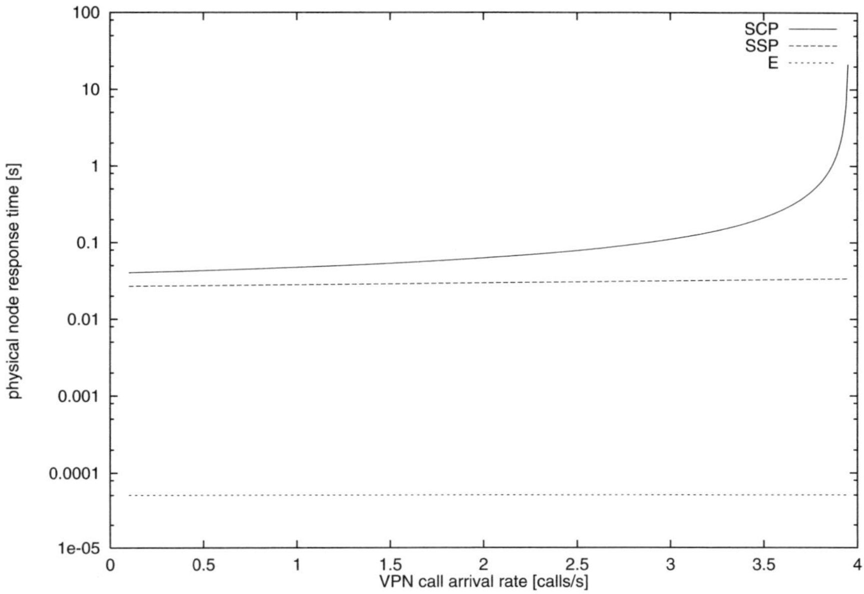


**Fig. 3.5.** Response time of physical nodes for different VPN call arrival rates.

3. Node utilization.
   In order to identify system bottlenecks, the utilization of physical network nodes is investigated.

**Case Study I: Bottleneck Identification and System Reconfiguration.** The VPN call arrival rate is varied between 0.1 and 3.95 calls per second (corresponding to 360 to 14220 calls/hour). Figure 3.5 illustrates the mean response times of the physical network nodes. It can be observed that the response time of the SCP reaches unacceptable values for call arrival rates larger than about 3 calls per second. The diagram also shows that the response time of the Ethernet component is negligible.

Considering the response times of the virtual service devices in Figure 3.6, the excessive response time of the SCP unit manifests again: calls to FEs which involve the SCP (i.e., SCF and SDF, see Figure 3.4) experience excessive delays for VPN call arrival rates larger than approximately 3.5 calls per second.

These results suggest that the SCP node represents a bottleneck of the system. Indeed, investigating the physical node utilization in Figure 3.7 shows that the SCP is quickly approaching 100% utilization for higher call arrival rates. For example, at 3.5 calls per second, the SCP utilization is 88.5%, while the SSP utilization is only 30.3%.

Noticing this large difference between SSP and SCP utilization, we propose to migrate the SDF functional entity to the SSP node in order to reduce

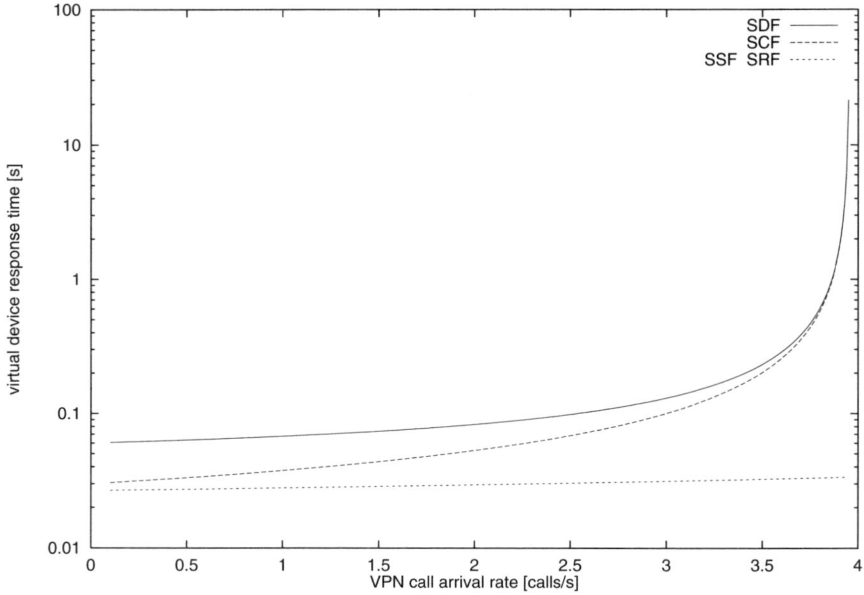


**Fig. 3.6.** SIB response times for different VPN call arrival rates.

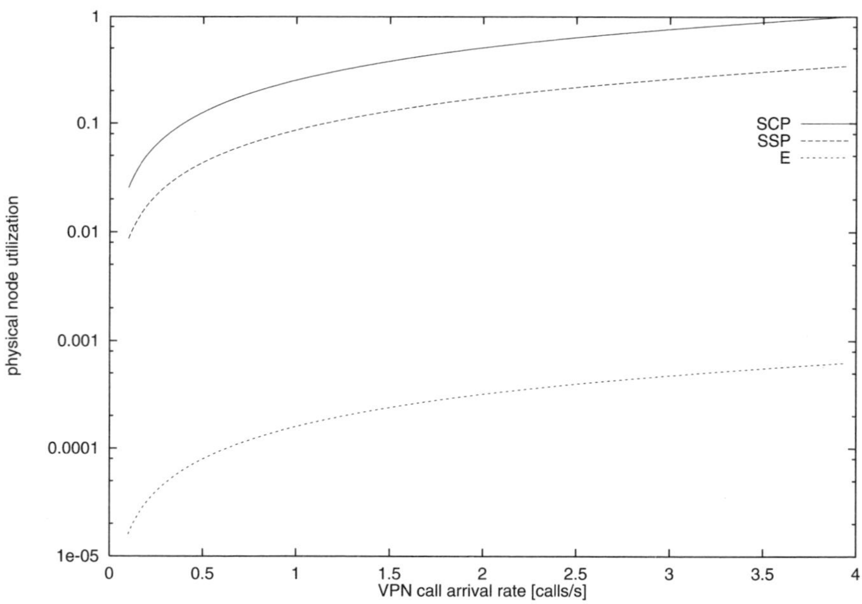


**Fig. 3.7.** Utilization of physical nodes for different VPN call arrival rates.

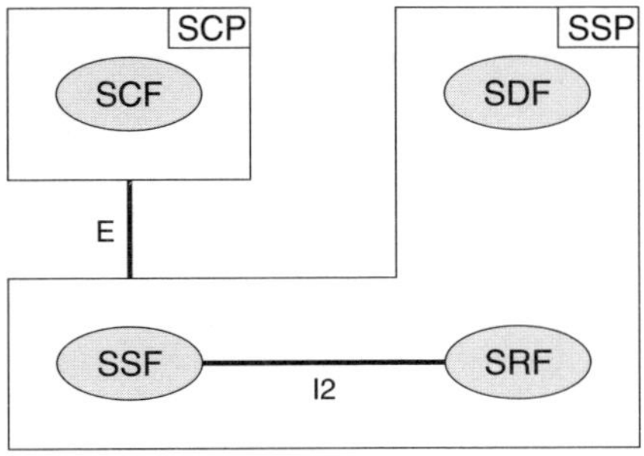


**Fig. 3.8.** Distributed functional plane view of the reconfigured system.

| physical node | functional entity SCF | SSF | SDF | SRF | IF1 | IF2 | IF3 | IF4 | node speed |
|---|---|---|---|---|---|---|---|---|---|
| SCP | 30 | 0 | 0 | 0 | 0 | 0 | 0 | 0 | 1000 [instr./s] |
| SSP | 0 | 40 | 60 | 40 | 0 | 0 | 0 | 0 | 1500 [instr./s] |
| E | 0 | 0 | 0 | 0 | 50 | 200 | 50 | 0 | $10^6$ [B/s] |
| I2 | 0 | 0 | 0 | 0 | 0 | 0 | 0 | 50 | $10^7$ [B/s] |

**Table 3.5.** Workload on physical nodes (and node speeds) after network reconfiguration.

the load on the SCP node (as illustrated in Figure 3.8). In the model specification, this is reflected by moving the SDF load of 60 instructions in Table 3.4 to the row belonging to the SSP. As the communication pattern changes as well, a mapping between virtual devices and physical nodes as shown in Table 3.5 is obtained.

Figure 3.9 illustrates the physical node utilization after reconfiguring the system in this manner. As opposed to the initial mapping, the SSP now shows the larger node utilization. However, its utilization is still much lower than the previous utilization of the SCP; at 3.5 calls per second its utilization amounts to 59.2% (SCP utilization now is 45.3%). As a result, the load is much better distributed among the physical nodes, leading to much lower response times.

The impact of this improvement is best illustrated by considering the user-perceived mean delay between entering a VPN short number and connection establishment. This delay value is illustrated in Figure 3.10, where values are provided for both the original and the reconfigured network. It can be seen that the reconfigured system yields connection setup delays of less than 0.3 seconds at arrival rates where the original system was far from being usable.

**Case Study II: Hardware Update.** As an alternative to reconfiguring the system by moving a virtual service device to another physical node, we might also choose to improve response times by using a faster SCP. Focusing on

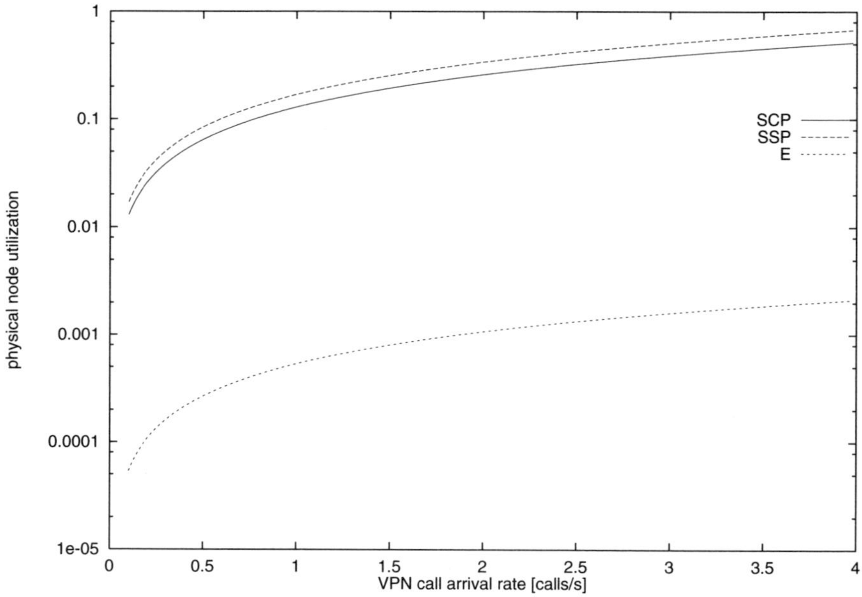


**Fig. 3.9.** Utilization of physical nodes after network reconfiguration.

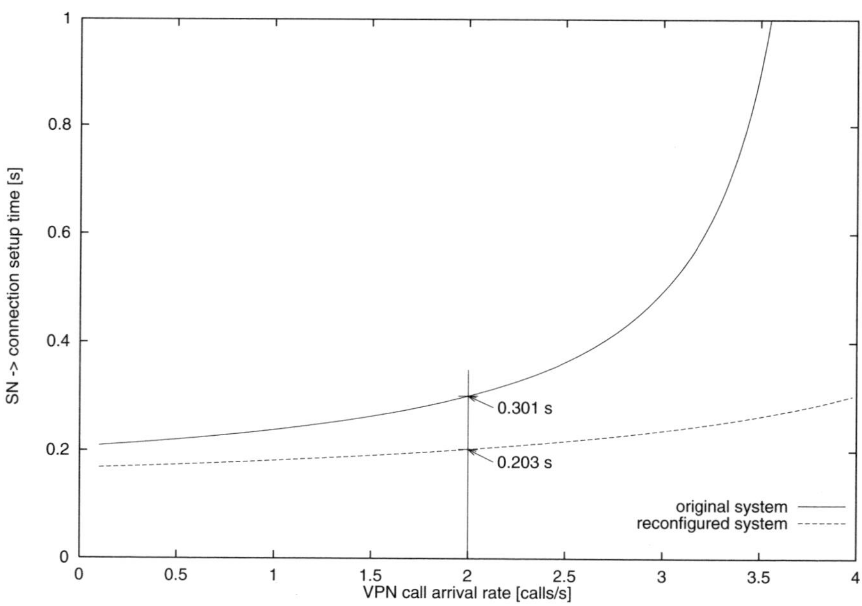


**Fig. 3.10.** Mean time between dialing a short number and connection establishment for different VPN call arrival rates.

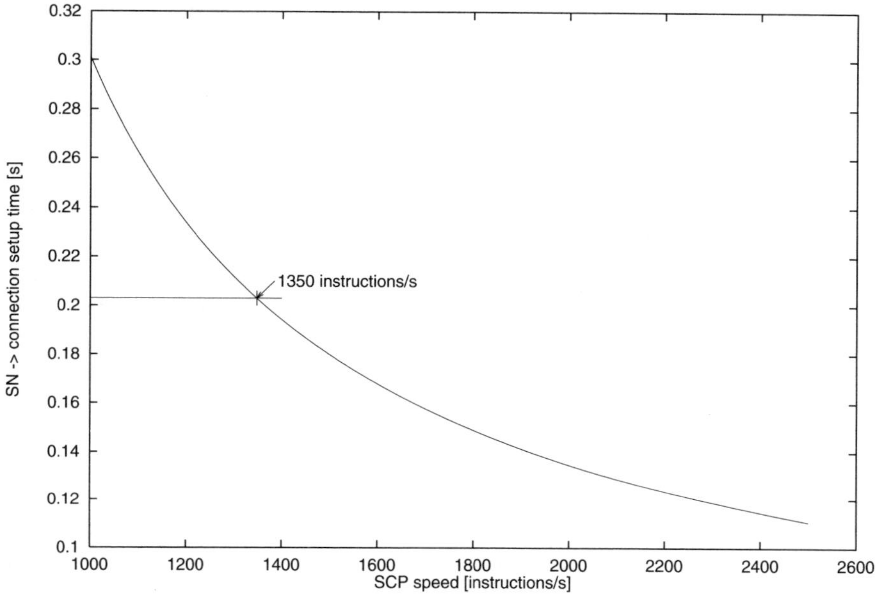


**Fig. 3.11.** Mean time between dialing a short number and connection establishment for different SCP speeds at a VPN call arrival rate of 2 calls per second.

user-perceived delays again, we consider the expected delay between entering a short number and connection establishment at different SCP speeds. The VPN call arrival rate has been fixed to 2 calls per second in this example, a value where the original system exhibits large yet tolerable connection setup delay of about 0.3 seconds (see vertical line in Figure 3.10).

In Figure 3.11, it can be observed that the delay time can be reduced by introducing faster SCP hardware. However, in order to obtain the same delay as in the reconfiguration case (which was 0.203 seconds, see Figure 3.10), the SCP speed has to be increased by 35% to 1350 instructions per second, which probably leads to higher costs than just reconfiguring the system.

In addition, at VPN call arrival rates larger than 2 calls per second, it becomes even more expensive to improve the system responsiveness by increasing the SCP speed. This is due to the fact that the relative gain of reconfiguring the system grows faster than linearly (see the increasing distance between the curves in Figure 3.10), while increasing the SCP speed only leads to a sub-linear improvement of the connection setup delay, as can be observed in Figure 3.11. Therefore, at increasing call arrival rates, more and more SCP power is needed in order to obtain the same improvement as gained by reconfiguration. However, it must be noted that if connection setup delays are required to be lower than about 0.2 seconds (depending on the system load), then this can not be reached by reconfiguration alone (see Figure 3.10), whereas this can be reached by introducing a faster SCP, as can be seen in Figure 3.11. Depending on the actual application case, a combination of both reconfiguration and hardware upgrade may be sensible.

## 3.4 Relation to Other Approaches

The importance of model-based performance evaluation frameworks for modern telecommunication networks, as illustrated in Chapter 2, has already been recognized before. Consequently, a wide range of approaches has been developed to cope with this complex issue. The purpose of this section is to summarize the most important results in this area, to illustrate the main properties of the individual approaches, and to relate them to the prototype framework presented in Section 3.2.

In Section 3.4.1, we first summarize approaches dealing with the model-based evaluation of a single system configuration. In Section 3.4.2, an overview is provided on those approaches which also tackle the mapping problem. Finally, we conclude with a comparison of the approaches to our protoype approach in Section 3.4.3.

### 3.4.1 Single System Evaluation Approaches

One of the first approaches specifically targeted towards the IN framework has been presented in [114]. The approach is based on a simulation environment, where the system workload is specified by message sequence charts of the signaling messages involved in IN calls, and by associating interarrival time distributions to these sequences. The simulation yields quality of service parameters like connection-setup and -release delays and call rejection probabilities. Due to the computationally very expensive simulation runs, the authors suggest to apply the approach to small networks only, where it is used to compare (for example) different load control algorithms where detailed models are a prerequisite. For evaluating larger networks, the authors propose a simple analytical approach, where the entire IN (as seen from the SSP at which quality of service measures are to be determined) is modeled by a single server queue with load-dependent delay. Evidently, this level of detail will not be sufficient to account for more complex system models and the derivation of more detailed performance measures.

In [202, 197, 116], a simulation-based approach for TINA networks is presented. Using a system description in the specification and description language (SDL), results are obtained for the TINA system throughput. While the authors employ a rather detailed description of the messages interchanged in the network, they assume a fully connected network and neglect propagation delays (a situation which will not be given as soon as the system under consideration is larger than a local Ethernet-based network). Also, external traffic is assumed to be Poisson, and all node service times are deterministic. The authors strongly emphasize the importance of the mapping problem to the overall system performance. Similar to our findings, the authors report the scarce availability on detailed system properties; to a large extent, model parameters are based on assumptions only. Concerning the granularity of node models, a simulation approach should easily be able to account for more detail; however, such an approach is likely to yield similarly large simulation times as reported by similar simulation approaches mentioned before.

A detailed IN simulation model has also been developed in [82]. The approach accounts for detailed node models, message-dependent processing times, different call type distributions and, in particular, for end-user mobility. In order to avoid the excessive simulation times, the authors also develop a simple analytical IN model. Similar to the prototype approach presented in Section 3.2, it is based on a network of M|G|1 queues and resorts to the Pollaczek-Khinchine formula for the analysis of the model. However, the framework presented in Section 3.2 also covers the complete trajectory from a SLP/SIB-oriented service specification to derive the traffic entering the physical nodes. Concerning the results presented in [82], the authors report a good accordance between the analytical results and those derived by the detailed simulation model. Given the close relationship between their analytical approach and our framework presented in Section 3.2, this also gives support to our prototype approach. Clearly, this approach also shares the level of detail of the node models with our approach, which may not be sufficient in many cases.

In [190], a simulation study of a SS7 signaling network is performed. Depending on the size of the network and the considered traffic intensity, simulation times of up to about 32 hours were needed for the evaluation of a single configuration; in addition, the memory requirements of the simulation environment were as large as 200 Mbytes. Even though faster execution is possible nowadays due to the availability of cheaper and faster computer hardware (the study has been performed in 1994), this clearly underlines that a simulation-based approach is not feasible for the evaluation of large telecommunication networks, especially if optimizations are to be performed on top of a basic performance evaluation environment.

In [199], a comprehensive analytical framework for the performance evaluation of SS7 networks has been presented. The solution approach is based on decomposition, and network-wide end-to-end measures are determined in three steps. First, a global traffic flow analysis is performed in order to determine the amount of load offered to the individual network nodes (this is accomplished by solving a set of linear equations). Then, individual nodes are analyzed in isolation by applying a result for M|G|1 priority queues with feedback [161]. Finally, per-node results are used to obtain network-wide delays, similar to the approach presented in Section 3.2.2. Recently, this framework for SS7 has also been extended to account for the traffic arising in INs [15, 17].

The analysis is computationally very attractive, and the results presented in [199] are closely matching those derived by simulation. However, it should be noted that due to the underlying M|G|1 formalism, it is not possible to employ more detailed node models or to account for advanced flow control mechanisms. Also, it should be noted that only mean values are derived, and no higher moments are provided by the analysis.

The author of [165] suggests an IN performance evaluation framework based on the software performance engineering approach. A key idea of this approach is a clear distinction between the software structure and the resource demands of the individual software modules, which fits very well in the modularized and decentralized framework of modern telecommunication networks. However, the approach is only formulated on a very high level, and the authors suggest to eventually employ the approach presented in [199] (see above) in order to actually compute the performance measures of interest.

### 3.4.2 Approaches Dealing with the Mapping Problem

In [176], an approach to tackle a special case of the mapping problem in the IN context is presented. As already suggested in Section 2.4.2, the authors employ an optimization algorithm on top of a basic model to estimate network deployment costs. While the reported results are promising, it must be said that the cost model is rather simple, as it does not consider performance issues, but only focuses on (linearly increasing) node and trunk utilization costs. Nevertheless, the combined usage of linear programming and simulated annealing for finding near-optimal solutions may also be of interest for solving the mapping problem in a performance-oriented context.

In a very similar spirit, linear programming has been applied in [115] to find an optimal placement of SCPs in an IN. Furthermore, also in [115], the authors present a different approach in order to compare various overload control algorithms in SSPs and SCPs, based on the simulation of a network of G|D|1|$k$ queues. However, no solution is presented to derive the incoming traffic at these queueing nodes (which is assumed to be Poisson) from a service-oriented system description, and no end-to-end performance measures are derived. Also, as mentioned above, simple deterministic service time distributions will often not be sufficient to model complex nodes.

An interesting approach in conjunction with the mapping problem has been developed in [74]. The authors present a system to automatically reconfigure a distributed service provisioning system in order to optimize the overall performance in case of node failures or changes of the system load. This is accomplished by automatically creating alternative system configurations and evaluating their performance under to the changed environment. Once better configurations have been found, the distributed system is automatically reconfigured accordingly. The analysis of candidate configurations is based on the numerical solution of a Markov chain model (which is derived from a modular high-level description based on stochastic Petri nets, see also

Chapter 5). Despite of the high modeling flexibility of this approach, these models need to be very small in practice, since many candidate models need to be evaluated quickly to choose a reconfiguration within reasonable time limits. While this problem has not been tackled by the authors, the study indicates both the feasibility of implementing an automatic reconfiguration system and the need for fast solution methods.

### 3.4.3 Discussion

Many of the existing approaches for assessing the performance of modern telecommunication systems are based on simulation. While they are suitable to investigate arbitrarily detailed system models, all approaches experience exceedingly large simulation times as the considered models become larger and as large volumes of traffic are involved. Since a fast solution trajectory represents one of the key requirements formulated in Section 2.4.2, simulation-based approaches cannot be considered a promising candidate in our context. This is especially true in view of the fact that the wealth of system parameters needed to build detailed system models are rarely available in practice, as witnessed both by the reported simulation studies and by our own cooperation project with Philips.

Concerning the analytical approaches, computation times are rarely an issue. However, care must be taken that the modeling formalism does not neglect too many of the system properties, and that it is possible to incorporate as much information as available (as long as this is sensible to the modeler). Clearly, many of the analytical approaches mentioned above do not match this requirement. The most flexible approaches in this area are those suggested in [82] and [199]. Interestingly, both approaches are based on M|G|1 node models, and especially the first one is very closely related to our prototype approach. However, even these approaches are not sufficient to reflect even slightly more complicated node models, or traffic flows other than Poisson. Especially the latter fact has recently be shown to be too strong an approximation in many cases (see Section 6.3 for details). Furthermore, these approaches are not suited to account for flow control mechanisms among nodes.

Concerning approaches directly aiming at the mapping problem, the situation is clearly not better, since the problem under consideration is even more complex than the evaluation of a single model. Here, only the approach presented in [74] proves to be feasible, under the condition that very simple system models are involved.

## 3.5 Summary and Concluding Remarks

In this section, we have first assessed the requirements to a model-based performance evaluation environment from a practical point of view. Compared to the theoretical findings of Chapter 2, our cooperation with an industrial partner led to very similar requirements.

Based on the validated requirements, a first modeling and evaluation approach has been developed; in a case study, it has been demonstrated that all evaluation requirements which were formulated by Philips can be met.

Subsequently, an overview on existing approaches in the area of model-based performance evaluation for modern telecommunication networks has been provided; interestingly, two of the most promising approaches are closely related to our prototype environment as they are also based on the analysis of M|G|1 node models. However, in practical modeling situations the expressiveness of these frameworks will not be sufficient, especially when it comes to modeling traffic with highly correlated and non-exponentially distributed interarrival times.

With purely analytical queueing-theoretic approaches this situation will be hard to tackle; however, simulative approaches are prohibitive from a computational point of view. We therefore suggest to employ a solution which retains the queueing-theoretic benefit of a fast solution, and adds additional modeling flexibility by incorporating a state-oriented Markov-chain based description of the individual queues. The class of *quasi-birth-and-death* processes represents a family of Markov chains which exactly fits these requirements; its description, analysis, and the easy specification of large systems is the subject of the following Part II of this work. As more detailed node models will also lead to more complex traffic between individual nodes, Part III of this work is dedicated to the analysis of networks of these nodes.

# Part II

# Node Analysis

# 4. Quasi-Birth-and-Death Processes

The focus of this chapter is a special class of stochastic processes, the class of so-called *quasi-birth-and-death* (QBD) processes. From a modeler's point of view, QBD processes are particularly interesting, as they combine a high degree of modeling expressiveness with efficient methods for the numerical analysis. On the one hand, QBD processes are closely related to conventional queueing systems and allow the specification of complex interarrival and service time distributions. Actually, their structure is general enough to reflect the underlying Markov chain of a stochastic Petri net, as will be shown later in Chapter 5. On the other hand, QBD processes are of Markovian nature, enabling the development of a number of very efficient solution algorithms. The first works on this class of stochastic processes appeared in the late 1960s, with the initial publications of Evans [66] and Wallace [194]. Since then, the interest in this class of stochastic processes has steadily increased, with the extensive treatment of Neuts [152] being one of the milestones which made them a widespread and fruitful research area.

The purpose of this chapter is twofold. First, we aim at providing an overview on the most recent results in this area with respect to numerical methods for the steady-state analysis of such processes, thereby also indicating interesting areas for further research. Secondly, based on this overview, an extensive comparison is performed both from a theoretical point of view and in a practical case study in order to highlight the strengths and weaknesses of the various approaches. As QBD processes will form the computational core of the high-level modeling approach to be presented in Chapter 5, this comparison provides guidance in the selection of appropriate numerical analysis methods.

This chapter opens with a characterization of QBD processes and their associated steady-state distribution in Section 4.1. We then review the most efficient algorithms for the state-state analysis in Sections 4.2 and 4.3, each section dealing with a different family of algorithms. In Section 4.4, the suitability of these algorithms for the analysis of systems which involve batch-arrivals and -departures is investigated, as this is a situation often encountered in practical modeling situations. After providing this theoretical overview, an application-oriented comparison of the various algorithms is presented in Section 4.5. In Section 4.6, we identify prospective areas for further research, concerning both the model class and the associated solution algorithms. We then conclude the chapter with a summary in Section 4.7.

## 4.1 Definition

The aim of this section is to provide a concise introduction to QBD processes. QBD processes (also termed *QBDs* in the following) represent a special class of Markov chains, where an infinite number of states and a particular transition structure between these states are assumed. In this chapter (and throughout this work), we focus on the steady-state analysis of *continuous-time* QBD processes; nevertheless, all algorithms and results are also applicable to discrete-time QBDs.

In Section 4.1.1, we first discuss the transition structure of QBD Markov chains in an informal manner. This special transition structure implies a particular form of the generator matrix of QBD Markov chains, which is presented in Section 4.1.2 in order to characterize the steady-state solution. There, we also comment on the relation of QBDs to the more general classes of M|G|1 and G|M|1 type Markov chains. Note that our definitions represent a slight extension of those provided in [152], as we take special care for efficiently dealing with non-regular behaviour at the "boundary" of the QBD state space.

### 4.1.1 State Space and Transition Structure

In a nutshell, the main idea of QBD Markov chains is that this class of stochastic processes represents a generalization of M|M|1 queueing systems by providing an additional dimension to account for the state of a stochastic environment. This environment may, for example, be used to reflect the state of a service station or the state of a complex arrival process. Thus, in contrast to the one-dimensional state space structure of an M|M|1 Markov chain, the state space of a QBD Markov chain can be considered as a semi-infinite strip, as illustrated in Figure 4.1. Individual states of the Markov chain are identified by pairs $(i, j)$, where the second component $j \in \{0, 1, \cdots\}$ is unbounded, whereas the first can only take a finite set of values. By grouping all states of the Markov chain with the same value $j$ in so-called *levels* with index $j$, the representation shown in Figure 4.1 is obtained.

Since $i$ has to be bounded, the number of states in any level is also bounded. For a QBD process, it has to be equal to $N$ states for all levels, except for the *boundary* level (where $j = 0$), where a different number of states $N_0$ is allowed. All levels which contain $N$ states (i.e., where $j \geq 1$) are referred to as *repeating* levels; for notational convenience (see Section 4.1.2), we count repeating levels starting from 0, so repeating level 0 corresponds to $j = 1$. The first repeating level 0 is also denoted as the *border* level. We will refer to the pair $(N_0, N)$ as the *order* of the QBD process.

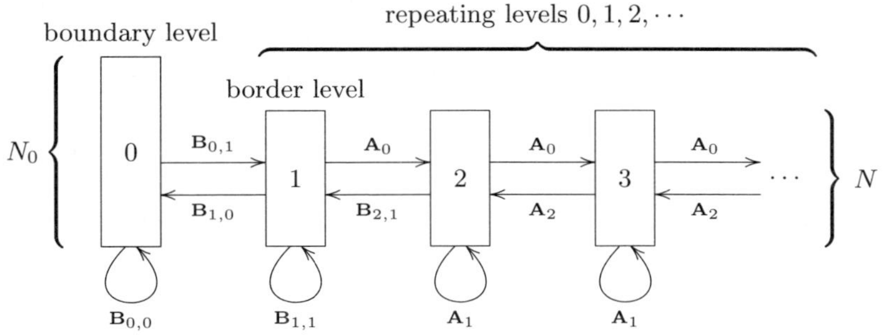


**Fig. 4.1.** States of a QBD process can be grouped in levels.

In a QBD Markov chain, positive transition rates are only allowed between states which belong to the same level, or which are located in adjacent levels.[1] Thus, transitions can only occur between states $(i_1, j)$ and $(i_2, j+d)$ with $d \in \{-1, 0, +1\}$. This enables us to describe all transitions by an (infinite) set of matrices $\mathbf{B}_{j,j+d}$ where the entry in the $i_1$-th row and $i_2$-th column contains the transition rate between states $(i_1, j)$ and $(i_2, j+d)$. Note that this is the reason for the similarity of Figure 4.1 to the transition structure of a simple M|M|1 queueing system, which coined the notion "quasi-birth-and-death".

As indicated in Figure 4.1, the transition rate matrices $\mathbf{B}_{j,j+d}$ must be constant for the repeating levels. More precisely, we have that

- $\forall j, j' \geq 1 : \mathbf{B}_{j,j+1} = \mathbf{B}_{j',j'+1} = \mathbf{A}_0$,
- $\forall j, j' \geq 2 : \mathbf{B}_{j,j} = \mathbf{B}_{j',j'} = \mathbf{A}_1$, and
- $\forall j, j' \geq 3 : \mathbf{B}_{j,j-1} = \mathbf{B}_{j',j'-1} = \mathbf{A}_2$.

Due to this fact, the entire (infinite) generator matrix of a QBD Markov chain can be described by the three matrices $\mathbf{A}_0$, $\mathbf{A}_1$ and $\mathbf{A}_2$ and the five matrices concerning the boundary and border levels occurring in Figure 4.1. In the next section, we will have a closer look at the structure of the generator matrix which results from the assumptions presented here. It will then be used to derive a concise description of the steady-state solution of QBD processes.

### 4.1.2 Generator Matrix and Steady-State Characterization

The assumption on the state space structure and the allowed transitions between states of QBD Markov chains which have been presented in the previous section imply a special structure of the (infinite) infinitesimal generator matrix of QBD processes. After describing it below, the stationary distribution can be characterized elegantly.

[1] We discuss how this requirement can be relaxed in Section 4.4.

**Generator Matrix.** Since QBD Markov chains can take an infinite number of states, its generator $\mathbf{Q}$ is a matrix of infinite size. However, due to the limitations on allowed transitions, this matrix exhibits a very regular structure.

As already indicated in Figure 4.1, all transition rates of a QBD process can be summarized in eight matrices. The matrices $\mathbf{B}_{0,0} \in \mathbb{R}^{N_0 \times N_0}$, $\mathbf{B}_{0,1} \in \mathbb{R}^{N_0 \times N}$, $\mathbf{B}_{1,0} \in \mathbb{R}^{N \times N_0}$ and $\mathbf{B}_{1,1} \in \mathbb{R}^{N \times N}$ contain the transitions rates within and between the boundary and the border level. The transitions within and between the repeating levels can be summarized in the matrices $\mathbf{A}_0, \mathbf{A}_1, \mathbf{A}_2 \in \mathbb{R}^{N \times N}$, and transitions between the border level and the second repeating level are described by matrix $\mathbf{B}_{2,1} \in \mathbb{R}^{N \times N}$.

If the states of the QBD process are sorted lexicographically, i.e., in a sequence $(0,0), \cdots, (N_0-1,0), (0,1), \cdots, (N-1,1), (0,2), \cdots$, then the generator $\mathbf{Q}$ exhibits the following block-tridiagonal structure:

$\mathbf{Q} =$

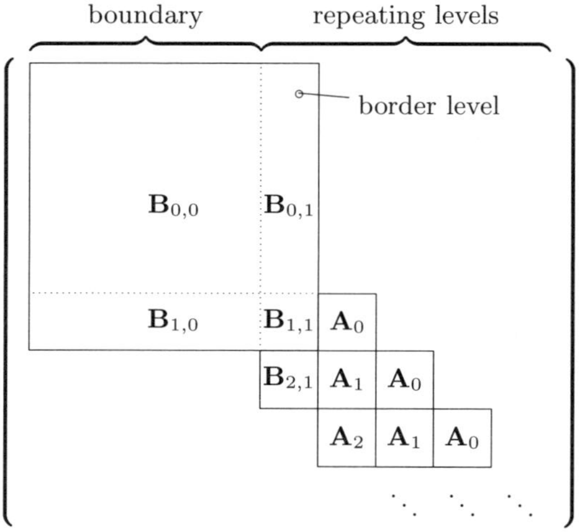
boundary
repeating levels
border level
B0,0
B0,1
B1,0
B1,1
A0
B2,1
A1
A0
A2
A1
A0

. (4.1)

It should be noted that for $N_0 = 0$ (i.e., if the QBD contains no boundary states), this definition is in accordance with the one provided in [152, p. 82]. However, by introducing an additional boundary level with $N_0$ states, our definition extends the one given there by allowing for a more complex boundary transition structure, which is needed to account for non-regular transitions at the beginning of the state space. Complex boundaries can also be accounted for by the definition given in [152, p. 82], however, only at the price of increasing $N$ to the value of $N_0$, which is computationally expensive for large boundaries. To summarize, we provide the following definition of what will be considered a QBD processes in this work:

**Definition 1.** A *quasi-birth-and-death (QBD) process* of order $(N_0, N)$ is a Markovian process on the state space $\{(i,j) \in \mathbb{N}_0^2 | (j = 0 \wedge 0 \leq i < N_0) \vee (j > 0 \wedge 0 \leq i \leq N)\}$ with a generator matrix conforming to Eq. (4.1). The tuple

$$\mathcal{Q} = (\mathbf{B}, \mathbf{B}_{2,1}, \mathbf{A}_0, \mathbf{A}_1, \mathbf{A}_2) \tag{4.2}$$

with $\mathbf{B} = \begin{pmatrix} \mathbf{B}_{0,0} & \mathbf{B}_{0,1} \\ \mathbf{B}_{1,0} & \mathbf{B}_{1,1} \end{pmatrix}$ and $\mathbf{B}_{0,0} \in \mathbb{R}^{N_0 \times N_0}, \mathbf{B}_{0,1} \in \mathbb{R}^{N_0 \times N}, \mathbf{B}_{1,0} \in \mathbb{R}^{N \times N_0}$ and $\mathbf{B}_{1,1}, \mathbf{B}_{2,1}, \mathbf{A}_0$, $\mathbf{A}_1, \mathbf{A}_2 \in \mathbb{R}^{N \times N}$ is called a *description* of a QBD process of order $(N_0, N)$.

Finally, we note that QBD processes represent a special case of two other classes of stochastic processes, namely, Markov chains of the M|G|1 and G|M|1 type. They can be considered as an extension of the embedded Markov chains of M|G|1 and G|M|1 queues, similar to the way in which QBD Markov chains represent an extension of M|M|1 queues. The (infinite) generator matrices of M|G|1 and G|M|1 type Markov chains have a structure similar to the one shown in (4.1), however, the upper or lower parts of the generator matrix may take non-zero (yet regular) entries, respectively. Thus, these matrices are of upper and lower block-Hessenberg form, respectively. While such Markov chains will not be dealt with in this chapter, many algorithms for the analysis of QBD processes are suitable for such systems as well. We will also comment on the role of such systems when dealing with non-skip-free QBDs in Section 4.4.

**Steady-State Solution.** With the generator matrix of a QBD process given in Eq. (4.1), it is possible to characterize the (infinite) steady-state probability vector $\boldsymbol{\pi} \in \mathbb{R}^{\mathbb{N}}$ as usual by the (infinite) set of global balance equations

$$\boldsymbol{\pi}\mathbf{Q} = \mathbf{0} \tag{4.3}$$

under the normalization condition

$$\boldsymbol{\pi}\mathbf{1} = 1. \tag{4.4}$$

Now consider a partitioning of the steady-state probability vector $\boldsymbol{\pi}$ according to the leveled state space structure into subvectors $\mathbf{b} \in \mathbb{R}^{N_0}$ and $\mathbf{v}_i \in \mathbb{R}^N, i = 0, 1, \cdots$, so that

$$\boldsymbol{\pi} = (\mathbf{b}, \mathbf{v}_0, \mathbf{v}_1, \cdots) \tag{4.5}$$

holds. Then $\mathbf{b}$ contains the steady-state probabilities of the states in the boundary level, and the vectors $\mathbf{v}_i$ hold the steady-state probabilities of repeating level $i$. Using this segmentation of $\boldsymbol{\pi}$, the global balance equations (4.3) can be rewritten as follows:

$$\begin{aligned} \mathbf{b}\mathbf{B}_{0,0} + \mathbf{v}_0\mathbf{B}_{1,0} &= \mathbf{0} && (4.6)\\ \mathbf{b}\mathbf{B}_{0,1} + \mathbf{v}_0\mathbf{B}_{1,1} + \mathbf{v}_1\mathbf{B}_{2,1} &= \mathbf{0} && (4.7)\\ \mathbf{v}_j\mathbf{A}_0 + \mathbf{v}_{j+1}\mathbf{A}_1 + \mathbf{v}_{j+2}\mathbf{A}_2 &= \mathbf{0} \qquad j = 0, 1, \cdots && (4.8) \end{aligned}$$

The normalization condition (4.4) can be expressed as

$$\mathbf{b1} + \sum_{j=0}^{\infty} \mathbf{v}_j \mathbf{1} = 1 \tag{4.9}$$

Evidently, the solution of Eq. (4.8) represents the main point in solving for the steady-state probability distribution, since Eqs. (4.6)–(4.7) represent a simple finite set of $N_0 + N$ linear equations. In fact, the regular structure of (4.8) is the reason why QBD processes can be solved efficiently. In the following two sections, we will discuss several approaches for solving these equations in order to derive the steady-state distribution.

## 4.2 Matrix-Geometric Solution Methods

Two different classes of solution techniques for deriving the steady-state solution of a QBD Markov chain can be distinguished. In both classes, the main focus is the solution of a matrix polynomial equation which is derived from Eq. (4.8). The first class of algorithms (which is considered in this section) directly operates on the matrices involved in this polynomial and applies probabilistic arguments in order to obtain a geometric solution form for the vectors $\mathbf{v}_j$. This solution is very similar to the (scalar) geometric solution $(1-\rho)\rho^j$ of a simple M|M|1 system, and we will thus refer to these approaches as *matrix-geometric* solution algorithms. All of these algorithms are of iterative nature: the computation ends after a sufficiently large probability mass has been considered. A second class of solution algorithms will be investigated later in Section 4.3.

After the review of some common properties which apply to all algorithms in Section 4.2.1, the individual algorithms are presented in Sections 4.2.2–4.2.4.

### 4.2.1 Preliminaries

**Solution Structure.** The main idea in the class of matrix-geometric solution approaches is that the probability vectors $\mathbf{v}_i$ have a so-called matrix-geometric form, i.e.,

$$\mathbf{v}_j = \mathbf{v}_0 \mathbf{R}^j, \quad \mathbf{R} \in \mathbb{R}^{N \times N}, \quad j = 0, 1, \cdots \tag{4.10}$$

Intuitively, this result resembles the steady-state solution of an M|M|1 queueing system, which is given as $\pi_i = (1-\rho)\rho^i$. Actually, the role of the matrix $\mathbf{R}$ is similar to that of the utilization $\rho$ in the scalar case: it determines how to derive the "next" probability (vector) from the previous one. Also, for stable systems, the spectral radius of the matrix $\mathbf{R}$ (i.e., its largest eigenvalue, denoted by $\mathrm{sp}(\mathbf{R})$) must be smaller than 1.

A formal discussion why this matrix-geometric solution form can be assumed can be found in [66, 152]; a very intuitive proof is also provided in [127].

In [152], it is also shown that $\mathbf{R}$ is the entry-wise smallest non-negative solution of the quadratic matrix equation

$$\mathbf{A}_0 + \mathbf{R}\mathbf{A}_1 + \mathbf{R}^2\mathbf{A}_2 = \mathbf{0}. \tag{4.11}$$

It is easily verified that a matrix $\mathbf{R}$ conforming to (4.11) yields a matrix-geometric solution which satisfies the global balance equations for the repeating levels (4.8). Using Eq. (4.10), they can be transformed to

$$\mathbf{v}_j\mathbf{A}_0 + \mathbf{v}_j\mathbf{R}\mathbf{A}_1 + \mathbf{v}_j\mathbf{R}^2\mathbf{A}_2 = \mathbf{0}, \qquad j = 0, 1, \cdots \tag{4.12}$$

$$\mathbf{v}_j(\mathbf{A}_0 + \mathbf{R}\mathbf{A}_1 + \mathbf{R}^2\mathbf{A}_2) = \mathbf{0}, \qquad j = 0, 1, \cdots \tag{4.13}$$

Obviously, the latter condition is fulfilled if $\mathbf{R}$ is chosen according to (4.11).

Thus, the core of the matrix-geometric solution algorithm is to solve (4.11) for an appropriate $\mathbf{R}$. The remainder of this section is devoted to the presentation of the most efficient approaches to perform this task.

**Boundary Equations.** In order to obtain the overall stationary distribution, also the boundary equations (4.6)–(4.7) have to be considered. If $\mathbf{R}$ has already been determined, $\mathbf{v}_1$ in (4.7) can be replaced by $\mathbf{v}_0\mathbf{R}$, yielding

$$\mathbf{b}\mathbf{B}_{0,1} + \mathbf{v}_0\mathbf{B}_{1,1} + \mathbf{v}_0\mathbf{R}\mathbf{A}_2 = \mathbf{0}. \tag{4.14}$$

Together with boundary equation (4.6), this equals in matrix form:

$$(\mathbf{b}, \mathbf{v}_0)\begin{pmatrix} \mathbf{B}_{0,0} & \mathbf{B}_{0,1} \\ \mathbf{B}_{1,0} & \mathbf{B}_{1,1} + \mathbf{R}\mathbf{A}_2 \end{pmatrix} = \mathbf{0} \tag{4.15}$$

The complete stationary distribution can be obtained by solving the set of equations (4.15) under the normalization condition (4.4). Using Eq. (4.10), the latter one can be transformed to

$$\mathbf{b}\mathbf{1} + \sum_{j=0}^{\infty} \mathbf{v}_0\mathbf{R}^j\mathbf{1} = 1 \tag{4.16}$$

$$\mathbf{b}\mathbf{1} + \mathbf{v}_0(\mathbf{I} - \mathbf{R})^{-1}\mathbf{1} = 1. \tag{4.17}$$

The last transformation takes advantage of the fact that for $\mathrm{sp}(\mathbf{R}) < 1$, the relation $\sum_{j=0}^{\infty} \mathbf{R}^j = (\mathbf{I} - \mathbf{R})^{-1}$ holds, similar to the scalar case.

The computational effort[2] which is needed for the solution of the boundary equation system amounts to $\mathcal{O}\left(\frac{1}{3}N^3\right)$ operations to derive $(\mathbf{I}-\mathbf{R})^{-1}\mathbf{1}$ in Eq. (4.17) (see next paragraph for details) and $\mathcal{O}\left(N^3\right)$ operations for computing $\mathbf{RA}_2$ in Eq. (4.15), plus $\mathcal{O}\left(\frac{1}{3}(N_0+N)^3\right)$ operations for solving Eq. (4.15), thus yielding a total of $\mathcal{O}\left(\frac{1}{3}(N_0+N)^3+\frac{4}{3}N^3\right)$ operations.

**An Efficient "Trick".** Many of the algorithms in this chapter involve the computation of a matrix product

$$\mathbf{X} := \mathbf{Y}^{-1}\mathbf{Z}. \tag{4.18}$$

A common approach for speeding up the computation of $\mathbf{X}$ is to avoid both the explicit inversion of $\mathbf{Y}$ and the following matrix multiplication, which would involve $\mathcal{O}\left(2\cdot N^3\right)$ operations. Instead, $\mathbf{X}$ can also be derived by solving the set of equations $\mathbf{YX}=\mathbf{Z}$ for $\mathbf{X}$. This is done by performing a LU decomposition of $\mathbf{Y}$ to obtain matrices $\mathbf{L}$ and $\mathbf{U}$ with $\mathbf{LU}=\mathbf{Y}$, and doing a forward/backward substitution step for each right hand side of $\mathbf{LUX}=\mathbf{Z}$. For each right hand side (i.e., column of $\mathbf{Z}$), forward and backward substitution involve $\mathcal{O}\left(\frac{1}{2}N^2\right)$ operations each, totalling to $\mathcal{O}\left(N^3\right)$ operations to derive $\mathbf{X}$. Together with $\mathcal{O}\left(\frac{1}{3}N^3\right)$ operations needed for the LU decomposition of $\mathbf{Y}$, this yields a total operation count of $\mathcal{O}\left(\frac{4}{3}N^3\right)$ (for a derivation of the mentioned operation counts, see e.g. [164]).

Furthermore, in the algorithms presented in the following, assignments of the form (4.18) have to be accomplished several times for different matrices $\mathbf{Z}$, but for the same matrix $\mathbf{Y}$. In this case, the LU decomposition of $\mathbf{Y}$ has only to be accomplished once, of course. For two such assignments, the overall effort is thus only $\mathcal{O}\left(\frac{7}{3}N^3\right)$ operations.

### 4.2.2 The Successive Substitution (SS) Method

This approach has initially been suggested in [152, p. 13]. A recursive expression for $\mathbf{R}$ is directly derived from Eq. (4.11) by transforming it to

$$\mathbf{R} = -(\mathbf{A}_0+\mathbf{R}^2\mathbf{A}_2)\mathbf{A}_1^{-1}. \tag{4.19}$$

The matrix $\mathbf{A}_1$ is always nonsingular for irreducible QBD processes [152]. Starting with $\mathbf{R}=\mathbf{0}$, it is shown in [152] that the iterative application of the right hand side of this equation monotonically converges to the desired matrix $\mathbf{R}$. As a stopping criterion, one may choose that the difference between successive values of $\mathbf{R}$ becomes sufficiently small. Alternatively, one could exploit the accuracy checks proposed in [152], i.e., $(\mathbf{B}_{1,1}+\mathbf{RB}_{1,0})\mathbf{1}=1$ or $(\mathbf{A}_0-\mathbf{RA}_2)\mathbf{1}=0$.

As each iteration involves three matrix multiplications, the complexity of the algorithm is $\mathcal{O}\left(\frac{9}{3}N^3\right)$ per iteration; if the "trick" for Eq. (4.18) is used,

[2] Here and in the rest of this work, the effort relates to the number of required floating point multiplications and divisions.

the additional effort outside the main iteration loop amounts to $\mathcal{O}\left(\frac{1}{3}N^3\right)$ operations for performing the LU decomposition of $\mathbf{A}_1$. An improved version of the SS procedure has been suggested as algorithm "U" in [128], which involves only $\mathcal{O}\left(\frac{7}{3}N^3\right)$ operations per iteration, plus the same number of operations outside the main iteration loop. Both algorithms can also be applied to G|M|1 type matrices, and analogous algorithms for M|G|1 type matrices have been proposed in [128, 153].

While the successive substitution algorithm was one of the first approaches available to compute $\mathbf{R}$, it is also one of the slowest ones. The reason for this is the low convergence rate of the algorithm, which decreases at higher traffic intensities, and which effectively eliminates the gain of a relatively low operation count per iteration. In many cases, several thousand iterations are necessary to satisfy a sensible accuracy condition. Even for moderately sized systems, this is clearly not feasible, and one should resort to the much more efficient approaches presented in the following.

### 4.2.3 The Logarithmic Reduction (LR) Approach

The LR algorithm has been proposed in [129]. Similar to the above-mentioned algorithm "U", it does not directly compute the matrix $\mathbf{R}$, but instead focuses on the related matrix $\mathbf{G}$, which is the solution of the following "dual" equation to (4.11):

$$\mathbf{A}_0 + \mathbf{A}_1\mathbf{G} + \mathbf{A}_2\mathbf{G}^2 = \mathbf{0}. \tag{4.20}$$

According to [126, 129], this matrix is related to $\mathbf{R}$ by the equation

$$\mathbf{R} = -\mathbf{A}_0(\mathbf{A}_1 + \mathbf{A}_0\mathbf{G})^{-1}. \tag{4.21}$$

The advantage of considering $\mathbf{G}$ instead of $\mathbf{R}$ is that its entries have a very intuitive probabilistic interpretation of $\mathbf{G}$,[3] which forms the basis for deriving the LR algorithm: starting from a state $(i, 1)$, the entry in row $i$ and column $i'$ denotes the probability that the first transition of the QBD process to level 0 is accomplished by entering state $(i', 0)$. The key idea of the LR algorithm is to successively approximate $\mathbf{G}$ under the condition that no levels above level $k = 1, 2, 4, 8, \cdots$ are visited. This exponential increase is in contrast to the SS and "U" algorithms, where the number of considered levels increases in a linear fashion. Consequently, the number of iterations required by the LR algorithm is only the logarithm of the number of iterations required by the SS and "U" algorithms. In [129], it has been shown that the rate of convergence to the actual matrix $\mathbf{G}$ is quadratic.

The LR algorithm doubles the number of considered levels in each iteration by performing the following steps (see [129] for details):

[3] Note that there also exists a stochastic interpretation of $\mathbf{R}$; however, especially in the continuous-time case, it is not as straightforward and involves the expected sojourn times in individual states. For details, see [152, p. 35].

1. **(Embedding)** First, the embedded QBD process is considered which is obtained by focusing on those time instants where a transition between levels occurs. This leads to an (infinite) block-tridiagonal transition probability matrix with a zero block diagonal, where transitions to upper and lower levels are described by the matrices $\mathbf{B}_0 = -\mathbf{A}_1^{-1}\mathbf{A}_0$ and $\mathbf{B}_2 = -\mathbf{A}_1^{-1}\mathbf{A}_2$, respectively. The $\mathbf{G}$ matrix of this discrete-time QBD is equivalent to the original $\mathbf{G}$.
2. **(Reduction)** Now, this QBD process is further restricted to only observing transitions which enter or leave the *even*-numbered levels. Here, the fact is used that the $\mathbf{G}$ matrix of this new process is given by $\mathbf{G}^2$. The transition probability matrix of this QBD process has again the usual block-tridiagonal structure, so that the first step can be applied again. Note that the restriction to the even-numbered levels is the key point which leads to the quadratic convergence rate of the LR approach.

After the reduction step, updated matrices $\mathbf{B}_0$ and $\mathbf{B}_2$ for the repeated application of the first step to the reduced process can be derived directly from their previous values by

$$\mathbf{B}_0' = \left(\mathbf{I} - \mathbf{B}_0\mathbf{B}_2 - \mathbf{B}_2\mathbf{B}_0\right)^{-1}\mathbf{B}_0^2, \qquad \text{and} \tag{4.22}$$

$$\mathbf{B}_2' = \left(\mathbf{I} - \mathbf{B}_0\mathbf{B}_2 - \mathbf{B}_2\mathbf{B}_0\right)^{-1}\mathbf{B}_2^2. \tag{4.23}$$

The quadratic relationship between the $\mathbf{G}$ matrices which appears in the second step is used to derive the main equation

$$\mathbf{G} = \sum_{k=0}^{\infty}\left(\prod_{i=0}^{k-1}\mathbf{B}_0^{(i)}\right)\mathbf{B}_2^{(k)}, \tag{4.24}$$

where the upper index $(i)$ of the matrices denotes their value after the $i$-th iteration. Note that the empty matrix product is assumed to be $\mathbf{I}$. The algorithm for implementing this procedure is given in Figure 4.2. The stopping criterion uses the probabilistic interpretation of $\mathbf{G}$ (i.e., the row sums must be equal to 1).

Each iteration of the LR algorithm involves two assignments of the form (4.18) and 6 additional matrix multiplications, thus, the total number of operations per iteration is given by $\mathcal{O}\left(\frac{25}{3}N^3\right)$. This is considerably more than what was needed for the SS algorithm, however, the number of iterations is drastically reduced. Usually, less than 10 iterations are needed to reach a sensible precision level (and, according to [129], models of practical relevance which need more than 40 iterations are hard to imagine). To reach the same precision, the "U" algorithm would take $\mathcal{O}\left(2^{10}\cdot\frac{7}{3}N^3\right) = \mathcal{O}\left(\frac{7168}{3}N^3\right)$ operations, as compared to $\mathcal{O}\left(\frac{250}{3}N^3\right)$ operations in the LR case (only accounting for the main iteration loop). Outside the main iteration loop, the LR approach involves $\mathcal{O}\left(\frac{14}{3}N^3\right)$ operations ($\mathcal{O}\left(\frac{7}{3}N^3\right)$ operations both before and after the main iteration loop).

| $\mathbf{B}_0 := -\mathbf{A}_1^{-1}\mathbf{A}_0, \mathbf{B}_2 := -\mathbf{A}_1^{-1}\mathbf{A}_2$ |
|---|
| $\mathbf{G} := \mathbf{B}_2,\ \mathbf{T} := \mathbf{B}_0$ |
| While $\|\|\mathbf{1} - \mathbf{G}\mathbf{1}\|\|_\infty \geq \epsilon$ |
|      $\mathbf{D} := \mathbf{I} - \mathbf{B}_0\mathbf{B}_2 - \mathbf{B}_2\mathbf{B}_0$ |
|      $\mathbf{B}_0 := \mathbf{D}^{-1}\mathbf{B}_0^2,\ \mathbf{B}_2 := \mathbf{D}^{-1}\mathbf{B}_2^2$ |
|      $\mathbf{G} := \mathbf{G} + \mathbf{T}\mathbf{B}_2,\ \mathbf{T} := \mathbf{T}\mathbf{B}_0$ |
| $\mathbf{R} := -\mathbf{A}_0(\mathbf{A}_1 + \mathbf{A}_0\mathbf{G})^{-1}$ |

**Fig. 4.2.** The logarithmic reduction algorithm of Latouche and Ramaswami.

In conjunction with its numerical stability, the efficiency of the LR algorithm made it the algorithm of choice for the solution of QBD models for a long time. It should be noted that the algorithm is still sensitive to the traffic intensity (as the probability mass "shifts" to the higher levels at higher intensities), however, the impact is reduced due to the ability of the approach to account for more levels for computing $\mathbf{G}$ in an exponential fashion.

### 4.2.4 Naoumov's Improved LR Algorithm

An improvement of the LR algorithm has recently been proposed in [148]. While retaining the ideas of the LR approach, the computational effort per iteration is reduced by employing a factorization of the matrices $\mathbf{B}_0^{(k)}$ and $\mathbf{B}_2^{(k)}$. To obtain this factorization, the authors define three matrices sequences

$$\mathbf{N}^{(0)} = \mathbf{A}_1, \quad \mathbf{N}^{(k+1)} = \mathbf{N}^{(k)} - \mathbf{M}^{(k)} \underbrace{\left(\mathbf{N}^{(k)}\right)^{-1}\mathbf{L}^{(k)}}_{=:\mathbf{X}^{(k)}} - \mathbf{L}^{(k)} \underbrace{\left(\mathbf{N}^{(k)}\right)^{-1}\mathbf{M}^{(k)}}_{=:\mathbf{Y}^{(k)}}, \tag{4.25}$$

$$\mathbf{L}^{(0)} = \mathbf{A}_0, \quad \mathbf{L}^{(k+1)} = -\mathbf{L}^{(k)}\left(\mathbf{N}^{(k)}\right)^{-1}\mathbf{L}^{(k)}, \tag{4.26}$$

$$\mathbf{M}^{(0)} = \mathbf{A}_2, \quad \mathbf{M}^{(k+1)} = -\mathbf{M}^{(k)}\left(\mathbf{N}^{(k)}\right)^{-1}\mathbf{M}^{(k)}, \tag{4.27}$$

and show that $\mathbf{B}_0^{(k)} = -\mathbf{X}^{(k)}$ and $\mathbf{B}_2^{(k)} = -\mathbf{Y}^{(k)}$ holds. This fact is used to derive the matrix

$$\mathbf{W} = \mathbf{A}_0 \sum_{k=0}^{\infty} \left( \prod_{i=0}^{k-1} \mathbf{B}_0^{(i)} \right) \mathbf{B}_2^{(k)} = \sum_{k=0}^{\infty} \mathbf{L}^{(k)}\left(\mathbf{N}^{(k)}\right)^{-1}\mathbf{M}^{(k)}. \tag{4.28}$$

According to Eq. (4.24), this is equivalent to

| $\mathbf{N} := \mathbf{A}_1$, $\mathbf{L} := \mathbf{A}_0$, $\mathbf{M} := \mathbf{A}_2$ |
|---|
| $\mathbf{X} := -\mathbf{N}^{-1}\mathbf{L}$, $\mathbf{Y} := -\mathbf{N}^{-1}\mathbf{M}$, $\mathbf{Z} := \mathbf{L}\mathbf{Y}$ |
| $\mathbf{W} := \mathbf{Z}$ |
| While $\|\|\mathbf{A}_0\mathbf{1} - \mathbf{W}\mathbf{1}\|\|_\infty \geq \epsilon$ |
|     $\mathbf{N} := \mathbf{N} + \mathbf{Z} + \mathbf{M}\mathbf{X}$, $\mathbf{L} := \mathbf{L}\mathbf{X}$, $\mathbf{M} := \mathbf{M}\mathbf{Y}$ |
|     $\mathbf{X} := -\mathbf{N}^{-1}\mathbf{L}$, $\mathbf{Y} := \mathbf{N}^{-1}\mathbf{M}$, $\mathbf{Z} := \mathbf{L}\mathbf{Y}$ |
|     $\mathbf{W} := \mathbf{W} + \mathbf{Z}$ |
| $\mathbf{R} := -\mathbf{A}_0(\mathbf{A}_1 + \mathbf{W})^{-1}$ |

**Fig. 4.3.** Naoumov's improved LR algorithm.

$$\mathbf{W} = \mathbf{A}_0\mathbf{G}, \tag{4.29}$$

which can be used in Eq. (4.21) to derive

$$\mathbf{R} = -\mathbf{A}_0(\mathbf{A}_1 + \mathbf{W})^{-1}. \tag{4.30}$$

Though the three matrix sequences (4.25)–(4.27) look complicated, the sum (4.28) can be computed in an efficient manner. First, note that the terms $\mathbf{L}^{(k)}\left(\mathbf{N}^{(k)}\right)^{-1}\mathbf{M}^{(k)}$ which form the infinite sum (4.28) are available "for free", as they are also used in the computation of $\mathbf{N}^{(k+1)}$. Secondly, the matrices $\mathbf{X}^{(k)}$ and $\mathbf{Y}^{(k)}$ can be computed efficiently with $\mathcal{O}\left(\frac{7}{3}N^3\right)$ operations. Also, these matrices can be reused in the computation of $\mathbf{L}^{(k+1)}$ and $\mathbf{M}^{(k+1)}$, so that merely 4 additional matrix multiplications are necessary in each iteration step, leading to a total number of $\mathcal{O}\left(\frac{19}{3}N^3\right)$ operations per iteration, which is about 24% less than what is required by the original LR algorithm. The corresponding algorithm is summarized in Figure 4.3. As can be observed, the operation count outside the main iteration loop equals $\mathcal{O}\left(\frac{16}{3}N^3\right)$ (i.e., $\mathcal{O}\left(\frac{10}{3}N^3\right)$ operations at the start of the algorithm and $\mathcal{O}\left(\frac{4}{3}N^3\right)$ at its end).

It should be emphasized that this algorithm no longer explicitly computes the matrix $\mathbf{G}$, but instead derives the product $\mathbf{W} = \mathbf{A}_0\mathbf{G}$. While this is irrelevant for the computation of $\mathbf{R}$ (as its derivation only involves the product $\mathbf{A}_0\mathbf{G}$, see Eq. (4.21)), this must be accounted for when formulating the stopping criterion for the iterative procedure. In our experiments, we used the fact that $\mathbf{A}_0\mathbf{G}\mathbf{1} = \mathbf{W}\mathbf{1}$ converges to $\mathbf{A}_0\mathbf{1}$ since $\mathbf{G}$ is stochastic. With this criterion, the algorithm took the same number of iterations as the LR algorithm in most of our experiments, which was desirable for the numerical comparisons presented in Section 4.5. However, this criterion clearly depends on $\mathbf{A}_0$, and thus one might prefer to take the change of $\mathbf{W}$ between iteration steps as a stopping criterion, as it has been proposed in [148]. It is worth noting that it is not feasible to "extract" the factor $\mathbf{A}_0$ from $\mathbf{W}$ by simply setting the initial factor $\mathbf{L}_0$ to $\mathbf{I}$, since this also changes the derivation of Eq. (4.28) to a form where an additional matrix multiplication with $\mathbf{A}_0$ is needed in each iteration.

Otherwise, the fact that $\mathbf{G}$ is not computed explicitly does not represent a shortcoming of this algorithm, since $\mathbf{G}$ can be derived at the end of the algorithm by using the identity (see [126])

$$\mathbf{G} = -(\mathbf{A}_1 + \mathbf{W})^{-1}\mathbf{A}_0. \tag{4.31}$$

## 4.3 Transform Methods

In this second section on numerical methods for deriving the steady-state solution of QBD Markov chains, we focus on another family of solution methods. In contrast to the matrix-geometric approaches discussed in Section 4.2, the algorithms in this section are not solely based on probabilistic arguments to directly solve the matrix polynomial equation (4.11), but instead employ a transformation to some other domain in order to derive a solution of the balance equations (4.8). In the following, we thus refer to this class of algorithms as *transform methods.*

In the subsequent sections, we consider the most powerful solution algorithms in this area: the cyclic reduction approach (Section 4.3.1), the invariant subspace algorithm (Section 4.3.2), and the spectral expansion method (Section 4.3.3). Special attention will be paid to the spectral expansion method, as it involves a different form of the boundary equation system. Furthermore, we clarify the relation of the spectral expansion solution to matrix-geometric solutions which are expressed in terms of the matrix $\mathbf{R}$.

### 4.3.1 The Cyclic Reduction Method

The cyclic reduction method proposed in [26] employs results from the theory of block Toeplitz matrices to efficiently derive the steady-state solution vector of a QBD process. Block Toeplitz matrices are matrices where the block entries may only depend on the difference of the column and the row index where a block is found. QBD generator matrices are a special case of block Toeplitz matrices where all block entries with an index difference other than $-1, 0$ or $1$ are equal to null matrices, i.e., they can be interpreted as block-tridiagonal block Toeplitz matrices. Also, generator matrices of M|G|1 and G|M|1 type are block Toeplitz (in this case, all entries with index differences smaller than $-1$ or larger than $1$ are null), thus, the cyclic reduction approach is suited for this type of Markov chains as well.

The cyclic reduction approach uses two key results. The first result yields a recursive algorithm to derive a series of LU decomposition of the generator matrix, where the matrices forming the decompositions become successively smaller. At the end, a final system of $N$ equations has to be solved, and the overall steady-state solution vector is obtained by successively performing backsubstitution for all of the preceding decomposition steps.

The second key result is used to derive the LU decomposition in each recursion step. The rather complex matrix operations in each step can be accomplished in a very efficient manner, since all matrix inversions and multiplications can be reduced to operations on triangular Toeplitz matrices. Such matrices have an elegant representation in the form of a power series, and operations on them can be performed efficiently by employing discrete Fourier transforms.

For QBD processes, however, this second result is not even necessary. In this case, the cyclic reduction algorithm leads to an iterative procedure which is identical to Naoumov's method [27]. We will thus not further comment on this approach, and instead refer the reader to the results for Naoumov's improved LR algorithm. It is, however, worth noting that cyclic reduction represents an efficient algorithm for more general stochastic processes than just QBD processes (see also our remarks in Section 4.4.2).

### 4.3.2 The Invariant Subspace Approach

The invariant subspace approach proposed in [8] is suitable for the analysis of M|G|1 and G|M|1 type Markov chains, and thus also applies to the class of QBD models considered here (see [7] for a discussion of the QBD case). The key idea of this approach is to formulate the matrix-geometric solution matrix $\mathbf{R}$ in terms of a particular invariant subspace $S \subseteq \mathbb{R}^{2N}$ of a matrix $\mathbf{E}_m^{\mathrm{T}} \in \mathbb{R}^{2N \times 2N}$.

A set $S \subseteq \mathbb{R}^{2N}$ is called an invariant subspace of $\mathbf{E}_m^{\mathrm{T}}$, if $\mathrm{Img}_S\, \mathbf{E}_m^{\mathrm{T}} \subseteq S$. In this case, there exists a matrix $\mathbf{E}'$ such that $\mathbf{E}_m^{\mathrm{T}}\mathbf{S} = \mathbf{S}\mathbf{E}'$, where the columns of the matrix $\mathbf{S}$ form a basis for the subspace $S$. The subspace $S$ of interest is the largest one for which the real part of all eigenvalues of $\mathbf{E}'$ is *negative*; therefore, it is called the *left* invariant subspace.

The computation of this left invariant subspace represents the computational main part of this approach. While there are several approaches to perform this task, the authors suggest to evaluate the *matrix sign function* [168]. The matrix sign function has the special property that if the matrix $\mathbf{Z}$ denotes the matrix sign of a matrix $\mathbf{M} \in \mathbb{R}^{N \times N}$, then $\mathrm{Img}_{\mathbb{R}^N}(\mathbf{Z} - \mathbf{I})$ provides the left invariant subspace of $\mathbf{M}$. To compute $\mathbf{R}$, a basis of the invariant subspace has to be derived; this is accomplished by QR decomposition of $\mathbf{Z} - \mathbf{I}$.

For the actual computation of the matrix sign of a matrix $\mathbf{M}$, the authors suggest a simple iterative procedure with quadratic convergence. Each iteration basically involves the inversion of the last iterate. Since the in-

| $\mathbf{A} := \mathbf{A}_0 + \mathbf{A}_1 + \mathbf{A}_2$ |
|---|
| Solve $\mathbf{xA} = \mathbf{x}$ for $\mathbf{x}$ |
| $\mathbf{H}_0 := \mathbf{A} - \mathbf{I}$, $\mathbf{H}_1 := 2(\mathbf{A}_2 - \mathbf{A}_1)$, $\mathbf{H}_2 := \mathbf{A}_0 - \mathbf{A}_1 + \mathbf{A}_2 + \mathbf{I}$ |
| $\mathbf{a} := (\mathbf{x}, \mathbf{0})$, $\mathbf{b} := ((\mathbf{H}_1\mathbf{1})^{\mathrm{T}}, (\mathbf{H}_2\mathbf{1})^{\mathrm{T}})$ |
| $\mathbf{E}_m := \begin{pmatrix} \mathbf{0} & -\mathbf{H}_0\mathbf{H}_2^{-1} \\ \mathbf{I} & -\mathbf{H}_1\mathbf{H}_2^{-1} \end{pmatrix} + \frac{1}{\mathbf{ab}^{\mathrm{T}}}\mathbf{b}^{\mathrm{T}}\mathbf{a}$ |
| Let $\mathbf{Z} := \mathbf{E}_m^{\mathrm{T}}$, and iteratively compute its matrix sign function: |
| $c := \lvert \det(\mathbf{Z}) \rvert^{1/N}$ |
| $\mathbf{Z} := \frac{1}{2c}\mathbf{Z} + \frac{c}{2}\mathbf{Z}^{-1}$ |
| Repeat until $\lVert\mathbf{Z}^{-1}\rVert_1 / \lVert\mathbf{Z}\rVert_1 < \epsilon$ |
| QR-factorize $\mathbf{Z} - \mathbf{I}$ to derive a basis for the left invariant subspace: $\mathbf{QRP} := \mathbf{Z} - \mathbf{I}$, with the orthogonal matrix $\mathbf{Q} = \begin{pmatrix} \mathbf{T}_1 & \cdot \\ \mathbf{T}_2 & \cdot \end{pmatrix}$ |
| $\mathbf{R} := ((\mathbf{T}_1 + \mathbf{T}_2)(\mathbf{T}_1 - \mathbf{T}_2)^{-1})^{\mathrm{T}}$ |

**Fig. 4.4.** The invariant subspace algorithm.

volved matrices have dimension $2N$, the complexity of each iteration step is $\mathcal{O}\left(\frac{24}{3}N^3\right)$. For an illustration of the overall algorithm, see Figure 4.4.

The computational effort outside the main iteration loop is larger than in the previous approaches, as it involves an additional QR decomposition of a $2N$ dimensional matrix. According to [164], QR decomposition involves about twice as many operations as LU decomposition, thus leading to an $\mathcal{O}\left(\frac{16}{3}N^3\right)$ operation count for a $2N$ dimensional matrix. Together with the remaining operations, this leads to a total effort outside the main iteration loop of $\mathcal{O}\left(\frac{28}{3}N^3\right)$ operations. The number of matrix sign function iterations for reaching a certain precision level does not depend on the traffic intensity, as opposed to the matrix-geometric approaches presented previously. Furthermore, it should be noted that the invariant subspace approach is also suitable to analyze *finite* M|G|1 and G|M|1 models at very low additional cost (this involves only an additional QR decomposition of $\mathbf{Z} + \mathbf{I}$ to derive the *right* invariant subspace). Also, it should be mentioned that a number of other approaches are available for the computation of the matrix sign function, both differing in the convergence rate and the complexity of an iteration. Recently, an alternative to the matrix sign function approach has been applied in [6], which proved to be faster for M|G|1 systems. As a thorough comparison of algorithms for the computation of invariant subspaces is beyond the scope of this work, we resort to computing the matrix sign function with the iterative approach taken by the authors of [8] in the tool telpack [2].

### 4.3.3 The Spectral Expansion Method

In contrast to the solution approaches presented so far, the spectral expansion approach does not lead to a geometric solution in the form $\mathbf{v}_j = \mathbf{v}_0\mathbf{R}^j$. Instead, the solution exhibits a component-wise geometric structure, which is more handy and more efficient for computing the steady-state probabilities of individual states of the QBD process.

Basically, the spectral expansion method can be considered as an approach for solving the matrix geometric main polynomial (4.11) in terms of the eigenvalues of the matrix $\mathbf{R}$. A first discussion of this approach can be found in [152, Section 1.6]. A more detailed description of the solution method and a first application is available in [71], where the method is termed *generalized eigenvalue approach.* For a thorough treatment of the spectral expansion method, including its extension to finite QBD systems and a comparison to the SS method, see [41].

In the next subsections, we will first introduce to the main idea of the spectral expansion approach, thereby also discussing computational issues. As the representation of the solution vectors also has implications on the solution of the boundary equations, this will be addressed as well. The presentation of the spectral expansion method is concluded with a discussion of the relation between the matrix-geometric solution form and its form in the spectral expansion case.

**Solution for the Repeating Levels.** The key idea of the spectral expansion approach is to consider the balance equations for the repeating levels, as given by Eq. (4.8), as a homogeneous (vector) difference equation with constant coefficients.

The solution of a vector difference equation (or, equivalently, a set of scalar difference equations) is accomplished by investigating its characteristic polynomial, which is given by

$$\mathbf{A}_0 + \lambda\mathbf{A}_1 + \lambda^2\mathbf{A}_2. \tag{4.32}$$

We now consider the quadratic eigenvalue problem[4] associated with this polynomial, defined as

[4] Polynomial eigenvalue problems represent a special class of eigenvalue problems, which can either be reduced to (standard) linear eigenvalue problems, or which can be solved directly by specialized methods. See Appendix A.1 for a further discussion of this topic.

$$\boldsymbol{\psi}(\mathbf{A}_0 + \lambda\mathbf{A}_1 + \lambda^2\mathbf{A}_2) = \mathbf{0}, \tag{4.33}$$

where $\lambda$ denotes an eigenvalue and $\boldsymbol{\psi} \neq \mathbf{0}$ the corresponding (left) eigenvector. As shown in Appendix A.1, this polynomial eigenvalue problem of dimension $N$ can be transformed to a $2N$-dimensional linear eigenvalue problem of the form

$$(\boldsymbol{\psi}, \boldsymbol{\phi}) \begin{pmatrix} \mathbf{0} & -\mathbf{A}_0\mathbf{A}_2^{-1} \\ \mathbf{I} & -\mathbf{A}_1\mathbf{A}_2^{-1} \end{pmatrix} = (\boldsymbol{\psi}, \boldsymbol{\phi})\lambda. \tag{4.34}$$

If an eigenvalue $\lambda$ and its corresponding eigenvector $\boldsymbol{\psi}$ have been found (both possibly being complex), then a solution of the original difference equation (4.8) has also been found. This is readily verified by the fact that (trivially)

$$\boldsymbol{\psi}\mathbf{A}_0 + \boldsymbol{\psi}\lambda\mathbf{A}_1 + \boldsymbol{\psi}\lambda^2\mathbf{A}_2 = \mathbf{0} \tag{4.35}$$

holds, *and* that this equation holds when multiplied by any power of the eigenvalue $\lambda$:

$$\boldsymbol{\psi}\lambda^j\mathbf{A}_0 + \boldsymbol{\psi}\lambda^{j+1}\mathbf{A}_1 + \boldsymbol{\psi}\lambda^{j+2}\mathbf{A}_2 = \mathbf{0} \qquad j = 0, 1, \cdots \tag{4.36}$$

When comparing the coefficients of (4.36) with those of the original set of balance equations (4.8), it becomes clear that

$$\mathbf{v}_j = \boldsymbol{\psi}\lambda^j \tag{4.37}$$

represents a valid solution to the set of balance equations.

Of course, *each* eigenvalue and associated eigenvector of (4.33) yields a valid solution. However, only those eigenvectors with $|\lambda| < 1$ need to be considered, since otherwise the normalization condition (4.4) cannot be satisfied, as the series (4.37) does not converge. Thus, if $\lambda_1, \cdots, \lambda_c$ are the eigenvectors of (4.33) strictly inside the unit disk (counting multiplicities), and $\boldsymbol{\psi}_1, \cdots, \boldsymbol{\psi}_c$ are their corresponding eigenvectors, then the overall solution space is given by linear combinations of the individual solutions:

$$\mathbf{v}_j = \sum_{k=1}^{c} x_k \boldsymbol{\psi}_k \lambda_k^j. \tag{4.38}$$

The vector $\mathbf{x} = (x_1, \cdots, x_c)$ holds the (complex) coefficients of a linear combination. If the rows of matrix $\mathbf{E}$ hold the eigenvectors $\boldsymbol{\psi}_1, \cdots, \boldsymbol{\psi}_c$, then the solution vectors can also be expressed in matrix form as

$$\mathbf{v}_j = \mathbf{x}\underbrace{\mathrm{diag}(\lambda_1^j, \cdots, \lambda_c^j)\mathbf{E}}_{\mathbf{E}_j}. \tag{4.39}$$

For notational convenience, we will refer to the product $\mathrm{diag}(\lambda_1^j, \cdots, \lambda_c^j)\mathbf{E}$ as $\mathbf{E}_j$ in the following. The role of the coefficient vector $\mathbf{x}$ is similar to the border probability vector $\mathbf{v}_0$ in the matrix geometric case. Similarly, it is derived from the solution of the boundary equation system, which is discussed below.

The numerical complexity of solving the eigenvalue problem clearly depends on the employed solution method. In our implementation, a standard QR composition based solution algorithm is employed to solve the linearized eigenvalue problem (4.34). This involves the following main steps:

1. Transform the coefficient matrix in Eq. (4.34) to upper Hessenberg form (zero below subdiagonal) by successive application of Householder transformations.
2. Use successive QR decompositions to transform the matrix to Schur form (only $1 \times 1$ and $2 \times 2$ matrices on the main diagonal).
3. Derive the eigenvalues and -vectors from the Schur matrix.

In [10], the overall complexity of these steps is estimated to $\mathcal{O}\left(\frac{79}{3}K^3\right)$ operations for a random coefficient matrix of dimension $K$, including a preprocessing step for improving the condition of the system. As the matrix appearing in Eq. (4.34) has dimension $2N$, this yields an operation count of $\mathcal{O}\left(\frac{632}{3}N^3\right)$ for the case considered here. This is equal to more than 24 iterations of the LR approach, however, the matrix which appears in Eq. (4.34) has a very special structure, which supposedly leads to much less effort than what is needed for a random matrix. For example, when considering the transformation of the coefficient matrix appearing in Eq. (4.34) to upper Hessenberg form, only the lower right $N \times N$ submatrix of the coefficient matrix plays a significant role, as its first $N$ columns are already very close to Hessenberg form. A thorough analysis of the complexity of the QR approach for the special coefficient matrix derived from polynomial eigenvalue problems is beyond the scope of this work, however, the numerical results presented in Section 4.5 confirm that the solution effort is actually much smaller than the one which has been estimated for random matrices in [10]. To conclude, we note that the matrix multiplications to derive the coefficient matrix in Eq. (4.34) involve another $\mathcal{O}\left(\frac{7}{3}N^3\right)$ operations.

**Boundary Equations.** Similar to the way the boundary vector $\mathbf{v}_0$ is determined in the matrix-geometric case, the unknown coefficients $x_i$ occurring in the linear combination (4.38) are determined by considering the boundary equations (4.6)–(4.7) under the normalization condition. Using Eq. (4.39) to express $\mathbf{v}_0$ and $\mathbf{v}_1$, the boundary equations yield (in matrix form):

$$(\mathbf{b}, \mathbf{x}) \left( \begin{array}{c:c} \mathbf{B}_{0,0} & \mathbf{B}_{0,1} \\ \hdashline \mathbf{E}_0\mathbf{B}_{1,0} & \begin{array}{l} \mathbf{E}_0\mathbf{B}_{1,1}+ \\ \mathbf{E}_1\mathbf{A}_2 \end{array} \end{array} \right) = \mathbf{0}. \tag{4.40}$$

The normalization condition can be transformed as follows:

$$\mathbf{b1} + \sum_{j=0}^{\infty} \mathbf{v}_j \mathbf{1} = 1 \tag{4.41}$$

$$\mathbf{b1} + \mathbf{x} \left( \sum_{j=0}^{\infty} \operatorname{diag}(\lambda_1^j, \cdots, \lambda_c^j) \right) \mathbf{E1} = 1 \tag{4.42}$$

$$\mathbf{b1} + \mathbf{x} \operatorname{diag} \left( \frac{1}{1-\lambda_1}, \cdots, \frac{1}{1-\lambda_c} \right) \mathbf{E1} = 1. \tag{4.43}$$

In [41] and [142], it has been conjectured that the number $c$ of eigenvalues inside the unit disk is always equal to $N$ for ergodic QBD processes, so that the system (4.40) under the normalization condition (4.43) has a unique non-zero solution. While no formal proof has been conducted to confirm this, the results of all experiments performed in [41, 142] and by ourself support this proposition.

Concerning the numerical complexity of the boundary equation solution, it should be noted that Eq. (4.40) represents a set of *complex* equations, due to the complex nature of the matrices $\mathbf{E}_j$. As the multiplication of two complex scalars involves 4 real multiplications, the solution effort amounts to $\mathcal{O}\left(\frac{4}{3}(N_0 + N)^3\right)$ operations. While the normalization condition (4.43) implies only effort of the order $\mathcal{O}\left(N^2\right)$, three complex/real matrix products have to be computed to determine the coefficient matrix in Eq. (4.40): $\mathbf{E}_0\mathbf{B}_{1,0}$ involves a $N \times N$ and a $N \times N_0$ matrix; $\mathbf{E}_0\mathbf{B}_{1,1}$ and $\mathbf{E}_1\mathbf{A}_2$ only involve matrices of order $N$. The effort for computing the individual products is thus $2N_0N^2$ and $2N^3$, respectively. Summarizing, this yields an overall effort for the solution of the boundary system of $\mathcal{O}\left(\frac{4}{3}(N_0 + N)^3 + 4N^3 + 2N_0N^2\right)$.

This is considerably more than what is required in the matrix-geometric case: the boundary solution effort is at least 3.2 as high in the spectral expansion case (for $N_0 = 0$). At $N_0 = 2N$, the *additional* effort for the set of boundary equations already equals more than 5 iterations of Naoumov's improved LR algorithm, which would practically render the spectral expansion approach useless for systems with large boundaries.

It is, however, possible to circumvent the complex nature of the boundary equation system [41, 42]. Since the coefficient matrix of the eigenvalue problem (4.34) is real, all eigenvalues and -vectors are either real, or they appear in complex conjugate pairs. This also holds for the associated coefficients $x_k$ in the linear combination (4.38), as the probability vectors $\mathbf{v}_j$ are real. Without loss of generality, we now assume that the first $r \leq c$ eigenvalues $\lambda_1, \ldots, \lambda_r$ (and the associated eigenvectors) are purely real. The remaining eigenvalues occur in $s$ complex conjugate pairs, so that $c = r + 2s$ holds. Now, Eq. (4.38) can be rewritten as

$$\mathbf{v}_j = \sum_{k=1}^{r} x_k \boldsymbol{\psi}_k \lambda_k^j + \sum_{k=r+1}^{r+s} x_k (\boldsymbol{\psi}_k \lambda_k^j) + x_k^* (\boldsymbol{\psi}_k \lambda_k^j)^*. \tag{4.44}$$

Using the fact that for complex values $a, b$ the property $ab + a^*b^* = 2\,\mathrm{Re}(a)\,\mathrm{Re}(b) - 2\,\mathrm{Im}(a)\,\mathrm{Im}(b)$ holds, and denoting $\mathrm{Re}(x_k)$ and $\mathrm{Im}(x_k)$ by $y_k$ and $z_k$, respectively, this leads to

$$\mathbf{v}_j = \sum_{k=1}^{r} y_k \boldsymbol{\psi}_k \lambda_k^j + \sum_{k=r+1}^{r+s} 2y_k \,\mathrm{Re}(\boldsymbol{\psi}_k \lambda_k^j) + \sum_{k=r+1}^{r+s} 2z_k \,\mathrm{Im}(\boldsymbol{\psi}_k \lambda_k^j). \tag{4.45}$$

Using this representation for $\mathbf{v}_0$ and $\mathbf{v}_1$ in the boundary equations, a real set of $N + N_0$ equations is obtained, which is equivalent to the matrix-geometric case. Note that if the normalization condition Eq. (4.43) shall be used to replace one of the boundary equations, then this can also easily be integrated in the real system of equations, as $\lambda_k = \lambda_{k+1}^*$ implies $1/\lambda_k = (1/\lambda_{k+1})^*$.

Using this real variant of the boundary equation system, no more complex arithmetics are required, and the overall solution effort reduces to $\mathcal{O}\left(\frac{1}{3}(N_0 + N)^3 + 2N^3 + N^2 N_0\right)$ operations, which is only slightly more than what is required in the matrix-geometric case.

**Relation to Matrix-Geometric Solution.** In contrast to the spectral expansion method, all solution approaches which have been presented so far yield a steady-state solution in terms of the matrix $\mathbf{R}$. The aim of this section is to clarify the relationship between this solution form and the spectral-expansion solution, thereby providing transformation steps to derive $\mathbf{R}$ from a spectral expansion solution, and vice versa.

In order to solve the polynomial matrix equation (4.11), consider a diagonalization of $\mathbf{R}$ which is given by the matrix $\mathbf{X}_\mathrm{L}$ of its left eigenvectors:

$$\mathbf{X}_\mathrm{L}\mathbf{R} = \mathrm{diag}(\lambda_1, \cdots, \lambda_N)\mathbf{X}_\mathrm{L} \tag{4.46}$$

$$\mathbf{R} = \mathbf{X}_\mathrm{L}^{-1}\,\mathrm{diag}(\lambda_1, \cdots, \lambda_N)\mathbf{X}_\mathrm{L} \tag{4.47}$$

Note that for stable QBD processes $\mathrm{sp}(\mathbf{R})$ is less than 1, and thus all $|\lambda_i|$ are less than one as well. Given this diagonalization, the original equation (4.11) can be rewritten as

$$\mathbf{A}_0 + \mathbf{X}_\mathrm{L}^{-1}\operatorname{diag}(\lambda_1, \cdots, \lambda_N)\mathbf{X}_\mathrm{L}\mathbf{A}_1 + \mathbf{X}_\mathrm{L}^{-1}\operatorname{diag}(\lambda_1^2, \cdots, \lambda_N^2)\mathbf{X}_\mathrm{L}\mathbf{A}_2 = \mathbf{0} \tag{4.48}$$

$$\mathbf{X}_\mathrm{L}\mathbf{A}_0 + \operatorname{diag}(\lambda_1, \cdots, \lambda_N)\mathbf{X}_\mathrm{L}\mathbf{A}_1 + \operatorname{diag}(\lambda_1^2, \cdots, \lambda_N^2)\mathbf{X}_\mathrm{L}\mathbf{A}_2 = \mathbf{0}. \tag{4.49}$$

When denoting the individual left eigenvectors of $\mathbf{R}$ by $\boldsymbol{\psi}_1, \cdots, \boldsymbol{\psi}_N$ we have $\mathbf{X}_\mathrm{L} = \begin{pmatrix} \boldsymbol{\psi}_1 \\ \vdots \\ \boldsymbol{\psi}_N \end{pmatrix}$, and thus derive

$$\begin{pmatrix} \boldsymbol{\psi}_1 \\ \vdots \\ \boldsymbol{\psi}_N \end{pmatrix} \mathbf{A}_0 + \begin{pmatrix} \lambda_1\boldsymbol{\psi}_1 \\ \vdots \\ \lambda_N\boldsymbol{\psi}_N \end{pmatrix} \mathbf{A}_1 + \begin{pmatrix} \lambda_1^2\boldsymbol{\psi}_1 \\ \vdots \\ \lambda_N^2\boldsymbol{\psi}_N \end{pmatrix} \mathbf{A}_2 = \mathbf{0}. \tag{4.50}$$

If we consider this equation row by row, we obtain

$$\boldsymbol{\psi}_i\mathbf{A}_0 + \lambda_i\boldsymbol{\psi}_i\mathbf{A}_1 + \lambda_i^2\boldsymbol{\psi}_i\mathbf{A}_2 = \mathbf{0} \quad i = 1, \cdots, N \tag{4.51}$$

$$\boldsymbol{\psi}_i(\mathbf{A}_0 + \lambda_i\mathbf{A}_1 + \lambda_i^2\mathbf{A}_2) = \mathbf{0} \quad i = 1 \cdots, N. \tag{4.52}$$

This, of course, is identical to the polynomial eigenvalue problem (4.33) which is solved in the spectral expansion method. Thus, up to row permutations and row scaling, $\mathbf{X}_\mathrm{L}$ is equal to the matrix $\mathbf{E}$ obtained in the spectral expansion method.

With this result it is easy to derive expressions for transforming a matrix-geometric solution to a spectral expansion solution and vice versa.

For the first case, we assume that a matrix-geometric solution is given by

$$\mathbf{v}_j = \mathbf{v}_0\mathbf{R}^j \quad j = 0, 1, \cdots \tag{4.53}$$

Now let $\lambda_1, \cdots, \lambda_N$ denote the eigenvalues of $\mathbf{R}$. The matrix $\mathbf{X}_\mathrm{L}$ holds in its rows the corresponding left eigenvectors. Then, according to (4.47), we obtain

$$\mathbf{v}_j = \mathbf{v}_0\mathbf{X}_\mathrm{L}^{-1}\operatorname{diag}(\lambda_1^j, \cdots, \lambda_N^j)\mathbf{X}_\mathrm{L} \quad j = 0, 1, \cdots \tag{4.54}$$

Since we did not assume any special sequence of the eigenvalues and eigenvectors occurring in the spectral expansion matrix $\mathbf{E}$, we can identify it with the matrix $\mathbf{X}_\mathrm{L}$ above. Furthermore, for stable systems all eigenvalues are inside the unit disk. Thus, by simply setting $\mathbf{x} := \mathbf{v}_0\mathbf{X}_\mathrm{L}^{-1}$, we obtain the spectral expansion solution form

$$\mathbf{v}_j = \mathbf{x}\operatorname{diag}(\lambda_1^j, \cdots, \lambda_N^j)\mathbf{E} \quad j = 0, 1, \cdots \tag{4.55}$$

To transform solutions in the other direction, assume a spectral expansion solution as given above. It can be rewritten as

$$\mathbf{v}_j = \mathbf{x}\mathbf{E}\mathbf{E}^{-1}\operatorname{diag}(\lambda_1^j, \cdots, \lambda_N^j)\mathbf{E} \quad j = 0, 1, \cdots \tag{4.56}$$

$$\mathbf{v}_j = \mathbf{x}\mathbf{E}(\mathbf{E}^{-1}\operatorname{diag}(\lambda_1, \cdots, \lambda_N)\mathbf{E})^j \quad j = 0, 1, \cdots \tag{4.57}$$

Setting $\mathbf{v}_0 := \mathbf{x}\mathbf{E}$ and

$$\mathbf{R} := \mathbf{E}^{-1}\operatorname{diag}(\lambda_1, \cdots, \lambda_N)\mathbf{E}, \tag{4.58}$$

the matrix-geometric solution form is obtained.

## 4.4 Non-Skip-Free QBDs

In computer and communication systems, it is often the case that jobs enter or leave a system not only in single instances, but also in form of platooned jobs (so-called *batches*). In the context of QBD Markov chains, platooned arrivals and services correspond to state transitions between *non-neighbouring* levels; in [77], the notion *non-skip-free* has been established for such systems.

The aim of this section is to investigate how the analysis of such systems can be accomplished. There are two conceivable ways to perform this task: transforming the non-skip-free process to a standard QBD, and applying one of the methods presented above, or the direct application of suitable algorithms to the non-skip-free system. The first alternative is described in Section 4.4.1, where we show how the transformation can be performed and what the computational implications of this approach are. In Section 4.4.2, we then discuss which of the presented methods for skip-free QBDs are suitable for the direct solution of non-skip-free QBDs and can take advantage of their particular structure.

### 4.4.1 Reduction to Standard QBD Processes

At first glance, it seems surprising that non-skip-free QBD processes represent a special case of standard QBDs, rather than a generalization. In the following, we thus illustrate first by a short example how non-skip-free QBDs can be transformed to skip-free QBDs. Subsequently, the general case is considered in terms of the generator matrix. Finally, we will also comment on computational issues for solving the transformed system.

**Example.** Consider a M|M|1-queue with batch arrivals. Arrivals as such take place as a Poisson process with rate $\lambda$. Upon an arrival, with probability $\alpha_1$ it turns out that exactly one customer just arrived, but with probability $\alpha_2$ two customers arrived at the same time ($\alpha_1 + \alpha_2 = 1$). We denote $\lambda_i = \alpha_i\lambda$ ($i = 1, 2$). The service rate for individual customers is $\mu$. For stability, we require that $\lambda_1 + 2\lambda_2 < \mu$. Figure 4.5 depicts the corresponding state-transition diagram. We clearly see that there are also transitions between non-neighbouring levels (note that every state exactly comprises a single level of the QBD process).

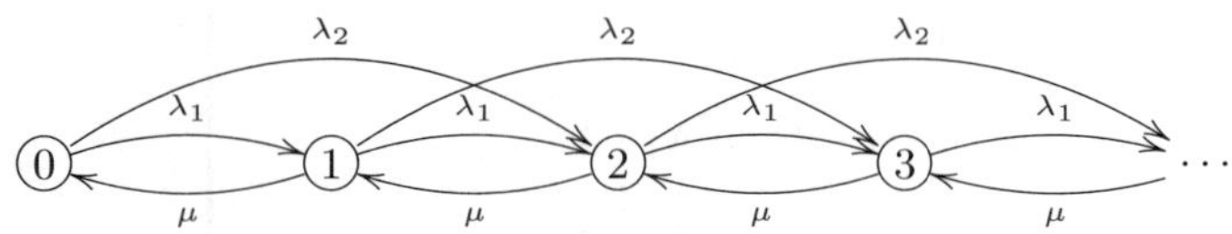


**Fig. 4.5.** State transition diagram of an M|M|1 queue with batch arrivals (maximum batch size is 2).

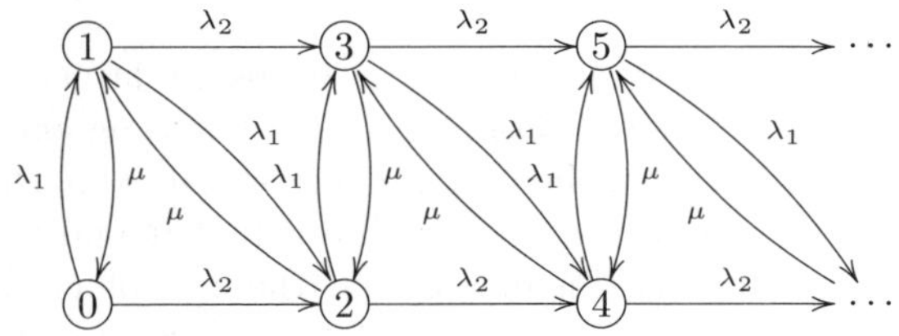


**Fig. 4.6.** Alternative representation of the state transition diagram of an M|M|1 queue with batch arrivals (maximum batch size is 2).

A skip-free QBD process can now be obtained by simply reordering the states in such a way that every two neighbour-states (starting from the left) are grouped into one level, leading to the state-transition diagram shown in Figure 4.6. In this figure, we see that there are only transitions between neighbouring levels, however, the size of the levels has doubled. We now have obtained a regular QBD again, and so the standard solution algorithms can be applied.

The basic idea of this re-organization of the state space is to group as many states into one level as necessary to guarantee that transitions entering or leaving a level are restricted to neighbouring levels only. Assuming that the maximum batch size is finite, this procedure will always work. The procedure is described more formally in the following section, where the generator matrices of non-skip-free QBDs are considered.

**General Reduction Procedure and Computational Issues.** When dealing with platooned job arrivals or departures, the generator matrix of a QBD no longer possesses the block-tridiagonal form shown in Eq. (4.1). For example, if the simultaneous arrival of two customers is allowed, a generator of the following form is obtained (block-diagonal entries are emphasized by shading):

$$
\mathbf{Q} = \begin{pmatrix}
\mathbf{B}_{0,0} & \mathbf{B}_{0,1} & & & & & \\
\mathbf{B}_{1,0} & \mathbf{B}_{1,1} & \mathbf{A}_1 & \mathbf{A}_0 & & & \\
 & \mathbf{B}_{2,1} & \mathbf{A}_2 & \mathbf{A}_1 & \mathbf{A}_0 & & \\
 & & \mathbf{A}_3 & \mathbf{A}_2 & \mathbf{A}_1 & \mathbf{A}_0 & \\
 & & & \ddots & \ddots & \ddots & \ddots
\end{pmatrix}. \tag{4.59}
$$

Now, by considering the same matrix in terms of larger submatrices, a block-tridiagonal form can be obtained again. Focusing on one block-row in Eq. (4.59), this is accomplished by making the diagonal blocks so large that only one block entry is "missed" both on the left and on the right side of the diagonal. In this case, this means that the new block diagonal entries must comprise $\mathbf{A}_2$ and $\mathbf{A}_1$, leading to the following representation:

$$
\mathbf{Q} = \left(\begin{array}{ccc|cc|cc}
\mathbf{B}_{0,0} & \mathbf{B}_{0,1} & & & & & \\
\mathbf{B}_{1,0} & \mathbf{B}_{1,1} & \mathbf{A}_1 & \mathbf{A}_0 & & & \\
 & \mathbf{B}_{2,1} & \mathbf{A}_2 & \mathbf{A}_1 & \mathbf{A}_0 & & \\
\hline
 & & \mathbf{A}_3 & \mathbf{A}_2 & \mathbf{A}_1 & \mathbf{A}_0 & \\
 & & & \mathbf{A}_3 & \mathbf{A}_2 & \mathbf{A}_1 & \mathbf{A}_0 \\
\hline
 & & & \ddots & & \ddots & \ddots
\end{array}\right). \tag{4.60}
$$

The fact that just one block-entry of the original system is "missed" leads to the skip-freeness of the "new" QBD process. In general, for a system with a maximum batch arrival size of $S_r$ jobs and $N$ states in each repeating level, the restructured QBD is formed by $S_r$ blocks of the original system, increasing the level size to $N \cdot S_r$.

For batch departures of at most $S_l$ jobs at once, the same procedure can be applied, leading to new block entries of the size $N \cdot S_l$. If a system involves both batch arrivals and batch departures, then the block size has to be large enough to cover both the maximum arrival and departure size, i.e., block entries of the size $N \cdot \max\{S_l, S_r\}$ are obtained.

While this restructuring step can easily be performed, the increased block size heavily impacts the performance of the solution algorithms. As the operation count of all algorithms is of order $\mathcal{O}\left(N^3\right)$, the computation time increases by a factor of 27 already for batch arrivals or departures of size 3; the memory requirements for storing the involved matrices increase by a factor of 9 (as they are of order $\mathcal{O}\left(N^2\right)$). For practically relevant systems, this approach is clearly not feasible. In the next section, we will thus discuss the suitability of solution algorithms for directly solving block-banded systems like the one appearing in Eq. (4.59).

### 4.4.2 Approaches for Direct Solution

As reducing non-skip-free QBDs to skip-free QBDs leads to a considerable increase of the QBD order, the aim of this section is to discuss alternative approaches which can be used to analyze non-skip free systems directly, and which are capable to take advantage of the special transition structure arising in non-skip-free QBDs. Depending on whether a QBD involves *both* batch arrivals and departures or not, two approaches can be distinguished.

**Algorithms for M|G|1 and G|M|1 Systems.** In the case that only batch arrivals (and no batch departures) are allowed, the generator of the non-skip-free QBD process has a structure as shown in Eq. (4.59). This structure is a special case of the upper block-Hessenberg generator matrices arising in M|G|1 type Markov chains. Therefore, algorithms which are suitable for the analysis of M|G|1 type systems can be applied to derive the steady-state distribution of QBD processes with batch arrivals. Conversely, QBD systems where only departures occur in batches can be analyzed by algorithms for G|M|1 type Markov chains.

In the case of the SS and "U" algorithms, this basically leads to a higher degree of the matrix polynomial equation (4.11), and therefore also to a larger number of terms in the recursion formula (4.19). The number of involved terms increases linearly with the maximum batch size $\max\{S_l, S_r\}$, which is more desirable than the exponential relationship in the reduction approach presented in Section 4.4.1. For large batch sizes, the experiments in [201] have shown that this approach is faster than applying the LR method to a reblocked QBD. However, as the SS and "U" algorithms still converge rather slowly, it is much more attractive to apply the invariant subspace approach or the cyclic reduction method, which are also suited for the analysis of M|G|1 and G|M|1 systems.

Unfortunately, the efficient LR method (and the related improvement by Naoumov) are limited to skip-free QBD systems, and so far no extensions have been presented for dealing with batch arrivals or departures.

It should be emphasized that considering the non-skip-free QBD as M|G|1 or G|M|1 system does not work if *both* batch arrivals and batch departures are allowed. In this case, the block-Hessenberg form is violated and a block-banded generator matrix is obtained. Thus, if reblocking shall be avoided, one of the approaches presented in the following section must be employed.

**The General Non-Skip-Free Case.** For the general case of non-skip-free QBDs where both batch arrivals and departures are allowed, only few algorithms are available if the transformation to a skip-free QBD shall be avoided.

In [28], the cyclic reduction approach has been extended to account for "displaced" M|G|1 type systems, i.e., systems where the block-diagonal is shifted to the left (as an example, if the generator in Eq. (4.59) involved additional non-zero block-subdiagonal entries in order to model batch departures, then the resulting matrix could be considered as a displaced M|G|1 type matrix). The approach is particularly efficient for QBD systems, and the displacement procedure increases the computation time per step only by a factor of $\mathcal{O}\left(k(\log k)^2\right)$, where $k$ denotes the shift size.

Another approach for solving non-skip free QBD process was suggested in [77], however, it has been shown in [28] that it is inferior to the cyclic reduction method both with respect to computational speed and numerical stability.

Similarly, variants of the SS algorithm for non-skip free Markov chains can be employed, such as Newton-related methods [183, Section 5.3.5]. Such methods, however, still suffer from the poor convergence properties of the original SS algorithm.

As a last alternative, the spectral expansion method can be employed to directly solve non-skip-free QBD processes. Here, the existence of transitions to non-neighbouring levels manifests in a higher degree of the considered vector difference equation (4.8), which increases from 3 to $S_l + S_r + 1$ coefficients. Similarly, the degree of the associated characteristic polynomial (4.32) changes from 2 to $S_l + S_r$. Using the linearization procedure outlined in Appendix A.1, this leads to an eigenvalue problem of dimension $(S_l + S_r) \cdot N$. Thus, compared to the skip-free QBD case (where the eigenvalue problem is of size $2N$), this represents an increase of the problem size by a factor of $[(S_l + S_r) \cdot N]/(2N) = \frac{S_l + S_r}{2}$. For the case $S_l = S_r = S$, this is equivalent to the factor $\max\{S_l, S_r\}$ which is obtained if the reblocking procedure of Section 4.4.1 is applied. However, it should be noted that the resulting eigenvalue problem becomes increasingly sparse (as only a fraction $1/S$ of the blocks of the coefficient matrix are non-null or non-unity), which effectively reduces the necessary computational effort. Furthermore, if $S_l \neq S_r$, the obtained coefficient matrix is smaller than in the reblocking case, as $\frac{S_l + S_r}{2} < \max\{S_l, S_r\}$. For $S_l = 1$ or $S_r = 1$, the difference approaches a factor of 2, which leads to a substantially reduced solution effort as the solution algorithm is of order $\mathcal{O}(N^3)$.

To summarize, it can be said that if a QBD system involves both batch arrivals and departures, then modified cyclic reduction and the spectral expansion method are the algorithms of choice, whereas the latter is especially advantageous if the difference between $S_l$ and $S_r$ is large. If only batch arrivals *or* batch departures are involved, then the invariant subspace approach (for M|G|1 and G|M|1 systems) should be considered as well.

## 4.5 Numerical Comparison of Solution Methods

The purpose of this section is to provide an experimental counterpart to the theoretical investigation of the solution methods in Sections 4.2 and 4.3. The motivation for this is two-fold: first, we so far only considered the leading coefficient of the highest order term $N^3$ in the complexity investigations.

In practical applications, the impact of lower order terms may still have significant impact on the necessary computation times. Secondly, many of the linear algebra computations performed in the various approaches might profit from the special structural properties of QBD Markov chains, and may thus require less effort than indicated by the theoretical complexity estimates. To the best of our knowledge, the extensive comparison results obtained in this section are the first which account for recent developments like the invariant subspace approach, Naoumov's improved LR algorithm, and the spectral expansion method; they represent a thorough extension of the results we presented in [92] and [160].

As the focus of this section is the suitability of the solution approaches in a practical performance evaluation study, the presented comparisons are based on the analysis of a large QBD Markov chain which models the behaviour of a checkpointing transaction processing system. Due to its largeness, we resorted to the high-level system description approach presented in Chapter 5.

Based on a summary of existing comparison studies, we open this section with a brief review of the main properties of the solution algorithms which have been presented so far in Section 4.5.1 in order to select the algorithms to be investigated subsequently. In Section 4.5.2, we then present the model under investigation and discuss its parameterization. Different approaches for its numerical solution will also be discussed there. Section 4.5.3 represents the main part of this section, where we present our numerical experiments and discuss the results. Finally, we conclude our investigations with a summary in Section 4.5.4.

### 4.5.1 Candidate Solution Algorithms

The investigations conducted in this section focus on a comparison of the most efficient solution algorithms which are currently available for the analysis of QBD Markov chains. For this reason, we will not consider matrix geometric solution algorithms with sub-quadratic convergence rates (i.e., the SS and "U" algorithms). Comprehensive comparisons of this class of algorithms can be found in [128] and [84]; the results of these studies indicate only marginal differences between the computational effort induced by algorithms in this class. In [129], it has been shown that the LR algorithm (and thus, also Naoumov's improved version) is clearly superior to these approaches. The same conclusion has been drawn in [142] for the relation of the spectral expansion method to the SS algorithm. Also, the cyclic reduction approach presented in Section 4.3.1 will not be considered here, as it coincides with Naoumov's method if QBD processes are considered.

In the subsequent studies, we will also analyze the impact of batch events on the computational effort. As we are interested in the behaviour of algorithms suited for the general case where both batch arrivals and departures are allowed, we employ the reblocking procedure of Section 4.4.1 in order to apply those algorithms which are not capable to deal with batch events

| | applicability | | | complexity for solving QBD systems | | |
|---|---|---|---|---|---|---|
| | finite systems | M\|G\|1, G\|M\|1 | batch events | boundary | repeating levels | typical #iters |
| SS | Yes | Yes | No | $\frac{1}{3}(N_0+N)^3+\frac{4}{3}N^3$ | $\frac{1}{3}N^3+k\cdot\frac{9}{3}N^3$ | $k \gg 100$ |
| "U" | Yes | Yes | No | $\frac{1}{3}(N_0+N)^3+\frac{4}{3}N^3$ | $\frac{1}{3}N^3+k\cdot\frac{7}{3}N^3$ | $k \gg 100$ |
| LR | Yes | No | No | $\frac{1}{3}(N_0+N)^3+\frac{4}{3}N^3$ | $\frac{14}{3}N^3+k\cdot\frac{25}{3}N^3$ | $k < 40$ |
| CR | Yes | Yes | Yes | *for QBD case, identical to NA* | | |
| NA | Yes | No | No | $\frac{1}{3}(N_0+N)^3+\frac{4}{3}N^3$ | $\frac{16}{3}N^3+k\cdot\frac{19}{3}N^3$ | $k < 40$ |
| IS | Yes | Yes | No | $\frac{1}{3}(N_0+N)^3+\frac{4}{3}N^3$ | $\frac{28}{3}N^3+k\cdot\frac{24}{3}N^3$ | $k < 40$ |
| SE | Yes | No | Yes | $\frac{1}{3}(N_0+N)^3+\frac{6}{3}N^3+N^2N_0$ | $\frac{7}{3}N^3+\frac{639}{3}N^3$ | – |

**Table 4.1.** Summary of the capabilities and numerical complexities of algorithms for QBD Markov chains.

directly. Note that the cyclic reduction approach with displacement has not been considered in our studies, as it represents a very recent development which involves elaborate numerical methods. To summarize, our investigations concentrate on the following four solution approaches:

- the logarithmic reduction (LR) method,
- Naoumov's improved LR algorithm (NA),
- the invariant subspace (IS) approach based on matrix-sign function iterations, and
- the spectral expansion (SE) method based on QR decomposition.

Table 4.1 summarizes the capabilities and numerical complexities of these algorithms from a theoretical point of view, as discussed in Sections 4.2 and 4.3. To provide a complete overview, the table also holds the properties of the SS and "U" methods and the cyclic reduction (CR) approach.

### 4.5.2 The Model under Investigation

As the aim of this section is to investigate the applicability of the different solution algorithms to real-world problems, we decided to focus on a QBD system arising from an actual application study which has also been considered in [89]. The model represents a transaction processing system with server breakdowns, which involves a checkpointing system for jobs which have not been completed yet. This rather complex system has been described using the high-level modeling approach which is developed in the next chapter; for a detailed description of the system and the corresponding model, refer to Section 5.1.4. The order $(N_0, N)$ of the corresponding QBD process depends on the size of the checkpointing interval, which is subsequently denoted by $K$ and which is related to $N_0$ and $N$ through $N_0 = N = 2K + 1$. Using this fact and the convenient high-level description of the model, we were able to

| Parameter | Model variable | Value |
|---|---|---|
| Service rate | $\mu$ | 10 |
| Failure rate | $f$ | 0.13 |
| Restart rate | $r$ | 11 |
| Checkpointing speed | $c$ | 21 |
| Checkpointing interval size | $K$ | variable |
| Maximum arrival batch size | $S$ | variable |

**Table 4.2.** The model parameters used in the experiments.

easily scale the QBD to be investigated by varying the value of $K$. In our experiments, the repeating level size will be increased to very large values, up to 201 states in the repeating levels. The actual values of the remaining model parameters are summarized in Table 4.2; for a detailed description of their meaning, please refer to the model description in Section 5.1.4. As we are also interested in the impact of batch events on the computation time, the same system has also been investigated for the case that jobs enter the system in uniformly distributed batches of size $1, \cdots, S$. For all algorithms except the spectral expansion method, the arising non-skip-free QBD has been reblocked to a standard QBD process. Note that this has not been performed at the QBD level, but also using the high-level modeling approach, as described in Section 5.3.4.

While the presented system has the nice property that the underlying QBD can be scaled to any desired order and that batch arrivals can easily be integrated, it should be pointed out that in the literature, several other models have been considered for comparing algorithms for QBD Markov chains. In [128], a set of 99 PH|PH|1 queues arising from different interarrival and service time distributions has been evaluated. A subset of these queues has also been investigated in [8] and [129]. Similarly, the teletraffic system which has been analyzed in [54] was also considered in [8] and [129]. In [28], random generator matrices were chosen to form an artificial benchmark for comparison reasons.

### 4.5.3 Numerical Results

In this section, the candidate solution algorithm are compared by applying them to the example model described in Section 4.5.2. Our studies do not emphasize the analysis of a wide range of different models (as, for example, the set of 99 queues considered in [128]), but instead focus on variations of one and the same model. While care must be taken when generalizing these results to different models, this approach has the advantage of giving insight in the sensitivity of the different algorithms to the varied parameters. The parameters have been chosen so that they represent fundamental properties of a QBD process, which occur in different applications as well.

After commenting on details of the used implementation and computing environment, we focus on four different system aspects: the traffic intensity, the order of the QBD process, the occurrence of batch arrivals, and the numerical stability and accuracy.

**Implementation and Computing Environment.** All solution algorithms except the IS approach were implemented in the C++ programming language, where we resorted to calling high-performance LAPACK [10] Fortran routines for all linear algebra operations. The IS computations were performed using original C code from the telpack [2] package (version 1.0.1a), which uses LAPACK functions for all linear algebra operations as well. In all routines, standard IEEE 754 double precision floating point arithmetic was used. The measurements were performed on a, otherwise idle, Pentium (P55C) system running at 233 MHz under the Linux operating system; the amount of physical memory was large enough to ensure that no paging occurred during the test runs. All runs for a specific configuration were performed six times, and the subsequent timing diagrams indicate the 95% confidence intervals of the consumed CPU time for obtaining the repeating level solution and solving the boundary equation system. Note that the discrete measurement points in all following diagrams are connected by lines to improve the readability.

The required accuracy (stopping criterion) for the iterative IS, LR and NA algorithms has been set to $10^{-8}$. In the few cases where this was necessary, the accuracy for the stopping criterion of Naoumov's approach has been changed slightly so that the number of iterations (and the obtained matrix $\mathbf{R}$) were identical to the LR method.

**Impact of Traffic Intensity.** To start with, we investigated the computational effort as a function of the traffic intensity (as also considered in [142]). To do so, we set the parameter $K$ to 80 (yielding a level size of $N = 161$), and varied the traffic intensity between $\rho = 0.1$ and $\rho = 0.97$ by appropriately changing the arrival rate $\lambda$. The resulting overall computation times for the algorithms under investigation are depicted in Figure 4.7.

Concerning the absolute computation times, it can be observed that the IS approach is clearly outperformed by all other considered algorithms, independent of the traffic intensity. The curves for the LR and NA methods nicely reflect the fact that these algorithms found on the same theory, and NA can be considered as a more efficient implementation of the LR method which reduces the computational effort by a constant percentage. The improvement observed in the experiments is very close to the theoretical value of 24% mentioned in Section 4.2.4.

Similar to the findings in [92, 142], it can be observed that the spectral expansion method is nearly insensitive to changes of the traffic intensity. This is in contrast to the LR and NA algorithms, where the computation time increases significantly for higher traffic intensities. However, these increases are much smaller than those reported in [142], where the inferior SS algorithm

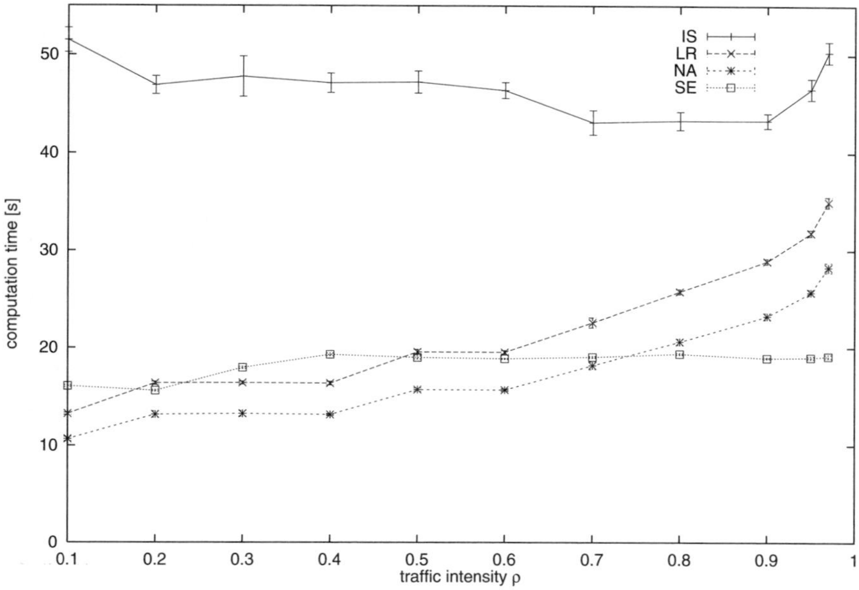


**Fig. 4.7.** The sensitivity of the overall computation time on the traffic intensity $\rho$, total solution time. The boundary and repeating level sizes have been set to $N_0 = N = 161$.

was used. Up to medium traffic intensities of about $\rho = 0.6$, the SE method performs about equally good as the LR method (and thus worse than the NA approach), however, at high traffic intensities, its insensitivity to the traffic intensity makes it the most efficient algorithm. At intensities above 0.5 it outperforms the LR method, and beyond $\rho = 0.8$ it is even faster than Naoumov's method.

Similar to the SE method, the IS approach is not very sensitive to the traffic intensity. However, it is worth to note the increased effort as the traffic intensity approaches saturation.

In addition to the overall computation times in Figure 4.7, it is interesting to distinguish the boundary solution effort and the solution effort for the repeating part of the state space, as illustrated by the diagrams in Figure 4.8. In the upper diagram, it can be observed that the boundary solution times for all considered algorithms except SE are very similar. This is due to the fact that the LR, NA and IS approaches all yield a solution of the repeating levels in terms of the matrix $\mathbf{R}$, which leads to identical boundary equation systems, and thus identical solution times. As described in Section 4.3.3, the SE method involves a different boundary equation system due to the different way of expressing the steady-state probability vectors. As $N_0 = N$ in our example model, the boundary solution effort in the matrix-geometric

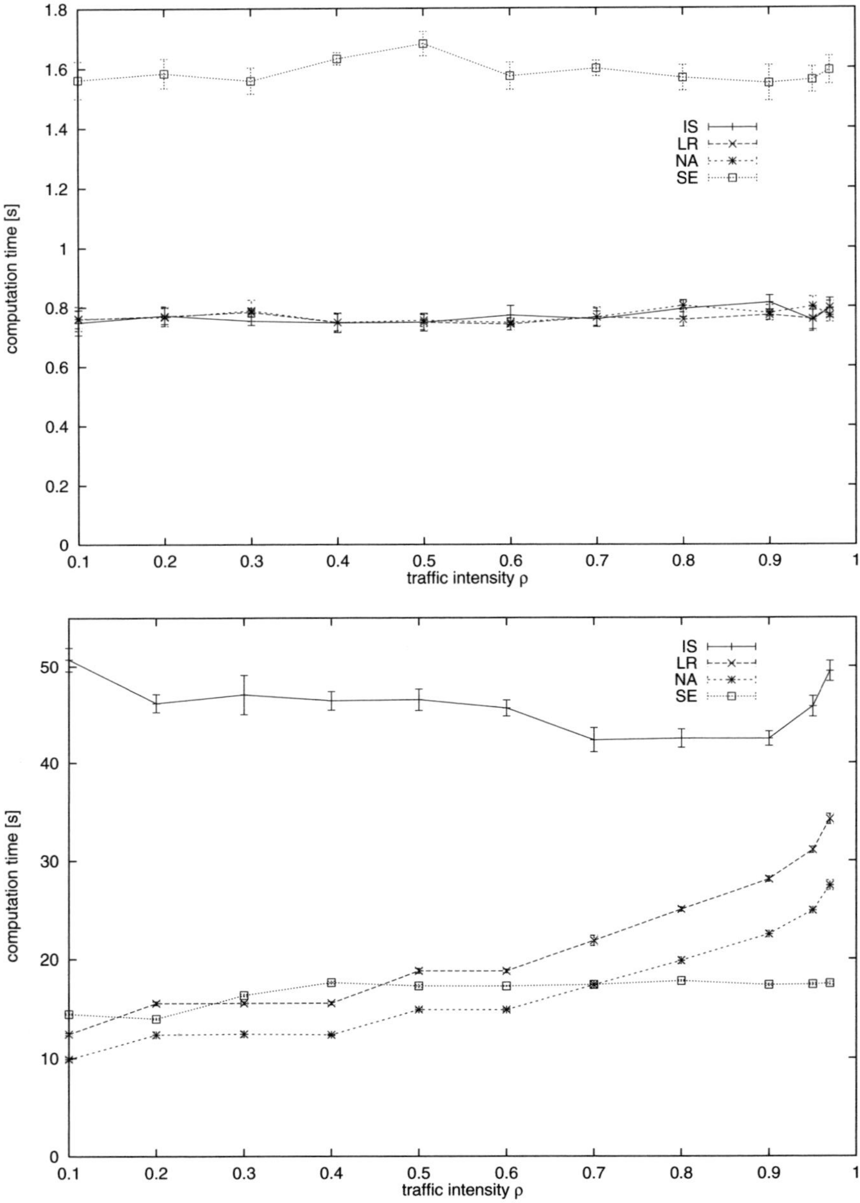


**Fig. 4.8.** The sensitivity of the computation time on the traffic intensity, boundary solution time (upper diagram) and repeating level solution time for $N_0 = N = 161$.

and the spectral expansion case are $\mathcal{O}\left(\frac{12}{3}N^3\right)$ and $\mathcal{O}\left(\frac{17}{3}N^3\right)$, respectively (see Table 4.1). The fact that the difference of the computation times in the upper diagram of Figure 4.8 is larger than the expected factor $\frac{17}{12}$ must be accounted to the fact that the coefficient matrix has more non-zero entries in the SE case, where the border probability vector $\mathbf{v}_0$ has to be expressed by $\mathbf{xE}_0$. Also, additional recalibration steps might be applied by the LAPACK routines if the condition of the boundary equation system in the SE case is worse than in the matrix-geometric case (see also our remarks on numerical stability below).

As the absolute solution times for the boundary equations are rather small compared to the repeating level solution effort depicted in the lower diagram in Figure 4.8, the latter diagram basically reflects the situation of the overall effort in Figure 4.7.

To conclude, it should be mentioned that the same experiments were also performed for other repeating level sizes $N$, indicating similar results. The impact of the order of the QBD on the computation time is considered in more detail in the next experiment.

**Impact of Repeating Level Size.** We now investigate the sensitivity of the computational effort on the size of the repeating levels. We changed the level size by varying the checkpointing interval $K$ between 10 and 100 (leading to a repeating level size of $N = 21, \cdots, 201$), and considered the computation times for traffic intensities $\rho = 0.2$ and $\rho = 0.9$. As can be observed in Figure 4.9, increasing the repeating level size clearly increases the required computational effort, reflecting the $\mathcal{O}\left(N^3\right)$ dependency of the computation time in all algorithms. Furthermore, the insensitivity of the IS approach to the traffic intensity can be observed again. The same is true for the spectral expansion method; the discrepancy between the corresponding two curves must be seen in view of the results of Figure 4.7, where this method performs slightly better at very low traffic intensities. Figure 4.9 also confirms that the IS approach is clearly outperformed by the other algorithms, and that the effort required by the NA algorithm takes a constant factor less than the LR approach. At a low traffic intensity ($\rho = 0.2$), NA performs slightly better than the spectral expansion method, and thus yields the best computation time of all considered algorithms. At the large traffic intensity (90%), the spectral expansion method outperforms the NA algorithm, taking about 10% less time.

**Impact of Batch Arrivals.** We continue with the investigation of the sensitivity of the computation time on the maximum arrival batch size $S$. Since for all algorithms except SE, the reblocking procedure of Section 4.4.1 had to be applied, this resulted in a reblocked QBD with a repeating level size of $N_{\mathrm{rb}} = S \cdot N$. In order to eliminate the influence of the level size (which has been investigated above), we kept the product $S \cdot N$ as close to a constant value as possible. In our experiments, the value 201 has been chosen, and thus $K$ was taken to be the integer closest to the solution of $S(2K + 1) = 201$,

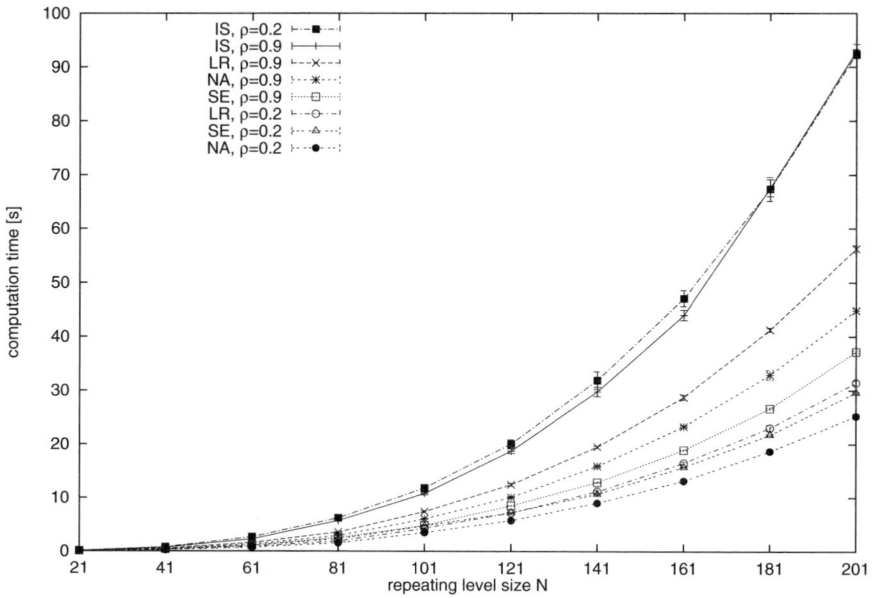


**Fig. 4.9.** The sensitivity of the overall computation time on the repeating level size $N$, total solution time.

where $S$ has been varied between 1 and 10; the actual values are indicated in Table 4.3. The experiments were again performed at a low traffic intensity level ($\rho = 0.2$) and at a high traffic intensity level ($\rho = 0.9$); the overall solution times are depicted in Figure 4.10.

As could be expected, the computation times for the LR and NA algorithms are fairly constant due to the (almost) constant effective repeating level size $N_{\text{rb}}$. It is, however, interesting to note that even these approaches experience an improvement for the first values of $S$; this is probably due to the fact that the involved transition matrices obtained from the reblocking procedure become increasingly sparse, which reduces the computational effort at least during the first iterations of these algorithms. This also explains the fact that the improvement is less emphasized for larger values of $S$. The IS method exhibits a rather irregular behaviour, especially for the low traffic intensity case. There, the computational effort reduces by about 30% up to $S = 6$, and then increases again. The absolute values are again much larger than those of all other methods.

For the spectral expansion method, the solution effort decreases significantly for larger values of $S$, and approaches less than 1/3 of the time required for the skip-free case ($S = 1$). From $S = 2$ on, it outperforms all other algorithms independent of the traffic intensity, including Naoumov's approach.

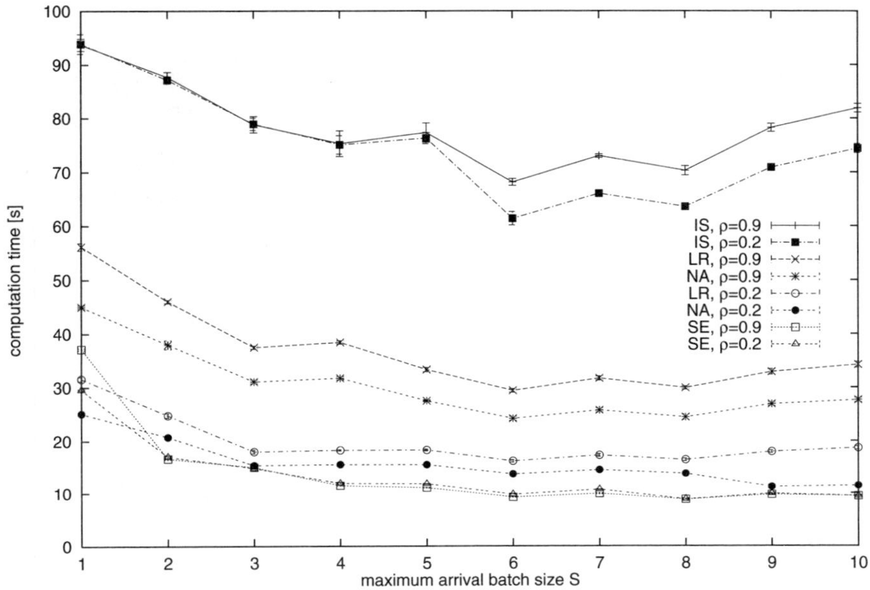


**Fig. 4.10.** The sensitivity of the computation time on the batch arrival size $S$, total solution time.

| $S$ | 1 | 2 | 3 | 4 | 5 | 6 | 7 | 8 | 9 | 10 |
|---|---|---|---|---|---|---|---|---|---|---|
| $K$ | 100 | 50 | 33 | 25 | 20 | 16 | 14 | 12 | 11 | 10 |
| $N_{\text{rb}}$ | 201 | 202 | 201 | 204 | 205 | 198 | 203 | 200 | 207 | 210 |

**Table 4.3.** The selected parameters for the batch arrival size experiment.

At $\rho = 0.9$ and $S \geq 2$, the SE algorithm is always more than twice as fast as Naoumov's method.

An interesting effect can be observed when distinguishing the boundary solution times and the repeating level solution effort, depicted in Figure 4.11. While the repeating level solution times exhibit the same behaviour as the overall solution effort due to the low absolute boundary solution times, a dramatic (relative) decrease of the boundary solution effort can be observed for the SE method. This is due to the fact that in the SE case, no reblocking has been performed. The size of the boundary equation system in the SE case thus depends on $N$, which is much smaller than the value $N_{\text{rb}}$ which is relevant for the other methods where reblocking had to be applied.

**Accuracy and Stability.** So far, we focused on the investigation of the computational effort induced by the different solution methods, i.e., we considered their computational *cost*. The aim of this section is to investigate the solution *quality* of the various algorithms and its dependence on other system properties. Key issues in these considerations are the behaviour of the

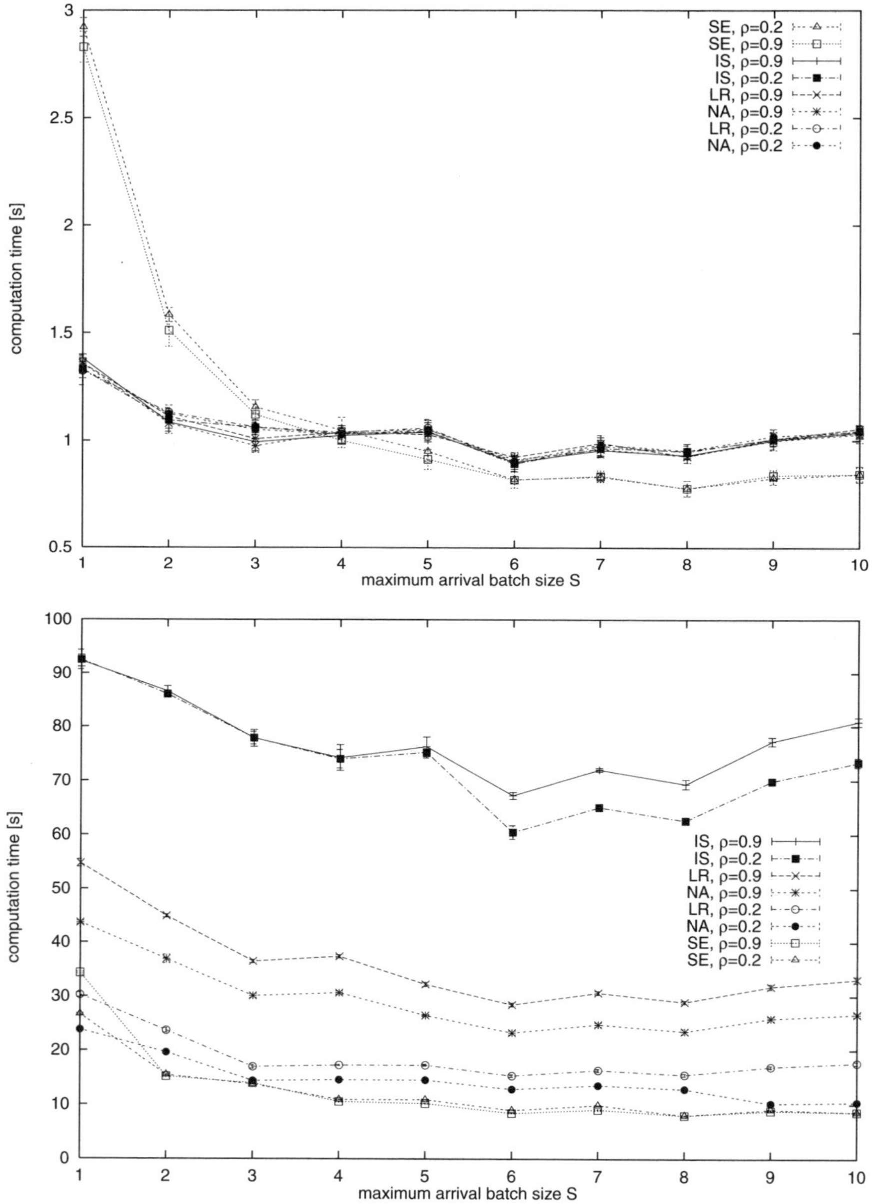


**Fig. 4.11.** The sensitivity of the computation time on the batch arrival size $S$, boundary solution time (upper diagram) and repeating level solution time (lower diagram).

IS method (where the stopping criterion $\epsilon$ has a different interpretation than in the LR and NA case), and the spectral expansion method, which does not allow the specification of an accuracy-oriented stopping criterion, but instead provides results on an "as is" basis.

For measuring the solution quality of those approaches which provide a solution in terms of the matrix $\mathbf{R}$, we resort to the approach taken in [129]. There, it has been suggested to consider the maximum absolute value of the matrix $\mathbf{A}_0+\mathbf{R}\mathbf{A}_1+\mathbf{R}^2\mathbf{A}_2$. According to Eq. (4.11), this value is ideally equal to zero. We will refer to this value as the *achieved* solution precision, where small values naturally indicate better solutions. Furthermore, we extend this notion also to the solution quality of the boundary equation system given by Eqs. (4.6) and (4.14), thereby considering the maximum absolute value of the vectors $\mathbf{b}\mathbf{B}_{0,0} + \mathbf{v}_0\mathbf{B}_{1,0}$ and $\mathbf{b}\mathbf{B}_{0,1} + \mathbf{v}_0\mathbf{B}_{1,1} + \mathbf{v}_0\mathbf{R}\mathbf{A}_2$.

Similar precision measures were derived for the spectral expansion case. Concerning the repeating level solution, one might suggest to apply Eq. (4.58) for deriving a solution in terms of the matrix $\mathbf{R}$, and then to apply the approach mentioned above. However, this conversion step introduces further numerical errors due to the inversion of $\mathbf{E}$, and we thus resorted to using Eq. (4.58) to transform Eq. (4.11) to

$$\mathbf{E}\mathbf{A}_0 + \mathbf{E}_1\mathbf{A}_1 + \mathbf{E}_2\mathbf{A}_2 = \mathbf{0}. \tag{4.61}$$

Again, the maximum residual of this equation has been computed. Apart from its relation to the previous measure, considering this equation corresponds to evaluating the solution quality of the polynomial eigenvector problem (4.33) for all used eigensolutions. Concerning the solution quality of the boundary equation system in the SE case, we took the straightforward approach of evaluating the residual of Eq. (4.40), being equivalent to the approach in the matrix-geometric case.

*Accuracy and Solution Effort.* In all experiments performed so far, the stopping criterion of the iterative LR, NA and IS algorithms has been set to a fixed value of $\epsilon = 10^{-8}$. To estimate the impact of this choice on the presented results, we first investigate the influence of the stopping criterion $\epsilon$ on the computation time. As the effort of the NA algorithm is directly related to the solution time of the LR algorithm and since these methods yield identical solutions, the NA algorithm is not considered here, and we solely focus on the LR and IS approaches here. The spectral expansion method is also not considered here, as the precision of the eigenvalue solution routines is fixed.

For low (20%) and high (90%) traffic intensity and repeating level size $N = 201$, we plot the computation times for deriving the repeating level solution (i.e., for computing the matrix $\mathbf{R}$), for $\epsilon = 10^{-l}$, $l = 1, \cdots, 12$, in Figure 4.12. As can be observed for low utilizations, the required times for the LR method are constant for $\epsilon \geq 10^{-4}$ and $\epsilon \leq 10^{-5}$. Clearly, increasing the required precision from $10^{-4}$ to $10^{-5}$ and beyond just increases the number of required iterations by 1. Similar behaviour can be observed at larger

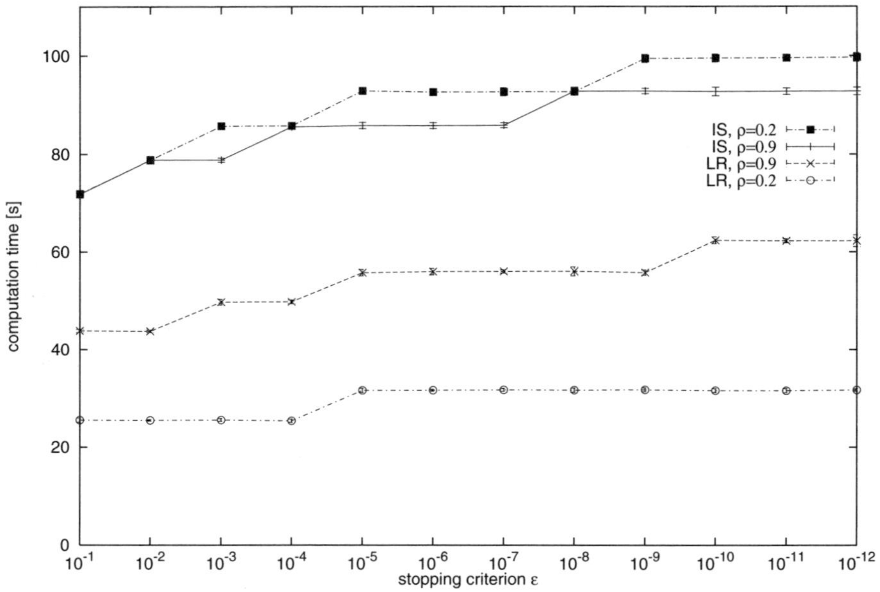


**Fig. 4.12.** The sensitivity of the computation time on the required accuracy $\epsilon$ for repeating level size $N = 201$.

utilizations, albeit that the number of 1-step increases is larger here; this is also true for the IS method. As a consequence of this relative insensitivity, our precision choice in the other experiments ($\epsilon = 10^{-8}$) does not have a large impact on the reported comparisons.

We note that similar experiments were also performed for lower values of $N$ and for non-skip free QBDs, showing similar behaviour for these cases as well.

*Stopping Criterion $\epsilon$ and Achieved Precision Level.* We now focus on the actual solution *quality* which is achieved for a given precision level $\epsilon$. This is particularly interesting in view of the fact that the stopping criterion for the IS method is not directly related to the matrix $\mathbf{G}$, but to the iterative routine for computing the matrix sign function. In [8], it has been stated that the IS approach usually provides more accurate solutions than the LR approach, as the value $||\mathbf{1} - \mathbf{G1}||_\infty$ is usually lower than in the LR case. It should, however, be noted that the LR algorithm is based on iteratively approaching the matrix $\mathbf{G}$, where (due to the probabilistic interpretation of $\mathbf{G}$) the matrix entries increase monotonically between iterates, which clearly justifies the quality measure $||\mathbf{1} - \mathbf{G1}||_\infty$ in this case. Such a probabilistic interpretation has not been given for the IS method (which does not operate directly on $\mathbf{G}$), and using the same measure is questionable here—for instance, any algorithm which simply yields the unity matrix $\mathbf{I}$ as a solution for $\mathbf{G}$ would be assigned

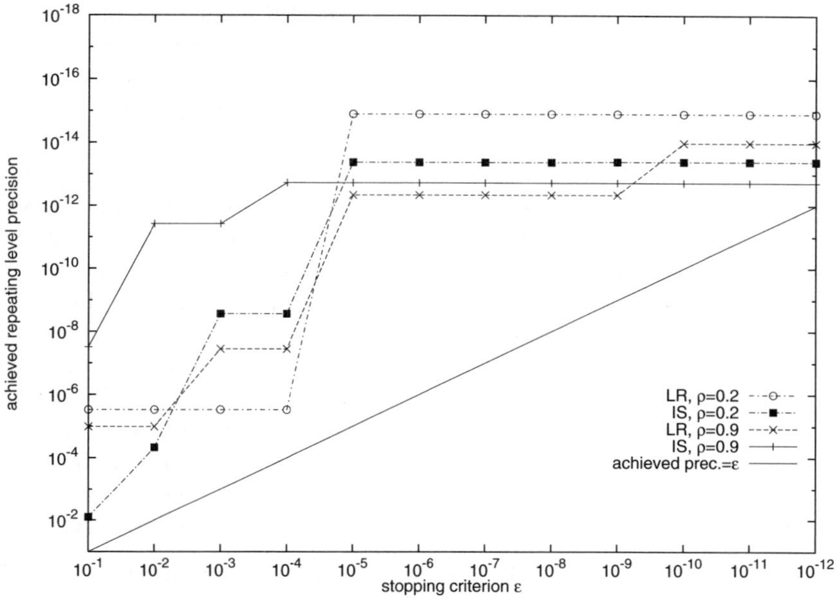


**Fig. 4.13.** The achieved repeating level accuracy versus the stopping criterion $\epsilon$.

maximum accuracy. See also [139] for further remarks on the accuracy of the IS method.

In Figure 4.13, we therefore depict the quality of the repeating level solution for the IS and LR algorithms (at high and low traffic intensities), for varying values of the stopping criterion $\epsilon$. The diagram also shows a line indicating the value of the stopping criterion itself. As users of the considered algorithms will expect the "input" precision $\epsilon$ to be associated to the achieved solution quality, it is desirable that the achieved precision is at least as large as $\epsilon$. As can be observed, this is clearly the case for both the LR and IS algorithm.

Comparing the solution quality of the LR and IS algorithms, it can be observed that at a low traffic intensity, the LR algorithm usually provides *better* results than the IS algorithm. The IS algorithm yields more accurate results only at precision levels $\epsilon = 10^{-3}$ and $\epsilon = 10^{-4}$; however, these low-precision cases are usually less important in practice. Concerning the LR method, it is interesting to note that the dramatic precision increase at $\epsilon = 10^{-5}$ is due to just one additional iteration of the LR algorithm, as it corresponds to the increase of the lower curve in Figure 4.12.

At a high traffic intensity, it can be observed that the achieved precision of the LR algorithm scales well with the value of $\epsilon$; the IS approach is not as sensitive to the value of the stopping criterion, and provides a significantly better accuracy for larger values of $\epsilon$ (i.e., at lower precision levels). If very accurate solution are required, the IS and LR algorithms provide about equally good solutions, with a slight advantage for the IS method at medium precision levels, and slightly better results for the LR method at very high precision levels. In any case, the precision reached by the LR method can easily be improved at the cost of a single iteration.

Summarizing, it can be said that the repeating level solution quality of both algorithms is about equally good. In contrast to the findings in [8], our experiments did not indicate a superior solution quality of the IS algorithm. Instead, the experiments show that the algorithms differ in the *dependence* of the achieved precision on the value of the stopping criterion $\epsilon$. The achieved precision of the LR method shows a higher dependence on $\epsilon$ than the IS method, thus providing a more fine-grained way of controlling the desired solution quality and the associated solution effort. Note that the boundary solution quality has been excluded form our considerations here, as both methods provide the same repeating level solution in terms of $\mathbf{R}$, and the achieved boundary precision only reflects the slightly different condition of the boundary equation system due to slightly different values of $\mathbf{R}$.

*Impact of the Model Size.* The precision of the solution algorithms for QBD Markov chains heavily depends on the order $(N_0, N)$ of the QBD system to be solved. For the LR, NA and IS algorithms the backsubstitution and matrix inversion steps become more inaccurate as the size of the involved matrices (which depends on $N$) increases; the same applies to the eigenvalue problem to be solved in the SE method. Furthermore, for large values of $N_0$ and $N$, the size of the boundary equation system which has to be solved in all algorithms can approach a size where the application of direct numerical solution methods becomes difficult. Especially the SE method is assumed to be sensitive to these issues; problems for large systems were both indicated from a theoretical point of view (cf. [152, pp. 29–30]) and by practical experiences [142].

As the quality of the repeating level solution is a prerequisite for good boundary solutions, we focus on this issue first. Figure 4.14 depicts the repeating level residuals of the LR, IS and SE methods for level sizes between 21 and 201 at high and low traffic intensities. In the experiments, the stopping criterion $\epsilon$ of the iterative LR and IS algorithms has been fixed to $10^{-8}$. As already observed in the previous experiment, the solution quality of the LR and IS algorithms always exceeds the value of $\epsilon$; these methods invariably provide a precision better than $10^{-10}$. It can also be observed that the achieved precision of the IS method is almost constant, independent of the system size. In view of the results of the previous experiment, the varying precision level of the LR method gives no reason to concern. The results of the SE approach are also encouraging, as they are insensitive to the system size and provide a good precision level of about $10^{-10}$. It should, however, be noted that the SE method should not be applied if very high precision levels are desired, as this can only be achieved by providing better algorithms for

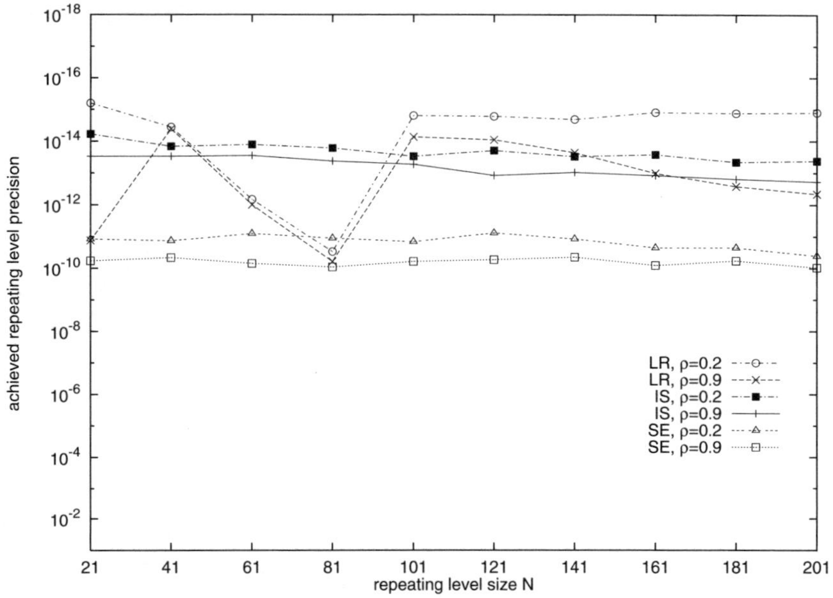


**Fig. 4.14.** The achieved repeating level accuracy for different repeating level sizes.

the eigenvalue problem which has to be solved. On the other hand, the fact that the quality of the SE solutions does not decrease even for large values of $N$ indicates that the numerical limits of the currently employed eigenvalue solution routine have not been reached yet.

Figure 4.15 depicts the boundary equation solution quality achieved in the experiments. First, note that the IS and LR algorithms again always provide accuracy levels which are better than the given value of $\epsilon = 10^{-8}$. Also, the relation of the corresponding curves to each other basically reflects the situation for the repeating level quality shown in Figure 4.14. Concerning the SE method, it can be observed that the achieved boundary solution precision severely deteriorates already at moderate level sizes, and in particular at low traffic intensities; at $N = 101$ and $\rho = 0.2$, the boundary equation residual is larger than $10^{-3}$. These results are due to the fact that the spectral expansion solution of the repeating levels leads to extremely ill-conditioned boundary equation systems. In several cases, the corresponding LAPACK solution routines reported the coefficient matrix being singular to floating point precision. We have been able to improve the boundary system condition by re-normalizing all eigenvectors to a certain, common length (which was determined experimentally), however, no general rule for this approach has been found so far, and it has yet to be investigated whether this approach can significantly improve the condition (the results in Figure 4.15 already account for this improvement). It should be noted that in spite of the apparently bad

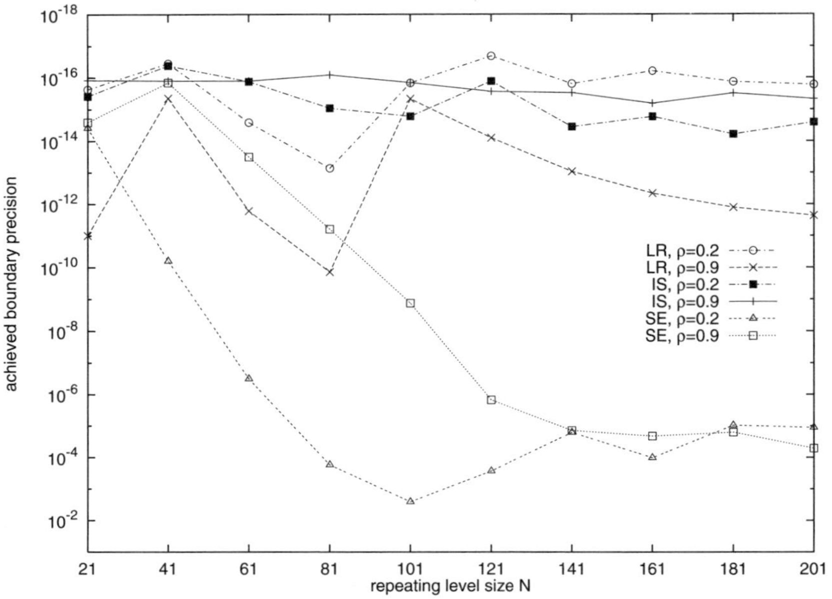


**Fig. 4.15.** The sensitivity of the achieved boundary accuracy on the repeating level size.

accuracy of the SE algorithm, we obtained both mean values (such as mean queue lengths) and individual steady-state probabilities which were very close to the corresponding values derived by the LR and IS solution. Nevertheless, it must be said that care must be taken if the SE approach is applied to QBD systems with repeating level sizes larger than 40.

*Impact of the Traffic Intensity.* In the investigation of the boundary solution accuracy of the SE method which has been presented above, the results indicated a strong dependency of the solution quality on the traffic intensity. We therefore conclude our numerical investigation study with an investigation of the impact of the traffic intensity on the achieved boundary solution quality.

In Figure 4.16, the boundary solution accuracy is given for the LR, IS and SE algorithms for a range of different traffic intensities. Note that two different repeating level sizes ($N = 41$ and $N = 161$) have been investigated. While no clear relationship between traffic intensity and boundary solution quality can be observed for the LR and IS methods, the precision achieved by these algorithms is again invariably better than the value of $\epsilon = 10^{-8}$. Concerning the SE method, it can be observed that the solution accuracy for the smaller system increases significantly for larger traffic intensities. This is in contrast to the larger system, where a good solution quality is only achieved at an extremely low traffic intensity. Thus, we conclude that the critical boundary system condition in the SE case in fact exhibits a strong

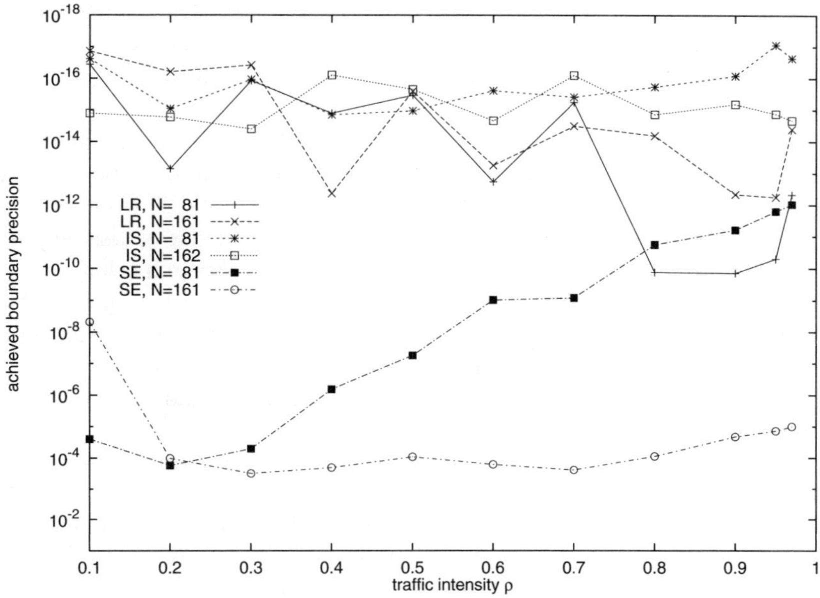


**Fig. 4.16.** The achieved boundary accuracy for different traffic intensities.

dependence on the traffic intensity, however, a general recommendation for "suitable" traffic intensities can not be given. Nevertheless, the fact that the SE method *can* provide accurate results at certain traffic intensities may provide a basis for future activities to improve the general applicability of the SE algorithm.

### 4.5.4 Conclusion

The results of our numerical comparison study can be divided in two areas, focusing on the computational speed of the algorithms on the one hand, and on the solution quality on the other hand.

Concerning the computational effort, it can be said that the LR algorithm in its "pure" form has been superseded by Naoumov's improved computation scheme. Both Naoumov's method and the spectral expansion approach can be considered as the fastest methods which are currently available for the analysis of QBDs. Naoumov's algorithm represents a general-purpose method which provides good results for skip-free QBDs up to reasonably high traffic intensities; the spectral expansion method performs particularly favourable at extreme traffic intensities, and for systems which involve batch arrivals. Especially in the latter case, the spectral expansion method completes significantly faster, as it can be applied directly to the non-skip-free QBD process, and no reblocking procedure has to be applied.

At least in its current form (i.e., based on the iterative evaluation of the matrix-sign function), the invariant subspace approach can not be considered as an alternative to these algorithms, as it takes significantly more time in *all* our experiments.

Regarding the solution quality and numerical stability of the algorithms, it must be said that the spectral expansion method can only be seen as an alternative to Naoumov's method if the order of the QBD process is relatively small. For repeating level sizes $N \geq 40$, the results of this algorithm should be subject to additional accuracy checks, as we experienced severe problems concerning the condition of the boundary equation system. While all other algorithms proved to be numerically stable, it is worth noting that the LR and NA algorithms scale better with the desired accuracy than the invariant subspace approach, which often yields much more precise results than actually desired.

To conclude, it must be said that Naoumov's method is currently the most efficient and numerically stable algorithm of those which have been considered in our study. The spectral expansion method represents a powerful alternative, however, until accuracy issues can be resolved, its application is restricted to smaller QBDs and non-skip-free processes, especially under high traffic intensities.

## 4.6 QBD Extensions

The aim of this Section is the suggestion and discussion of possible extensions of standard QBD processes. The following sections cover four main areas which have been identified. In Section 4.6.1, we discuss the application of approximation techniques for the solution of QBD systems. The focus of Section 4.6.2 is an approach to deal with immediate returns from any repeating level to the boundary and border levels. Section 4.6.3 presents some thoughts on applying the concept of *quasi-stationarity* to QBD processes, and in Section 4.6.4, the treatment of multi-dimensional QBD processes is discussed.

### 4.6.1 Approximate Analysis

Approximations can be considered as a key approach in situations where the exact solution of a problem is not possible due to its hard theoretical tractability or due to computational constraints in the implementation of a solution algorithm. However, as of today, no approaches exist for the approximate solution of QBD Markov chains. Especially in the field of model-based performance evaluation, such solutions would prove useful, as exact system characteristics are often not available, and the qualitative system behaviour is of more interest than exact state probabilities.

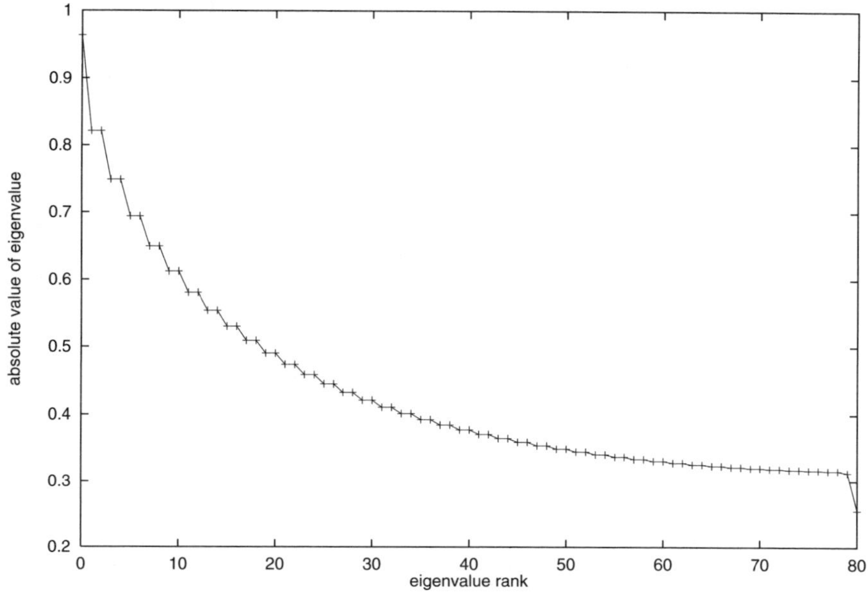


**Fig. 4.17.** The magnitude of the eigenvalues occurring in the spectral expansion solution of the example model considered in Section 4.5 ($\rho = 0.9, N = N_0 = 81$).

In this context, especially the spectral expansion method has an interesting property which may lead to an approximate solution algorithm. The fact that the steady-state solution vectors are expressed in terms of powers of eigenvalues inside the unit disk (see Eq. (4.38)) implies that the impact of small eigenvalues on the steady-state probabilities becomes negligible as states not too close to the border of the state space are considered. Approximate solutions may thus be derived by omitting the computation of small eigenvalues by specialized numerical methods for eigenvalue problems.

We investigated the eigenvalues which appear in the example model considered in Section 4.5 in order to estimate the magnitude of large and small eigenvalues; the absolute values of the eigenvalues (in descending order) are illustrated in Figure 4.17. As can be observed, the absolute values of the largest and smallest eigenvalues differ substantially; furthermore, the absolute values of the smaller eigenvalues are much closer to each other than those of the larger eigenvalues, except for the smallest eigenvalue. It thus seems promising to omit the solution of all smaller eigenvalues, or to just compute them partly and interpolate missing eigenvalues. If the number of eigenvalues which are actually used is reduced, then the problem has to be tackled that the linear combination (4.38) consists of fewer eigenvalue/eigenvector pairs and coefficients $x_i$, leading to an over-specified boundary equation system.

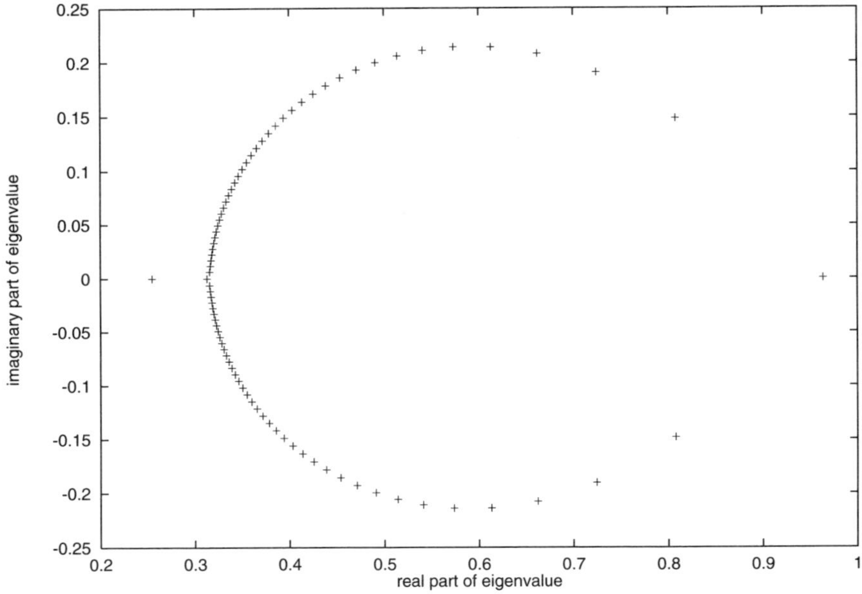


**Fig. 4.18.** A plot in the complex plane of the eigenvalues occurring in the spectral expansion solution of the example model considered in Section 4.5 ($\rho = 0.9, N = N_0 = 81$).

For completeness, we also provide a plot of the computed eigenvalues in the complex plane, see Figure 4.18. A special structure can also be recognized here. In addition to approximation purposes, this particular structure may be used to derive specialized eigenvalue solution routines. Note that similar results have been obtained for other system sizes and traffic intensities.

### 4.6.2 Buffer Resets

In standard QBD processes (as presented in Section 4.1), non-zero transition rates are only allowed between states in identical or adjacent levels. This leads to the typical block-tridiagonal transition matrix structure which yields the regular structure of the global balance equations for the repeating levels (see Eq. (4.8)), which can than be solved efficiently.

As the coefficients of each global balance equation correspond to the individual *columns* of the generator matrix, a more general form can be allowed in those columns of the generator matrix which do not affect the balance equations for the repeating levels (i.e., the first $N_0 + N$ columns). In matrix form, this corresponds to the following structure:

$$
\mathbf{Q} = \begin{pmatrix}
\mathbf{B}_{0,0} & \mathbf{B}_{0,1} & & & & & \\
\mathbf{B}_{1,0} & \mathbf{B}_{1,1} & \mathbf{A}_0 & & & & \\
\mathbf{B}_{2,0} & \mathbf{B}_{2,1} & \mathbf{A}_1 & \mathbf{A}_0 & & & \\
\mathbf{B}_{3,0} & \mathbf{B}_{3,1} & \mathbf{A}_2 & \mathbf{A}_1 & \mathbf{A}_0 & & \\
\mathbf{B}_{4,0} & \mathbf{B}_{4,1} & & \mathbf{A}_2 & \mathbf{A}_1 & \mathbf{A}_0 & \\
\vdots & \vdots & & & \ddots & \ddots & \ddots
\end{pmatrix}. \tag{4.62}
$$

The new matrices $\mathbf{B}_{j,0}$ and $\mathbf{B}_{j,1}$ contain the transition rates from *any* repeating level $i$ to the border and boundary levels, respectively.

From a modeling point of view, generators of the form (4.62) are useful to account for the often encountered situation that all jobs in a system are lost due to some special event (like, for example, a server breakdown). Structurally, such generator matrices can be considered as a special case of G|M|1 type matrices due to their lower block-tridiagonal form. The numerical solution of such models can thus easily be accomplished with methods for G|M|1 type Markov chains, however, many of the much more efficient methods for QBD processes may be applicable as well, since the form of the balance equations for the repeating levels remains unchanged. These methods must, however, be suited to deal with the fact that the row sums of $\mathbf{A}_0 + \mathbf{A}_1 + \mathbf{A}_2$ are no longer equal to one. Due to its probabilistic interpretation, this is intuitively true for the LR method and Naoumov's improved LR algorithm. Also the spectral expansion method is applicable, as its solution approach imposes no special restrictions on the considered set of vector difference equations.

No matter what approach is taken to solve the repeating part of the state space, the boundary equations (4.6) and (4.7) must now account for the newly introduced matrices $\mathbf{B}_{j,0}, j \geq 2$ and $\mathbf{B}_{j,1}, j \geq 3$, yielding

$$
\mathbf{b}\mathbf{B}_{0,0} + \sum_{j=0}^{\infty} \mathbf{v}_j \mathbf{B}_{j+1,0} = \mathbf{0} \tag{4.63}
$$

$$
\mathbf{b}\mathbf{B}_{0,1} + \sum_{j=0}^{\infty} \mathbf{v}_j \mathbf{B}_{j+1,1} = \mathbf{0}. \tag{4.64}
$$

If there exists a limit $j'$ from which on all matrices $\mathbf{B}_{j,0}, j > j'$ and all matrices $\mathbf{B}_{j,1}, j > j'$ are identical, then these equation can easily be solved due to the geometric structure of the solution vectors $v_j$. This is, for example, the case if a constant system failure rate shall be modeled. While more elaborate methods may be needed for more complicated cases, the presented approach presents a feasible solution to account for many situations where immediate resets of the buffer contents shall be modeled.

### 4.6.3 Quasi-Stationary Solution

Quasi-stationary distributions can be used to describe the long-term behaviour of *transient* (non-ergodic) Markov chains which seem to settle down to an equilibrium before the transient behaviour eventually manifests. An illustrative example in this context is a Markov chain which models a system with an extremely low failure rate, leading to eventual absorption in a "broken" state. Clearly, this Markov chain is non-ergodic, nevertheless, the state probabilities of the quasi-equilibrium which settles down before absorption takes place is of high practical interest. Recently, the quasi-stationary distribution has also been proposed to account for the dependencies between subsystems in decomposition approaches [146].

The consideration on quasi-stationary distributions emerged in the early sixties with the initial works of Vere-Jones [191] and Kingman [117]. Formally, quasi-stationary distributions are closely related to the *limiting conditional distribution* $\boldsymbol{\rho}$ of a Markov chain, which is given by the limit

$$\lim_{t\to\infty} \Pr(X(t) = j | \text{ absorption eventually takes place}) = \rho_j. \tag{4.65}$$

Under the assumption that absorption is certain, the above probabilities are independent of $t$, and

$$\Pr(X(t) = j | \text{ absorption eventually takes place}) = \rho_j. \tag{4.66}$$

is a quasi-stationary distribution. For finite Markov chains, the limiting conditional (and thus, quasi-stationary) distribution is given by the solution of the eigenvalue problem

$$\boldsymbol{\rho}\mathbf{Q} = -\lambda\boldsymbol{\rho}, \tag{4.67}$$

where $\lambda$ denotes the smallest eigenvalue of $\mathbf{Q}$, and $\boldsymbol{\rho}$ is the corresponding eigenvector (normalized to yield a proper distribution). $\lambda$ is then referred to as the *decay rate* of the Markov chain.

Recently, quasi-stationary solutions were also considered in the context of absorbing quasi-birth-and-death processes, however, no general solution is available yet. For the special case where absorption is only possible from the border states of a QBD process, a generalization of the LR approach has been derived in [21]. We found [147] that the eigenvalues which appear in the spectral expansion solution of an absorbing QBD processes are related to $\lambda$, but the introduction of absorption rates from arbitrary states of a QBD process still needs considerable work. In some cases, the quasi-stationary distribution does not exhibit a geometric structure, even when considering simple birth-and-death processes. It is also worth mentioning that the intuitive solution of restricting the state space of an absorbing QBD to a finite subset of states is difficult (see [50] for a discussion).

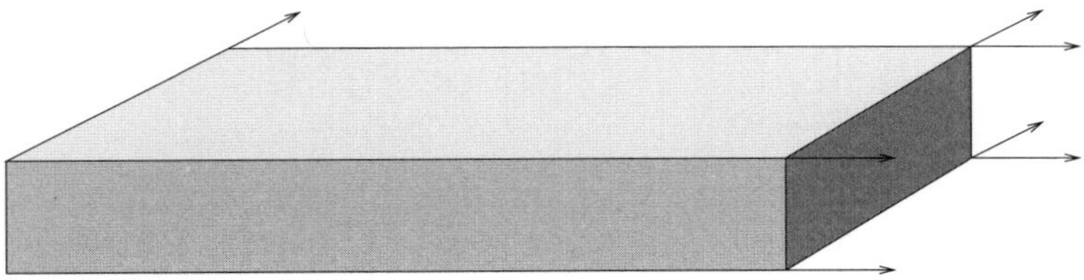

**Fig. 4.19.** State space structure of a multi-dimensional QBD process which involves two unbounded buffers.

### 4.6.4 Multi-Dimensional QBD Processes

In this chapter, QBD process were introduced as an efficient approach to model and analyze systems with complex interarrival and service time distributions which involve a very large or unbounded buffer. A natural question in this context is whether this can be extended to systems which involve *multiple* buffers. In theory, additional finite buffers can be introduced by accounting for them in the QBD's finite dimension, however, for any reasonable buffer size, this quickly leads to QBD processes of an order which is hardly tractable in practice. Furthermore, this approach can not be taken if additional infinite buffers are to be considered. Thus, the question arises whether it is possible to introduce an additional dimension to the QBD state space for each additional buffer, leading to a multi-dimensional QBD process.

As an example, consider the state space structure arising from the simultaneous consideration of two unbounded buffers, as depicted in Figure 4.19. The (non-boundary) transition rates of such a system can easily be described by nine matrices, which, in addition to the conventional repeating level transition rate matrices $\mathbf{A}_0, \mathbf{A}_1$ and $\mathbf{A}_2$, describe transition rates between levels in the second infinite dimension, and simultaneous transitions in both unbounded dimensions. By looking at the state space "from the top", the transition behaviour can be described as shown in Figure 4.20.

The extension of conventional solution approaches for QBDs to multi-dimensional QBDs is, however, a formidable task. As the overall state space of any Markovian queueing *network* can be described by a multi-dimensional QBD, a generic solution approach also provides an efficient approach for the analysis of arbitrary queueing networks.

Even for the simple case of two unbounded buffers and a repeating level size of 1 (i.e., the class of random walks on a two-dimensional grid), only few results are available. Clearly, if there are no transitions which simultaneously change both state variables, then we are dealing with a system of two independent M|M|1 queues, which has a simple product form solution (for other conditions for a product form solution, see e.g. [19]). For general random walks, the solution becomes much more difficult. There exist approaches based on generating functions (see e.g. [24, 49]), however, they often do not lead to explicit expressions for the steady-state probabilities. The *compen-*

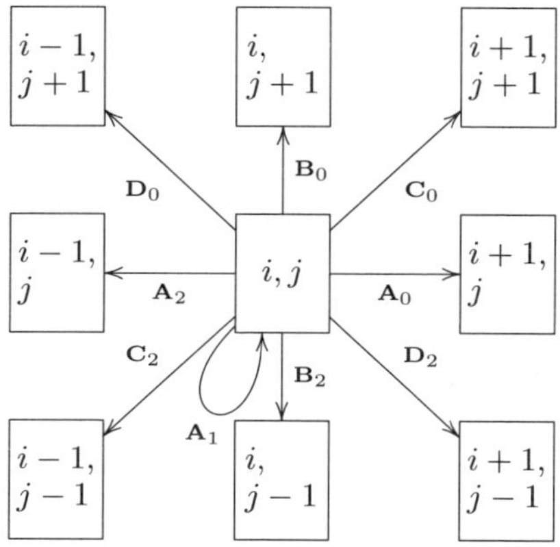


**Fig. 4.20.** The matrices needed to describe the transition rates of a two-dimensional QBD process.

*sation approach*, proposed in [1] and further extended in [103], is based on deriving product-form solutions for the non-boundary states, and then superposing them to satisfy the boundary conditions. The approach is capable to account for more than two unbounded dimensions, however, a rather strict limitation concerning the "drift" of the random walk is assumed (basically, a positive step in one dimension is only permitted in parallel with negative steps in all other dimensions). In [30], a recursive algorithm for expressing the steady-state probabilities in terms of a power-series has been presented. This approach is particularly interesting in view of its ability to deal with the complete class of multi-dimensional QBDs, including the consideration of more than two unbounded dimensions. The recursive algorithm, however, still lacks a thorough formal foundation, and it does not provide error bounds on the computed solutions. Furthermore, the practical applicability of this approach may be restricted due to the fact that its memory requirements increase exponentially with the number of involved queues, and that the computation time depends on the traffic intensity and the number, the variation coefficients, and the similarity of the involved queues.

## 4.7 Summary and Concluding Remarks

In this chapter, we focused on a thorough treatment of QBD processes from a modeler's point of view, in order to employ them for modeling and analyzing complex queueing systems. Therefore, the investigations concentrated on two main issues: first, a comprehensive survey on available methods for the steady-state analysis of QBD processes has been provided, and secondly, an extensive comparison of these algorithms has been performed.

In our overview of solution methods, we included standard methods like the successive substitution method and the well-established LR approach as well as more recent developments like Naoumov's algorithm, the spectral expansion method, and the invariant subspace approach. Without going too much into details of the solution methods, we aimed at providing enough information to understand the basic idea of all algorithms, to relate them to each other, and to provide guidance in the implementation of the various approaches. Furthermore, in our definition of QBD processes, special care has been taken to efficiently deal with the non-regular transition behaviour at the boundary of the QBD state space; this has also been accounted for in the computational complexity estimates which were provided for all presented algorithms. Finally, the suitability of the algorithms for analyzing systems with batch-arrivals has been discussed.

In the second part of this chapter, we conducted a thorough numerical comparison of the different approaches in an application-oriented case study. While we concentrated on one base model only, the sensitivity of the solution algorithms on basic properties like model size and traffic intensity has been investigated, providing insight in the general behaviour of the algorithms as well. Our study extended the results of previously published comparison studies with respect to the following aspects:

- The presented comparison accounts for recent developments like the spectral expansion method, the invariant subspace approach, and Naoumov's improved LR algorithm, which have so far not been compared in other studies.
- We investigated the suitability of the various algorithms to analyze non-skip-free QBD processes.
- In addition to the computational cost of the algorithms, we also considered the achieved solution quality and the numerical stability of the approaches.

In our study, Naoumov's improved LR method was found to be the most robust and efficient solution approach in most cases. The spectral expansion method represents a very attractive alternative for smaller systems and non-skip-free systems, especially under high traffic intensities; however, for larger QBDs, this method suffers from numerical instabilities.

Our considerations concluded with a discussion of further extensions to the class of QBD processes. Here, especially accounting for additional transitions from arbitrary states to the border and boundary levels represents a feasible extension which can easily be integrated in existing modeling environments.

To summarize, this chapter provides a thorough guide to selecting and applying suitable algorithms from a range of the most recent and efficient approaches which are available for the steady-state solution of QBD Markov chains. From an *analytical* point of view, this provides a basis to realize an efficient performance evaluation environment allowing for much more detailed node models than those provided by the approaches investigated in Chapter 3.

Now, the question has to be tackled how detailed QBD node models can be specified in a user-friendly way; the direct specification of a QBD node on the Markov chain level is certainly not an approach many modelers are willing to take. Therefore, a high-level *modeling* approach which specifically fits the QBD-based environment will be developed in the next chapter; it enables the modeler to avoid the cumbersome and error-prone model specification directly on Markov chain level.

# 5. High-Level System Specification with iSPNs

In the previous chapter, QBD processes have been introduced as a powerful class of stochastic processes which is amenable to very efficient solution algorithms. Furthermore, the QBD framework provides a concise and memory-efficient description of the infinitely large generator matrix. Thus, from a computational point of view, it is desirable to employ QBD processes as a low-level mechanism for the numerical analysis. From a user perspective, however, the development of QBD models directly at the Markov chain level suffers from the same drawbacks as the direct specification of large finite Markov chain models: the model development is cumbersome and error-prone, as is the derivation of the performance measures of interest.

Therefore, the purpose of this chapter is to develop a high-level model description formalism for the easy specification of performance models which can be analyzed by efficient techniques for QBD processes. The approach developed here is a natural extension of the stochastic Petri net (SPN) formalism, which already proved to be very powerful for the specification of finite Markov chains.

In Section 5.1, we motivate our decision for an SPN-based environment, and then formally introduce our proposed *infinite-state* SPN (iSPN) framework. The model class will also be illustrated by a small application example. The aim of Section 5.2 is to proof that *any* iSPN model specification is amenable to QBD-based solution algorithms, and, conversely, that any QBD can be described by an iSPN, thus ensuring that we can take advantage of the entire class of QBD processes. While the proofs conducted in this section already provide thorough guidelines for the design of an iSPN-based performance evaluation tool, we will comment on some additional implementation issues in Section 5.3. In Section 5.4, we briefly discuss extensions to our iSPN model class for dealing with buffer resets and the specification of quasi-stationary models. Finally, the results of this chapter are summarized in Section 5.5.

## 5.1 The iSPN Modeling Environment

In this section we develop a class of SPNs which is suitable to be analyzed by efficient numerical techniques for QBD processes. After motivating our choice of a SPN-based approach in Section 5.1.1, we will present the key ideas in Section 5.1.2, where related approaches in this area will also be discussed. In Section 5.1.3, we formally introduce the SPN class which forms the basis for our own class, and then present the extensions for its integration in a QBD-based evaluation framework. To illustrate the modeling formalism and to familiarize with our graphical notation, we conclude with the presentation of an example in Section 5.1.4.

### 5.1.1 High-Level Modeling Approaches

With the availability of fast computers and large amounts of memory at reasonable prices, it becomes possible to design more and more detailed system models which involve Markov chains with very large state spaces. However, whereas the numerical treatment of Markov chains has become more and more easy, system designers faced the problem that the direct construction of Markov chain models becomes cumbersome and error-prone due to the enormous amount of states. Thus, several approaches have been suggested for a high-level specification of large Markov chains.

Queueing models may be considered as one of the first approaches for a concise description formalism, however, their focus is a rather particular kind of systems which involve buffers with jobs which arrive and are served according to certain distributions.

Another prominent class of approaches stems from the Petri net formalism, initially suggested by Petri in the mid-60s [163]. By extending this formalism to account for timed behavior (see e.g. [5, 145] and [20] for a good introduction), it became possible to specify large Markov chains in an intuitive graphical way.

A third important approach for the specification of system models is derived from the theory of process algebras (SPAs) like CSP [99], CCS [141] or LOTOS [33] by extending it so that timing information can also be dealt with (see [97] for an overview). Here, one of the main advantages is the fact that large systems can easily be constructed from smaller subsystems (a property known as *compositionality*). There also exists a first approach for employing SPAs for the specification of QBDs, see [143, 156].

The aim of the work in this chapter is to bring QBD processes and the corresponding solution methods from Markov chain level to the end user. Here, we favour the intuitive graphical specification of stochastic Petri nets to reach this task. The stochastic Petri net formalism is closer to the idea of a state-based system description (and thus, the underlying Markov chain) than SPAs are. On the one hand, this makes it easier to develop appropriate SPNs for the specification of QBD processes, on the other hand, this will also enable the system designer to decide which specification alternative is better from a computational point of view. Last but not least, we consider the popularity of SPN models in the performance evaluation field as a good basis for the proliferation of QBD-based modeling in this area.

### 5.1.2 Basic Idea and Related Approaches

In conventional SPNs, it is usually assumed that the Markov chain which is obtained from an SPN specification is finite. Thus, the steady-state solution of the model can be obtained by solving a finite system of global balance equations. The finiteness of the underlying Markov chain implies that the SPN is *bounded*, i.e., there is some maximum number of tokens which can be found in any place of the SPN.

The main idea of the SPN class to be introduced in this section is that we allow for *one* place to be unbounded. We will then associate the number of tokens in this place to the level of the underlying QBD process. As all other places are still bounded, the number of markings of the SPN which correspond to one QBD level is finite. By introducing some slight limitations on the SPN class, we are able to guarantee that, except for some unrestricted boundary behaviour, the level size is constant, and that all transitions within and between levels comply with the restrictions of QBD processes. This approach has initially been suggested in [88, 89], with additional implementation work accomplished in [118] and [73]. In this chapter, we will provide a thorough formal basis for this approach to prove its main properties, which so far were understood in an intuitive manner only.

There also exist other approaches which try to account for unbounded SPNs. In particular, the class of *one-place unbounded* SPNs (also referred to as *complex-queue* nets) which has been introduced in [70, 71] is related to our approach. As in our approach, the authors allow for one place of unbounded capacity, and derive a QBD description of the underlying Markov chain. Then, the Markov chain is analyzed with either the successive substitution method or by spectral expansion (in [71] termed as *generalized Eigenvalue approach*). The approach differs from the one presented here in that the considered class of SPNs is much more restricted, as it does not allow for enabling functions or immediate transitions. Instead, the authors focus on this restricted SPN class to prove ergodicity conditions of the underlying stochastic process.

Another related approach has been proposed in [204, 205]. The considered class of SPNs also allows one place of unbounded capacity, however, the author assumes that the considered SPNs are of *free-choice* type.[1] Furthermore, the approach does neither allow for immediate transitions and arc multiplicities, nor does it consider marking-dependencies or enabling functions. The analysis of the underlying Markov chain is accomplished by building sets of equivalent markings and assuming a geometric relationship between the steady-state probabilities of states in these classes. Thus, though the model class is less expressive, the performance analysis can be accomplished without the application of computationally more expensive matrix-geometric procedures.

[1] In a free-choice SPN, it is required that if a place has more than one output transition, then all of them must have the same set of predecessor places.

### 5.1.3 Formal Definition of iSPNs

In this section we formally define our class of *infinite-state* SPNs. We will first summarize the general modeling features of this SPN class in an initial definition. In a second step, the class will be extended specifically for a seamless integration into a QBD-based analysis framework. Note that the definitions given here represent a strict formalization of the SPN classes suggested in [88], [89] and [90]. The aim of this section is to build a strict formal basis for the proofs which are conducted in the following sections.

Our model class is based on the well-known class of *generalized* SPNs (GSPNs), initially suggested in [5]. It incorporates many of the GSPN extensions proposed in [3] and can be seen as a simplified version of the class of *stochastic reward nets* (SRNs) [47]:

- Transition priorities can be used to establish firing preferences. The lowest priority is assigned to timed transitions, so that immediate transitions always have precedence.[2]
- Similar to [3] and [47], weights are used to probabilistically select among concurrently enabled immediate transitions of identical priority.
- Arc multiplicities allow the simultaneous removal or generation of several tokens.

Compared to extended GSPNs and SRNs, the restrictions of our class concern the marking dependency of rates, weights, and arc multiplicities:

- SRNs allow for marking-dependent arc multiplicities. In our SPN class, constant multiplicity values are assumed.
- Rates and weights of transitions can be assigned different values in each marking of an extended GSPN or an SRN. We allow for a weaker marking-dependency by introducing *enabling functions* to explicitly disable transitions depending on the number of tokens in a place, or by allowing the rate or weight of a transition to be multiplied by the number of tokens in a place.

These restrictions have been taken to simplify the definition of marking dependencies. The explicit and completely free definition of marking-dependent rates and weights would quickly lead to confusing model specifications. So far, these restrictions did not impose any problems in all our application studies.

The properties of our SPN class are formally summarized in the following definition:

**Definition 2.** A (generalized) *stochastic Petri net (SPN)* is a tuple

$$\mathcal{P} = (P, T, Pr, I, O, W, R, E, \mathbf{m}_0),$$

where the individual components are defined as follows:

[2] Note that SRNs not only allow to assign priority values to transitions, but the definition of a more general transitive priority relation ">" among transitions. This way, the SPN-wide impact of transition priorities can be avoided.

- $P$ and $T$ denote the finite sets of places and transitions.
- $Pr: T \to \mathbb{N}_0$ assigns a priority to each transition. The lowest priority 0 is reserved for timed transitions.
- $I: P \times T \to \mathbb{N}_0$ and $O: T \times P \to \mathbb{N}_0$ are functions mapping multiplicities to the input and output arcs of the Petri net. Multiplicities of 0 indicate non-existing arcs.
- $W : T \to \mathbb{R}^+$ and $R : T \to P \cup \{\perp\}$ assign a firing rate to each timed transition. Thus, the firing time of a transition $t \in T$ is exponentially distributed with parameter $W(t)$. For immediate transitions, $W(t)$ indicates the transition's *weight.* In both cases, if $R(t) = p$, then the values $W(t)$ are to be multiplied by the number of tokens in place $p$. The symbol "$\perp$" is used to indicate that there is no place-dependence for a transition.
- Each distribution of tokens in the SPN is a called a *marking* and is represented as a vector in $M = {\mathbb{N}_0}^{|P|}$; $\mathbf{m}_0$ is the initial marking of the Petri net.
- $E : T \to 2^F$ assigns a set of enabling functions $f \in F$, $f : M \to \{0, 1\}$ to each transition. We restrict to enabling functions $f_{p<n}, f_{p=n}, f_{p>n}, n \in \mathbb{N}_0, p \in P$, which yield 1 if the number of tokens in place $p$ is less than, equal to, or larger than a value $n$, and which equal 0 otherwise. It is a necessary condition that all enabling functions of a transition evaluate to 1 before it can fire.

Note that this definition does not explicitly account for inhibitor arcs. As they represent a special case of an enabling function ($f_{p<1}$), they are not treated separately here.

Later, we will extend this SPN class to ensure that the underlying Markov chain is of QBD type. In order to show this, we need to clearly specify a relation which identifies possible transitions between markings of the SPN, since these correspond to transitions in the underlying Markov chain. We use the notation $\mathbf{m}_p$ to denote the number of tokens in place $p \in P$ in a particular marking $\mathbf{m} \in M$.

Using the terminology of [3], a transition $t \in T$ is called *firable* in a marking $\mathbf{m} \in M$, if

- all places contain more tokens than the corresponding input arc multiplicities, and
- all its enabling functions evaluate to 1, and
- the transition's rate (or weight) is non-null (thus, we allow for the *implicit disabling* of transitions, which occurs when the marking dependent rate/weight multiplier function $R$ yields 0).

A transition is called *enabled* in a marking $\mathbf{m}$, if it is firable and there are no other firable transitions of higher priority. All transitions which are enabled in a marking $\mathbf{m}$ are summarized in the set $En(\mathbf{m})$. More formally, the set $Fir(\mathbf{m})$ of firable transitions is defined as

$$\begin{aligned} Fir(\mathbf{m}) = \{t \in T | \forall p \in P : \mathbf{m}_p \geq I(p,t) \wedge \\ \forall f \in E(t) : f(\mathbf{m}) = 1 \wedge \\ R(t) \neq \perp \Rightarrow m_{R(t)} > 0\}. \end{aligned} \tag{5.1}$$

Then $Em(\mathbf{m})$ contains the firable transitions with the highest priority:

$$En(\mathbf{m}) = \{t \in Fir(\mathbf{m}) | \nexists t' \in Fir(\mathbf{m}) : Pr(t') > Pr(t)\}. \tag{5.2}$$

With this definition, several relations can be defined to specify whether a marking can be reached from some other marking by the firing of one or more transitions. With these relations, we will derive the reachability set of all markings which can occur in the specified SPN if we start with $\mathbf{m}_0$. This is summarized in the following definition:

**Definition 3.** The one-step transition-relation $\stackrel{\cdot}{\longrightarrow} \subseteq M \times T \times M$ indicates whether a marking $\mathbf{m}_2$ can be reached from a marking $\mathbf{m}_1$ by firing a transition $t \in T$. It is given by

$$\begin{aligned} \mathbf{m}_1 \stackrel{t}{\longrightarrow} \mathbf{m}_2 \iff t \in En(\mathbf{m}_1) \wedge \\ \forall p \in P : \mathbf{m}_{2,p} = \mathbf{m}_{1,p} - I(p,t) + O(t,p). \end{aligned} \tag{5.3}$$

The relation $\longrightarrow \subseteq M \times M$, defined as

$$\mathbf{m}_1 \longrightarrow \mathbf{m}_2 \iff \exists t \in T : \mathbf{m}_1 \stackrel{t}{\longrightarrow} \mathbf{m}_2, \tag{5.4}$$

indicates that $\mathbf{m}_2$ is reachable from $\mathbf{m}_1$ by firing any transition. By considering the reflexive and transitive closure $\stackrel{*}{\longrightarrow}$ of $\longrightarrow$, we can define the *reachability set* $Rs(\mathcal{P}) \subseteq M$ of all markings which are reachable from the initial marking $\mathbf{m}_0$:

$$Rs(\mathcal{P}) = \{\mathbf{m} \in M | \mathbf{m}_0 \stackrel{*}{\longrightarrow} \mathbf{m}\}. \tag{5.5}$$

In conventional Petri nets, it is assumed that the reachability set is finite, so that well-known techniques for solving finite systems of linear equations can be applied. As the markings in the reachability set correspond to the states of the underlying Markov chain of the SPN, this guarantees the finiteness of the underlying Markov chain. $Rs(\mathcal{P})$ being finite implies that there is some upper limit on the number of tokens in any place. In the class of SPNs we define below, this requirement will be dropped. Infinite reachability sets will be obtained by allowing one place in the SPN to take any integral, non-negative number of tokens. However, in order to guarantee that the underlying (infinite) Markov chain is of QBD type, some additional assumptions have to be made:

**Definition 4.** An *infinite-state stochastic Petri net (iSPN)* is a stochastic Petri net $\mathcal{P} = (P, T, Pr, I, O, W, R, E, \mathbf{m}_0)$ which complies with the following conditions:

– $Rs(\mathcal{P})$ is *infinite*.
– Except for one designated place $\mathsf{P}_0 \in P$, the number of tokens in all places is bounded:

$$\exists n_{\max} \in \mathbb{N} : \forall \mathbf{m} \in Rs(\mathcal{P}), p \in P : p \neq \mathsf{P}_0 \Rightarrow \mathbf{m}_p < n_{\max}. \tag{5.6}$$

– In the initial marking, the number of tokens in $\mathsf{P}_0$ is 0: $\mathbf{m}_{0,\mathsf{P}_0} = 0$.
– The multiplicity of all arcs involving $\mathsf{P}_0$ is equal to one:

$$\forall t \in T : I(\mathsf{P}_0, t) \leq 1 \wedge O(t, \mathsf{P}_0) \leq 1. \tag{5.7}$$

– $\mathsf{P}_0$ must not occur in rate multiplier functions:

$$\forall t \in T : R(t) \neq \mathsf{P}_0. \tag{5.8}$$

– Connections between immediate transitions and $\mathsf{P}_0$ are limited to "test" arcs. That is, the firing of an immediate transition must not alter the number of tokens in $\mathsf{P}_0$:

$$\forall t \in T : Pr(t) > 0 \Rightarrow I(\mathsf{P}_0, t) = O(t, \mathsf{P}_0). \tag{5.9}$$

Note that the additional requirements in this definition only affect those parts of the iSPN which are associated to $\mathsf{P}_0$. Since the reachability set of conventional SPNs is usually assumed to be finite, iSPNs can be considered as a superset of the SPN class that has been specified in Definition 2. Also note that the last requirement (5.9) will be relaxed in Section 5.3.3.

The intuition behind the above definition is that the number of tokens in $\mathsf{P}_0$ in a particular marking will be associated to the level index of the corresponding state in the underlying QBD Markov chain. The limited multiplicity of arcs which involve $\mathsf{P}_0$ reflects the fact that only transitions within a level or between neighbouring levels are allowed. Finally, the absence of $\mathsf{P}_0$ in rate multiplier functions assures that transition rates are independent of the level to which a marking will be mapped. We will formally proof that the underlying Markov chain is of QBD type in Section 5.2.

### 5.1.4 An Example iSPN Model

In this section, we briefly present a smaller example iSPN model to illustrate the graphical representation of iSPNs as well as a typical application of how the unbounded place can be used in a modeling study.

The system under consideration is a transaction processing system as also discussed in [89]. Jobs (e.g., database transactions) enter a system and are served by a server which is subject to breakdowns. The server performs a "checkpointing" operation every time $K$ transactions have been completed (thus, $K$ is referred to as *checkpointing interval*). If the server breaks down, all transactions which have not been checkpointed yet have to be served again.

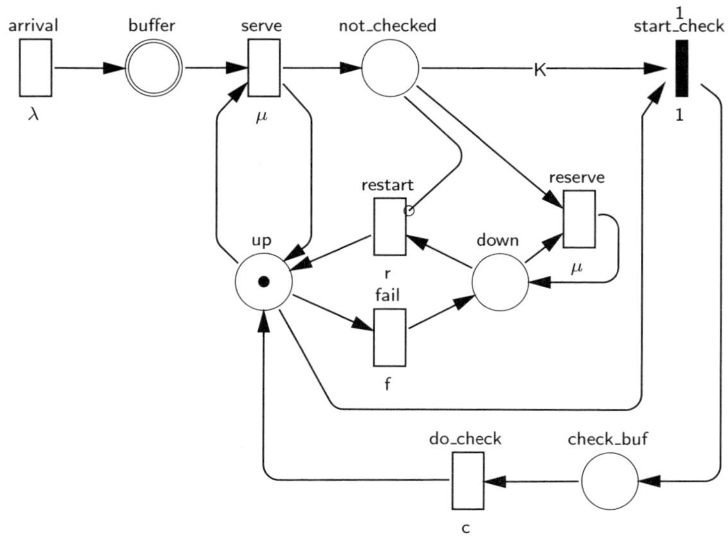


**Fig. 5.1.** Example iSPN model of a checkpointing transaction processing system.

The corresponding iSPN model is illustrated in Figure 5.1. Concerning notation, places are as usual represented by circles; the iSPN-specific unbounded place is distinguished by a double circle. Timed and immediate transitions are illustrated by hollow rectangles and thin bars, respectively. The rates and weights of transitions are denoted below their graphical representation, while the priority of immediate transitions is denoted above them. Arc multiplicities larger than 1 are denoted amidst the arcs. For convenience, we also denote inhibitor arcs (which are terminated by a circle) as a shortcut for the corresponding enabling function.

Transactions enter the system in a Poissonian manner by firings of transition arrival. The transactions are then awaiting service and queue up in a buffer, which is modeled by the unbounded place buffer. If the server is operational (as indicated by a token in place up), jobs are taken out of the buffer and served by transition serve. The served transactions are then stored in place not_checked since they have yet to be checkpointed. Due to the arc multiplicity $K$ of the arc to transition start_check, the checkpointing procedure is initiated immediately as soon as $K$ transactions have been served. Note that the server is allocated to perform the checkpointing operation, as indicated by an arc from up to transition start_check. After the checkpointing has been accomplished (as indicated by the firing of transition do_check), the server is available for normal servicing again.

The firing of transition fail indicates a server breakdown. In that case, all transactions which have not been checkpointed yet have to be re-served, as indicated by transition reserve. Due to the inhibitor arc to transition restart, the server can return to normal operation only after all transactions in not_checked have been re-served.[3] For simplicity, our model assumes that the server cannot fail when it is checkpointing or recovering.

Note that using the iSPN's unbounded place to model some kind of large or unbounded buffer (as in this example) is a quite common application (as also illustrated by the case studies in Chapter 6). Often, the policy of how jobs enter and leave the buffer is independent of the buffer filling, thus, restriction (5.8) is not of practical importance in many cases. The same holds for condition (5.7) as long as no platooned customer arrivals or services have to be modeled. If this is the case, a "folding" procedure like the one suggested in [92] can be applied to ensure that all iSPN restrictions are still fulfilled.

## 5.2 Equivalence to QBD Markov chains

The aim of this section is to proof that the stochastic processes underlying iSPNs are always QBDs, and (vice versa) that any QBD can be formulated as an iSPN. Thus, we show that iSPNs represent a high-level description equivalent for QBD processes.

After briefly dealing with notational issues and deriving some basic results in Section 5.2.1, we first show in Section 5.2.2 that QBDs are obtained from iSPNs if a special condition holds. Afterwards, this result will be extended in Section 5.2.3 to hold for arbitrary iSPNs, without any additional assumptions. We conclude with a proof that any QBD can be formulated at the iSPN level in Section 5.2.4, thus leading to the main result: the equivalence of the specification of iSPNs to the direct specification of a QBD process. In the remainder of this chapter, we consider an iSPN $\mathcal{P}$ as defined in Definition 4.

### 5.2.1 Preliminaries

Concerning notation, we will often refer to a marking $\mathbf{m}$ as $(j, \bar{\mathbf{m}})$, where $j = \mathbf{m}_{\mathsf{P}_0}$ and $\bar{\mathbf{m}}$ is referred to as *submarking* which holds the number of tokens in all places except $\mathsf{P}_0$. Furthermore, we partition the reachability set in so-called $j$-sets $S_j$ with identical number of tokens in place $\mathsf{P}_0$: $S_j = \{(j, \bar{\mathbf{m}}) \in Rs(\mathcal{P})\}$. To solely focus on the submarkings in a $j$-set, we will also use the sets $\bar{S}_j = \{\bar{\mathbf{m}} | (j, \bar{\mathbf{m}}) \in S_j\}$.

The notion of a $j$-set forms the basis for mapping markings of an iSPN to levels of the underlying QBD process. In the following two propositions, we show that transitions can only occur within a $j$-set, or between adjacent $j$-sets. We then prove that there is a number of tokens in $\mathsf{P}_0$ from which point onwards the transitions between markings exhibit a repetitive structure.

[3] Note that according to the iSPN model, jobs are actually re-served *before* the server restarts. While this does not affect buffer-oriented performance measures, this simplification must be considered if one is e.g. interested in the server availability characteristics.

**Proposition 1.** For all transitions between markings in an iSPN, the following condition holds:

$$(j, \bar{\mathbf{m}}) \longrightarrow (k, \bar{\mathbf{m}}') \Rightarrow |j - k| \leq 1. \tag{5.10}$$

*Proof.* If there is no arc connecting transition $t$ to place $\mathsf{P}_0$, then the number of tokens in $\mathsf{P}_0$ remains constant if $t$ fires. Thus, the destination marking is in the same $j$-set as the originating marking and the proposition is fulfilled, as $j - k = 0$. If transition $t$ is connected to $\mathsf{P}_0$, then the arc multiplicity is 1 according to (5.7). Thus, the number of tokens in $\mathsf{P}_0$ increases or decreases by one token or remains constant, leading to a destination marking in $j$-set $S_{j+1}, S_{j-1}$ or $S_j$, respectively. Hence, the proposition holds in this case as well. □

**Proposition 2.** Let

$$\begin{aligned} j_{\min} = \max(\{1\} \cup \{j \in \mathbb{N} | \exists t : & f_{\mathsf{P}_0 < j} \in E(t) \\ & \vee f_{\mathsf{P}_0 = j-1} \in E(t) \\ & \vee f_{\mathsf{P}_0 > j-1} \in E(t)\}) \end{aligned} \tag{5.11}$$

Then, the enabling of any transition in a marking $(j, \bar{\mathbf{m}})$ does not depend on $j$ for all $j \geq j_{\min}$:

$$\forall (j, \bar{\mathbf{m}}), j \geq j_{\min} : \; (j_{\min}, \bar{\mathbf{m}}) \xrightarrow{t} (j_{\min} + 1, \bar{\mathbf{m}}') \iff (j, \bar{\mathbf{m}}) \xrightarrow{t} (j + 1, \bar{\mathbf{m}}') \tag{5.12}$$

$$\wedge (j_{\min}, \bar{\mathbf{m}}) \xrightarrow{t} (j_{\min} - 1, \bar{\mathbf{m}}') \iff (j, \bar{\mathbf{m}}) \xrightarrow{t} (j - 1, \bar{\mathbf{m}}') \tag{5.13}$$

$$\wedge (j_{\min}, \bar{\mathbf{m}}) \xrightarrow{t} (j_{\min}, \bar{\mathbf{m}}') \iff (j, \bar{\mathbf{m}}) \xrightarrow{t} (j, \bar{\mathbf{m}}'). \tag{5.14}$$

Furthermore, the firing rate of transition $t$ which is enabled in marking $(j, \bar{\mathbf{m}})$ does not depend on $j$.

*Proof.* A transition in the iSPN is enabled if and only if all its enabling functions evaluate to 1, if all its predecessor places have a sufficient number of tokens, and if there are no firable transitions of higher-priority. Since the submarking $\bar{\mathbf{m}}$ is identical on both sides of (5.12)-(5.14), the number of tokens in all places except $\mathsf{P}_0$ is constant, and the enabling functions which do not involve $\mathsf{P}_0$ yield identical values. The enabling functions $f \in E(t)$ which do involve $\mathsf{P}_0$ are also constant, since

- if $f = f_{\mathsf{P}_0 < k} \in E(t)$, we have $j_{\min} \geq k$ according to Eq. (5.11). Thus, $f_{\mathsf{P}_0 < k}(j, \bar{\mathbf{m}})$ is 0 for all $j \geq j_{\min}$.
- If $f = f_{\mathsf{P}_0 = k} \in E(t)$ or $f = f_{\mathsf{P}_0 > k} \in E(t)$, we have that $j_{\min} - 1 \geq k$ and thus $j_{\min} > k$. Hence, $f_{\mathsf{P}_0 = k}(j, \bar{\mathbf{m}}) = 0$ and $f_{\mathsf{P}_0 > k}(j, \bar{\mathbf{m}}) = 1$ for all $j \geq j_{\min}$.

If the number of tokens in the input places of $t$ is sufficient in a marking $(j_{\min}, \bar{\mathbf{m}})$, then this is also the case for all markings with $j > j_{\min}$, since the number of tokens in all places except $\mathsf{P}_0$ is constant. Dependencies on place $\mathsf{P}_0$ are always satisfied since $j_{\min} \geq 1$ and the multiplicity of arcs leaving $\mathsf{P}_0$ is per definition at most 1.

Due to the constant nature of the transition priorities, a transition which is enabled in marking $(j_{\min}, \bar{\mathbf{m}})$ is also enabled for all values $j \geq j_{\min}$.

The fact that the firing rate does not depend on the number of tokens follows directly from (5.8) in Definition 4. As rate multipliers must not involve place $\mathsf{P}_0$, and all other places have the same number of tokens in the originating markings, the transition rates are constant. □

Summarizing the two preceding propositions, it has been shown that transitions can only occur between markings which belong to adjacent $j$-sets, and that these transitions exhibit a repeating structure for all markings where the number of tokens in place $\mathsf{P}_0$ is larger than or equal to $j_{\min}$. Now, it remains to show that starting with some index $r$, all $j$-sets contain the same submarkings. As the submarking is used to map the different markings in a $j$-set to states in a level of the underlying Markov chain, this implies that the Markov chain is of QBD type.

### 5.2.2 The Simple Case: Two Successive Submarking-Equivalent $j$-Sets

In this section we show that it is sufficient to find two successive $j$-sets with identical submarkings to ensure that all further $j$-sets are also identical, and we thus have an underlying QBD structure. In Section 5.2.3, we will extend the results of this section to show that the underlying Markov chain is always of QBD type, even if equivalent submarkings can not be found in *successive* $j$-sets.

**Proposition 3.** If the submarkings in two successive $j$-sets $S_r$ and $S_{r+1}$ with $r \geq j_{\min}$ are identical, then they are also identical to the submarkings in all $S_j$ with $j \geq r$.

*Proof.* We first show that if $\bar{S}_r = \bar{S}_{r+1}$, then also $\bar{S}_{r+2} = \bar{S}_r$ holds. Since the initial marking of $\mathsf{P}_0$ in an iSPN is 0 (see Definition 4), and due to the bounded jump size (Proposition 1), any reachable marking $(j, \bar{\mathbf{m}}')$ with $j \geq 1$ must be reachable by a finite sequence of transitions which starts with a transition from a marking in $S_{j-1}$ to a marking in $S_j$, and subsequently only reaches markings in $j$-sets $S_j$ or above. Formally, for $(r+2, \bar{\mathbf{m}}') \in S_{r+2}$, the following must hold:

$$
\begin{aligned}
(r+2, \bar{\mathbf{m}}') \in S_{r+2} \iff & \exists (r+1, \bar{\mathbf{m}}) \in S_{r+1}, \exists n \in \mathbb{N}, \\
& \exists l_1, \cdots, l_n, t_1, \cdots, t_n, \bar{\mathbf{m}}_1, \cdots, \bar{\mathbf{m}}_n : \\
& \forall i : l_i \geq r+2 \wedge l_n = r+2 \wedge \bar{\mathbf{m}}_n = \bar{\mathbf{m}}' \wedge \\
& (r+1, \bar{\mathbf{m}}) \xrightarrow{t_1} (l_1, \bar{\mathbf{m}}_1) \xrightarrow{t_2} \cdots \xrightarrow{t_n} (l_n, \bar{\mathbf{m}}_n).
\end{aligned}
$$

According to Proposition 2, the transitions occurring above are independent of the number of tokens in $\mathsf{P}_0$; they can be "shifted" to the left by one $j$-set since $r+1 \geq j_{\min}+1$. Furthermore, the submarkings in $S_{r+1}$ and $S_r$ are identical. Thus, the right-hand side of the relation can be transformed to obtain

$$
\begin{aligned}
(r+2, \bar{\mathbf{m}}') \in S_{r+2} \iff & \exists (r, \bar{\mathbf{m}}) \in S_r, \exists n \in \mathbb{N}, \\
& \exists l_1, \cdots, l_n, t_1, \cdots, t_n, \bar{\mathbf{m}}_1, \cdots, \bar{\mathbf{m}}_n : \\
& \forall i : l_i \geq r+1 \wedge l_n = r+1 \wedge \bar{\mathbf{m}}_n = \bar{\mathbf{m}}' \wedge \\
& (r, \bar{\mathbf{m}}) \xrightarrow{t_1} (l_1, \bar{\mathbf{m}}_1) \xrightarrow{t_2} \cdots \xrightarrow{t_n} (l_n, \bar{\mathbf{m}}_n),
\end{aligned}
$$

which is equivalent to

$$(r+2, \bar{\mathbf{m}}') \in S_{r+2} \iff (r+1, \bar{\mathbf{m}}') \in S_{r+1}.$$

Hence, the submarkings in $S_{r+2}$ are identical to those in $S_{r+1}$ and $S_r$. The repeated application of this result for larger values of $r$ proves the proposition. □

Using this result, we are able to prove the following theorem:

**Theorem 1.** If the reachability set of an iSPN contains two successive $j$-sets $S_r$ and $S_{r+1}$ with $r \geq j_{\min}$, which contain the same submarkings, then the underlying Markov chain is a QBD Markov chain.

*Proof.* Assume the following mapping of iSPN markings to states of the underlying Markov chain (refer to Figure 5.2 for an illustration):

- All markings in $S_0 \cup \cdots \cup S_{r-1}$ correspond to boundary states of the QBD process;
- the markings in $S_{r+k}, k \in \mathbb{N}_0$ are mapped to repeating level $k$.

Without loss of generality, we assume that in each level, all markings are arranged identically in the lexicographical order of the submarkings.

According to Proposition 3, the submarkings of markings associated to repeating levels $0, 1, \cdots$ are identical. Thus, those levels contain the same number of states. Furthermore, the following statements hold:

- According to Proposition 1, transitions originating from markings which correspond to repeating levels $0, 1, \cdots$ can only reach states within the same level or which are in adjacent levels.
- The transition rates between and within these levels are identical (Proposition 2).

Thus, all transitions within and between the repeating levels, and from border level 0 to the first repeating level, can be grouped in three matrices $\mathbf{A}_0, \mathbf{A}_1$ and $\mathbf{A}_2$. Together with a description of the transitions within and between the boundary states and the border level in a matrix $\mathbf{B}$, a valid QBD process description is obtained. □

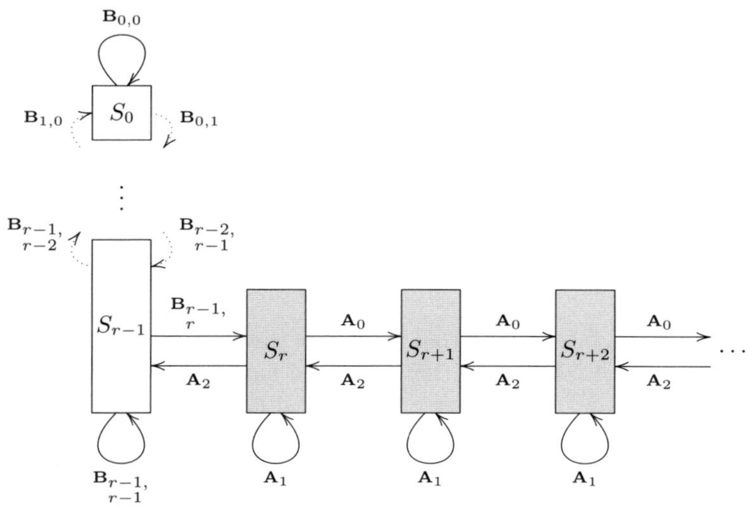


**Fig. 5.2.** Mapping markings of an iSPN to states of a QBD Markov chain. The shaded areas represent submarking-equivalent $j$-sets.

Based on the mapping described in the proof above, there are a few interesting facts worth noting. They are best illustrated by looking at the generator matrix of the QBD process, which is given by[4]

$$\mathbf{Q} = \begin{pmatrix} \mathbf{B}_{0,0} & \mathbf{B}_{0,1} & & & & & & & \\ \mathbf{B}_{1,0} & \mathbf{B}_{1,1} & \mathbf{B}_{1,2} & & & & & & \\ & \ddots & \ddots & \ddots & & & & & \\ & & \mathbf{B}_{r-1,r-2} & \mathbf{B}_{r-1,r-1} & \mathbf{B}_{r-1,r} & & & & \\ & & & \mathbf{B}_{r,r-1} & \mathbf{A}_1 & \mathbf{A}_0 & & & \\ & & & & \mathbf{A}_2 & \mathbf{A}_1 & \mathbf{A}_0 & & \\ & & & & & \mathbf{A}_2 & \mathbf{A}_1 & \mathbf{A}_0 & \\ & & & & & & \ddots & \ddots & \ddots \end{pmatrix} . \tag{5.15}$$

[4] Note that the size of the boundary transition matrices $\mathbf{B}_{i,j}, 0 \le i, j < r$ is not necessarily identical to the dimension of the matrices $\mathbf{A}_0, \mathbf{A}_1, \mathbf{A}_2$, as Eq. (5.15) might suggest. In many cases, these matrices may even be non-square for $i \ne j$.

The structure of $\mathbf{Q}$ is not as general as the form allowed by Eq. (4.1). First, due to the mapping of $j$-sets to levels of the QBD Markov chain and the limit of transitions to adjacent levels only, the QBD-structure also manifests in the boundary part of $\mathbf{Q}$, leading to a block-tridiagonal structure of the system of boundary equations. Tools for the numerical analysis of iSPNs could therefore exploit this fact by applying efficient specialized linear system solvers. However, it should be noted that the number of $j$-sets falling in the boundary level heavily depends on the modeling problem at hand. For problems with a boundary which corresponds to just one single $j$-set nothing can be gained.

Secondly, it can be observed that the size of the boundary can be decreased in case $\mathbf{B}_{r-1,r} = \mathbf{A}_0$ holds. In that case, the current border level becomes repeating level 1, and we still have that transitions to repeating level 1 are described by matrix $\mathbf{A}_0$. This is due to the fact that the selection of $j_{\min}$ according to Eq. (5.11) in Proposition 2 is not as tight as possible. For the proofs which are conducted in this section, the current choice is more handy. However, from a computational point of view, as many states as possible should be attributed to the repeating part of the state space. We will thus elaborate on a more sophisticated selection of $j_{\min}$ in Section 5.3, where implementation issues are discussed.

Finally, it should be noted that up till here, timed and immediate transitions have been treated identically to clarify the presentation. However, before a valid QBD Markov chain is obtained, all *vanishing* markings in the reachability graph (i.e., those markings where immediate transitions are enabled) must be eliminated. We will also comment on this in Section 5.3.

### 5.2.3 The General Case: All iSPNs Lead to QBD Processes

In the previous section, it has been shown that if two successive $j$-sets with identical submarkings can be found, then all successive $j$-sets contain the same submarkings, and thus the underlying Markov chain is of QBD type (Theorem 1). It is tempting to suppose that such a pair of $j$-sets can always be found due to the repetitive transition structure above $j_{\min}$. However, this is not the case.

The reason for this is that there are in fact no *direct* dependencies of transitions on the number of tokens in $\mathsf{P}_0$ for $j$ sets above $j_{\min}$. However, the enabling of transitions in marking $(j, \bar{\mathbf{m}})$ still depends on the submarking $\bar{\mathbf{m}}$. If there is a dependency between the submarking $\bar{\mathbf{m}}$ and the value $j$, then there exists an *indirect* dependency on $\mathsf{P}_0$. As an example, consider the iSPN given in Figure 5.3(a). This example model represents a system where jobs arrive at a rate $\lambda$ in the unbounded place. Jobs can only be served if there is an even number of jobs in the queue (for example, because the server imposes some alignment restriction on the jobs to be served). The place $\mathsf{parity}$ keeps track of whether the number of jobs in $\mathsf{P}_0$ is odd or even since it is toggled each time a job arrives at the queue or a job is removed.

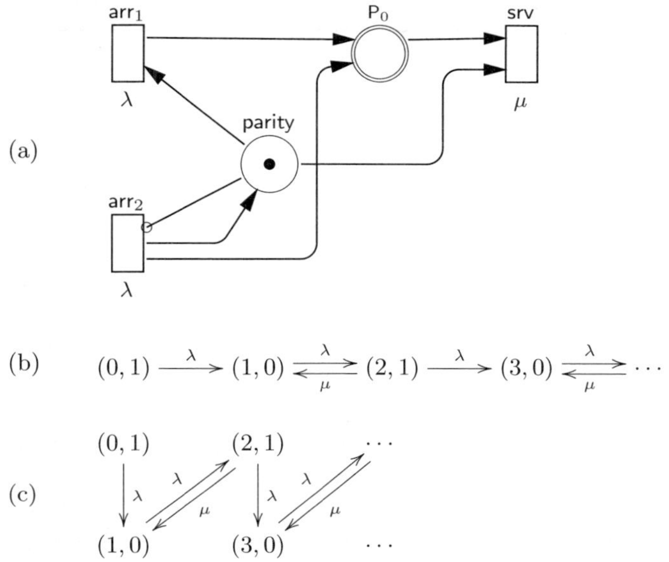


**Fig. 5.3.** (a) Example iSPN with indirect dependency on $\mathsf{P}_0$. The underlying Markov chain with states labeled with their corresponding iSPN-marking is depicted in (b); (c) shows the re-ordered state space for obtaining a QBD structure.

Figure 5.3(b) illustrates the underlying Markov chain of this iSPN, as obtained from the mapping suggested in Figure 5.2. As intended, services can only take place if the number of jobs in the queue is even. Note that the submarkings in the $j$-sets of this reachability graph constantly change between (0) and (1), thus, two successive $j$-sets with identical markings will never occur. Since services (backward transitions) can only occur in levels with submarking 1, the Markov chain is not of QBD type. We can, however, obtain a QBD process by re-arranging the state space, as shown in Figure 5.3(c). By putting the markings of several successive $j$-sets into *one* repeating level of the QBD state space, irregularities in successive $j$-sets are eliminated. Note that this is only possible if the sequence of differing sets of submarkings in successive $j$-sets exhibits a repetitive structure (like $(1), (0), (1), (0), \cdots$ in the example).

In the following two propositions, it will first be shown that such a repetitive structure can always be found. Using this fact and a formalization of the above-mentioned mapping approach, we then prove that *any* iSPN has an underlying Markov chain of QBD type.

**Proposition 4.** For each iSPN, there exist values $r \geq j_{min}$ and $k \in \mathbb{N}$ so that the $j$-sets $S_r$ and $S_{r+k}$ are submarking-equivalent: $\bar{S}_r = \bar{S}_{r+k}$.

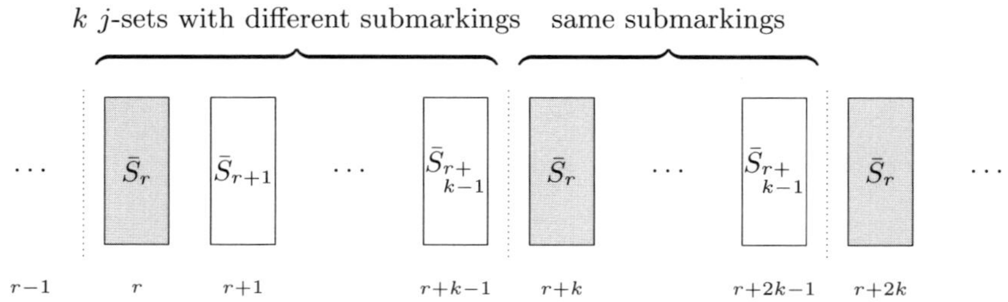


**Fig. 5.4.** Illustration of the repetitive occurrence of submarkings once two identical $j$-sets $\bar{S}_r = \bar{S}_{r+k}$ have been found.

*Proof.* By Definition 4, all places except $\mathsf{P}_0$ are bounded. Let $K$ denote the maximum number of tokens which can be found in any of these places. Then, the set of all submarkings in a $j$-set, $\bar{S}_j$, must be a subset of the finite set $M = \{0, \cdots, K\}^{|P|-1}$. Consequently, there exists only a finite number of different submarking sets $\bar{S}_j$. Now, by definition, the reachability set of an iSPN is infinite, which implies an infinite number of $j$-sets and corresponding submarking sets. Thus, there must be sets of submarkings which occur infinitely often, and which thus can also be found at least two times in $j$-sets $S_{r_1}$ and $S_{r_2}$ with $r_1, r_2 \geq j_{\min}$. □

Now, we will show that if two $j$-sets with identical submarkings have been found, then the submarkings of all $j$-sets between them repeat infinitely often, as shown in Figure 5.4.

**Proposition 5.** If the submarkings in two $j$-sets $S_r$ and $S_{r+k}$ with $r \geq j_{\min}$ and $k \in \mathbb{N}$ are identical, then the submarkings of the $j$-sets $S_r, \cdots, S_{r+k-1}$ are identical to the submarkings in all corresponding $j$-sets which are shifted by any multiple of $k$:

$$\forall n \in \mathbb{N}, 0 \leq l < k : \bar{S}_{r+l} = \bar{S}_{r+kn+l}. \tag{5.16}$$

*Proof.* The proof is similar to the one conducted for Proposition 3. We first show that if $\bar{S}_r = \bar{S}_{r+k}$, then also $\forall 0 \leq l \leq k : \bar{S}_{r+l} = \bar{S}_{r+k+l}$ holds (i.e., (5.16) holds for $n = 1$). Note that we include $l = k$, thus also showing that $\bar{S}_{r+k} = \bar{S}_{r+2k}$. We assume $l \geq 1$, as $\bar{S}_r$ and $\bar{S}_{r+k}$ are already identical by assumption. Since the initial marking of $\mathsf{P}_0$ in an iSPN is zero according to Definition 4, and due to the bounded jump size (Proposition 1), any reachable marking $(j + l, \bar{\mathbf{m}}')$ with $j + l \geq 1, l \geq 1$ must be reachable by a finite sequence of transitions which starts with a transition from a marking in $S_j$ to a marking in $S_{j+1}$, and subsequently only reaches markings in $j$-sets $S_{j+1}$ or above. Thus, for a marking $(r + k + l, \bar{\mathbf{m}}') \in S_{r+k+l}, l \geq 1$, the following must hold:

$$
\begin{aligned}
(r+k+l,\bar{\mathbf{m}}') \in S_{r+k+l} \iff & \exists (r+k,\bar{\mathbf{m}}) \in S_{r+k}, \exists n \in \mathbb{N}, \\
& \exists l_1, \cdots, l_n, t_1, \cdots, t_n, \bar{\mathbf{m}}_1, \cdots, \bar{\mathbf{m}}_n : \\
& \forall i : l_i \geq r+k+1 \wedge l_n = r+k+l \\
& \wedge \bar{\mathbf{m}}_n = \bar{\mathbf{m}}' \wedge \\
& (r+k,\bar{\mathbf{m}}) \xrightarrow{t_1} (l_1,\bar{\mathbf{m}}_1) \xrightarrow{t_2} \cdots \xrightarrow{t_n} (l_n,\bar{\mathbf{m}}_n).
\end{aligned}
$$

According to Proposition 2, the transitions occurring above are independent of the number of tokens in $\mathsf{P}_0$; they can be "shifted" to the left by $k$ $j$-sets since $r+k \geq j_{\min}+1$ (note that $k \geq 1$). Furthermore, the submarkings in $S_r$ and $S_{r+k}$ are identical. Thus, the right-hand side of the relation can be transformed to obtain

$$
\begin{aligned}
(r+k+l,\bar{\mathbf{m}}') \in S_{r+k+l} \iff & \exists (r,\bar{\mathbf{m}}) \in S_r, \exists n \in \mathbb{N}, \\
& \exists l_1, \cdots, l_n, t_1, \cdots, t_n, \bar{\mathbf{m}}_1, \cdots, \bar{\mathbf{m}}_n : \\
& \forall i : l_i \geq r+1 \wedge l_n = r+l \wedge \bar{\mathbf{m}}_n = \bar{\mathbf{m}}' \wedge \\
& (r,\bar{\mathbf{m}}) \xrightarrow{t_1} (l_1,\bar{\mathbf{m}}_1) \xrightarrow{t_2} \cdots \xrightarrow{t_n} (l_n,\bar{\mathbf{m}}_n),
\end{aligned}
$$

which is equivalent to

$$(r+k+l,\bar{\mathbf{m}}') \in S_{r+k+l} \iff (r+l,\bar{\mathbf{m}}') \in S_{r+k}.$$

Consequently, the submarkings in $S_r, \cdots, S_{r+k}$ are identical to those in $S_{r+k}, \cdots, S_{r+2k}$. By the repeated application of this result the proposition is derived. □

Note that Proposition 3 represents a special case of this result, as it is obtained for $k=1$.

With the aid of the preceding propositions, Theorem 1 can be extended to apply to arbitrary iSPNs, thus showing that any iSPN leads to an underlying QBD process:

**Theorem 2.** The underlying Markov chain of an iSPN is always a QBD Markov chain.

*Proof.* According to Proposition 4 we are able to find a $j$-set $S_r$ and an offset $k \geq 1$ so that $\bar{S}_r = \bar{S}_{r+k}$. Due to Proposition 5 the submarkings in $\bar{S}_r, \cdots, \bar{S}_{r+k-1}$ repeat infinitely in all further $j$-sets.

We now assume the following mapping of iSPN markings to states of the underlying Markov chain (see Figure 5.5 for an illustration):

- For $n = 0, 1, \cdots$, all markings in $j$-sets $S_{r+nk+1} \cup \cdots \cup S_{r+(n+1)k}$ are mapped to repeating level $n+1$. Let $N = |S_{r+nk+1} \cup \cdots \cup S_{r+(n+1)k}|$ denote the number of states in a repeating level (note that $N$ is independent of $n$ according to Proposition 5).

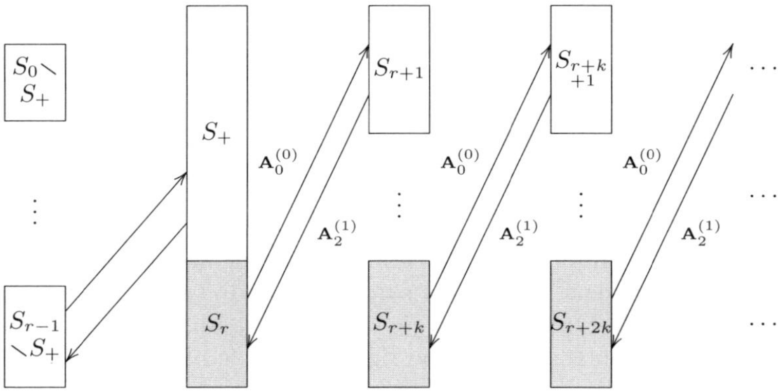


**Fig. 5.5.** Mapping markings of an iSPN to states of a QBD process, general case. Note that transitions within levels of the QBD have been omitted for better readability.

- Without loss of generality, we assume that $|S_0 \cup \cdots \cup S_r| \geq N$. If this is not the case, $r$ can be increased appropriately to satisfy this condition. The border level of the QBD now contains all markings in $S_r$ *and* a number of markings contained in $S_+ \subseteq (S_0 \cup \cdots \cup S_{r-1})$ so that $|S_r \cup S_+| = N$. This way, the border level contains the same number of states as the repeating levels.
- All remaining states in $(S_0 \cup \cdots \cup S_{r-1}) \setminus S_+$ are mapped to the boundary level of the QBD.

We again assume that in each level, all markings are arranged according to their lexicographical order.

According to Proposition 5, the markings in $S_{r+nk+1} \cup \cdots \cup S_{r+(n+1)k}$ associated to repeating levels $1, 2, \cdots$ are identical up to a shift in the number of tokens in place $\mathsf{P}_0$. Due to Proposition 1, the transitions corresponding to a $j$-set $S_{r+nk+l}$ can thus be described by three matrices $\mathbf{A}_0^{(l)}, \mathbf{A}_1^{(l)}$ and $\mathbf{A}_2^{(l)}$, with $0 \leq l < k$. These matrices are independent of $n$ and describe the rates between states which belong to the same or an adjacent $j$-set.

Note that with the suggested mapping, this independence of $n$ corresponds to the fact that transitions within the repeating levels $1, 2, \cdots$ are identical. The rates can be summarized in a matrix $\mathbf{A}_1$ which is built from the submatrices $\mathbf{A}_0^{(l)}, \mathbf{A}_1^{(l)}, \mathbf{A}_2^{(l)}$ as follows:

$$\mathbf{A}_1 = \begin{pmatrix} \mathbf{A}_1^{(0)} & \mathbf{A}_0^{(0)} & & & \\ \mathbf{A}_2^{(1)} & \mathbf{A}_1^{(1)} & \mathbf{A}_0^{(1)} & & \\ & \ddots & \ddots & \ddots & \\ & & \mathbf{A}_2^{(k-2)} & \mathbf{A}_1^{(k-2)} & \mathbf{A}_0^{(k-2)} \\ & & & \mathbf{A}_2^{(k-1)} & \mathbf{A}_1^{(k-1)} \end{pmatrix}. \tag{5.17}$$

Note that inter-level transitions can only occur between markings in $j$-sets $S_{r+nk}$ and $S_{r+nk+1}$ (see Figure 5.5). Thus, matrices $\mathbf{A}_0$ and $\mathbf{A}_2$ can be built similar to $\mathbf{A}_1$, however, they only contain the submatrix $\mathbf{A}_0^{(k-1)}$ in their lower left corner, or the submatrix $\mathbf{A}_2^{(0)}$ in their upper right corner, respectively.

To summarize, we have that

- border level and repeating levels contain the same number of states,
- transitions between the repeating levels are limited to a jump size of at most one level, and
- the rates of these transitions are identical, as given by $\mathbf{A}_0, \mathbf{A}_1, \mathbf{A}_2$.

It remains to show that transitions from the border level to repeating level 1 also occur according to matrix $\mathbf{A}_0$. This follows from the fact that transitions from states which belong to $j$-set $S_r$ are identical to those captured by $\mathbf{A}_0^{(k-1)}$ in matrix $\mathbf{A}_1$, and that there are no transitions from states in the boundary level which belong to $S_0 \cup \cdots \cup S_{r-1}$ due to Proposition 1. Thus, by summarizing all transitions within and between boundary and border states in a matrix $\mathbf{B}$, a valid QBD description has been found. □

### 5.2.4 Coverage of all QBDs by iSPNs

The aim of this section is to show that there is an iSPN description for *any* QBD process. This ensures that nothing of the expressiveness of QBDs is lost when the high-level iSPN-based description is used. In conjunction with Theorem 2, this result implies the expressive equivalence of the iSPN-based system description to the direct specification of a QBD process.

**Construction of equivalent iSPNs.** We constructively show that any QBD can be represented as iSPN by developing an iSPN specification from the QBD, where the underlying Markov chain equals the original QBD process. This is achieved by closely coupling transitions of the iSPN to transitions between states in the original QBD process.

The basic idea is to represent states $(i, j)$ of the QBD process by iSPN markings where $j$ is reflected by the number of tokens in the unbounded place, and $i$ corresponds to the fact that the $i$th bounded place contains one token (and all other bounded places are empty). Then, for each positive rate entry in the QBD description matrices, a transition is added to the iSPN. We assure that these transitions are only firable in the correct QBD level by assigning an enabling function which depends on the unbounded place.

In the following, we consider a QBD process $\mathcal{Q} = (\mathbf{B}, \mathbf{B}_{2,1}, \mathbf{A}_0, \mathbf{A}_1, \mathbf{A}_2)$ of order $(N_0, N)$. As usual, the submatrices of $\mathbf{B}$ are denoted by $\mathbf{B}_{0,0}, \mathbf{B}_{0,1}, \mathbf{B}_{1,1}$ and $\mathbf{B}_{1,0}$. Without loss of generality, we assume that $N_0 \geq 1$ and thus $\mathbf{B}_{0,0}, \mathbf{B}_{0,1}$ and $\mathbf{B}_{1,0}$ have positive dimensions. If $N_0 = 0$, the QBD process can be shifted appropriately to obtain $N_0 = N$.

We now define an iSPN $\mathcal{P} = (P, T, Pr, I, O, W, R, E, \mathbf{m}_0)$ where each state $(i, j)$ of the QBD process directly corresponds to a marking $(j, \bar{\mathbf{m}})$ where

$\bar{\mathbf{m}}_{\mathsf{P}_i} = 1$ and all other bounded places are empty. To account for all values of $i$ in the boundary level as well as in the repeating levels, $N' = \max\{B, N\}$ bounded places are needed. Thus, besides an unbounded place $\mathsf{P}_0$, $P$ contains places $\mathsf{P}_1, \cdots, \mathsf{P}_{N'}$. The initial marking is $\mathbf{m}_0$ with $\mathbf{m}_{0,\mathsf{P}_1} = 1$ and $\mathbf{m}_{0,\mathsf{P}_i} = 0$ for all other places.

We ensure that in any marking, there is exactly one token in all places $\mathsf{P}_1, \cdots, \mathsf{P}_{N'}$ by only employing transitions with exactly one input place and one output place in $\{\mathsf{P}_1, \cdots, \mathsf{P}_{N'}\}$. An additional input or output arc to $\mathsf{P}_0$ is used to reflect transitions between adjacent levels of the QBD process.

The transitions, transition rates, and enabling functions of $\mathcal{P}$ are defined as follows:

- For each positive entry $\lambda$ in the $i$-th column and $j$-th row of the matrices

$$\mathbf{B}_{0,0}, \mathbf{B}_{0,1}, \mathbf{B}_{1,0}, \mathbf{B}_{1,1}, \mathbf{B}_{2,1}, \mathbf{A}_0, \mathbf{A}_1 \quad \text{and} \quad \mathbf{A}_2,$$

  the set $T$ of the iSPN contains a unique transition

$$\mathsf{T}_{0,0}^{i,j}, \mathsf{T}_{0,1}^{i,j}, \mathsf{T}_{1,0}^{i,j}, \mathsf{T}_{1,1}^{i,j}, \mathsf{T}_{2,1}^{i,j}, \mathsf{T}_{0}^{i,j}, \mathsf{T}_{1}^{i,j} \quad \text{and} \quad \mathsf{T}_{2}^{i,j},$$

  respectively. For convenience, we refer to this transition as $\mathsf{T}$ in the following.
- The firing rate of a transition equals the transition rate in the QBD process:

$$R(\mathsf{T}) = \lambda.$$

- Input and output arcs of the transition are selected to reflect the change in the QBD's finite dimension from state $i$ to state $j$:

$$I(\mathsf{P}_i, \mathsf{T}) = 1 = O(\mathsf{T}, \mathsf{P}_j).$$

- If the QBD transition under consideration involves a change between levels, an additional arc from or to place $\mathsf{P}_0$ is needed. We have that
  - $O(\mathsf{T}, \mathsf{P}_0) = 1$ for transitions $\mathsf{T}_{0,1}^{i,j}$ and $\mathsf{T}_{0}^{i,j}$, as the matrices $\mathbf{B}_{0,1}$ and $\mathbf{A}_0$ correspond to transitions to upward levels.
  - $I(\mathsf{P}_0, \mathsf{T}) = 1$ for transitions $\mathsf{T}_{1,0}^{i,j}, \mathsf{T}_{2,1}^{i,j}$ and $\mathsf{T}_{2}^{i,j}$, as they correspond to the matrices $\mathbf{B}_{1,0}, \mathbf{B}_{2,1}$ or $\mathbf{A}_2$ which contain the rates to lower levels of the QBD.
- The dependence of a QBD transition on $i$ is already accounted for by the input arc of $\mathsf{T}$. Dependencies on $j$ at the QBD level are reflected at the iSPN level by the definition of an appropriate enabling function that depends on $\mathsf{P}_0$:

$$E(\mathsf{T}) = \begin{cases} f_{\mathsf{P}_0=0}, & \text{for transitions } \mathsf{T}_{0,0}^{i,j} \text{ and } \mathsf{T}_{0,1}^{i,j}, \\ f_{\mathsf{P}_0=1}, & \text{for transitions } \mathsf{T}_{1,0}^{i,j} \text{ and } \mathsf{T}_{1,1}^{i,j}, \\ f_{\mathsf{P}_0=2}, & \text{for transitions } \mathsf{T}_{2,1}^{i,j}, \\ f_{\mathsf{P}_0>0}, & \text{for transitions } \mathsf{T}_{0}^{i,j}, \\ f_{\mathsf{P}_0>1}, & \text{for transitions } \mathsf{T}_{1}^{i,j}, \\ f_{\mathsf{P}_0>2}, & \text{for transitions } \mathsf{T}_{2}^{i,j}. \end{cases}$$

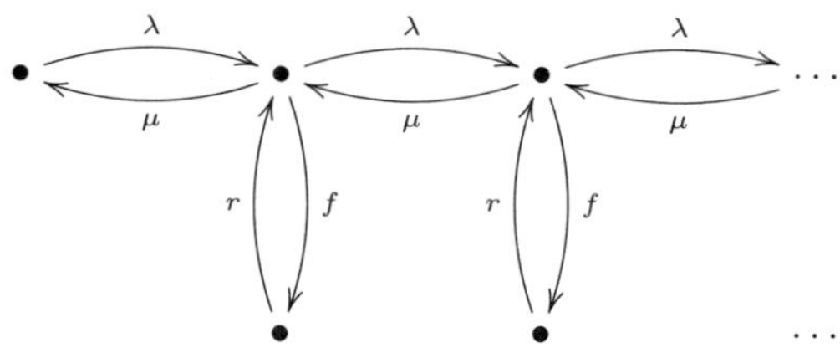


**Fig. 5.6.** Example QBD process for illustrating the mapping to iSPNs.

To summarize, the underlying Markov chain of $\mathcal{P}$ is a QBD process where markings $(j, \bar{\mathbf{m}})$ with $\bar{\mathbf{m}}_{\mathsf{P}_i} = 1$ correspond to states $(i, j)$ of $\mathcal{Q}$. Due to our definition of transitions and enabling functions, transitions between markings can only occur if they are possible in $\mathcal{Q}$ as well. The irreducibility of a QBD ensures that all markings corresponding to QBD states are reachable, and that the initial marking does not impact the reachability set as long as exactly one place $\mathsf{P}_k, k \geq 1$ has initially one token. The following theorem summarizes these results:

**Theorem 3.** For any QBD process, there exists an iSPN with equivalent underlying Markov chain.

*Proof.* This follows from the construction scheme and the arguments presented above. □

In conjunction with the results of Section 5.2.3, this leads to the following main conclusion:

**Corollary 1.** The class of stochastic processes underlying iSPNs is identical to the class of QBD processes.

*Proof.* This follows directly from the fact that the stochastic process underlying an iSPN is always a QBD (Theorem 2), and that any QBD can be represented as iSPN (Theorem 3). □

**Example.** To illustrate the iSPN construction procedure presented above, we develop an iSPN description for an example QBD process. We consider the QBD depicted in Figure 5.6. The QBD may represent a system where jobs arrive at rate $\lambda$ and are served at rate $\mu$. If the server is non-idle, it is subject to breakdowns at rate $f$. Once the server breaks down, both arrivals and services are suspended. Server repairs take place at rate $r$.

The transition rate matrices of this QBD are as follows (negative entries are denoted by a dot):

$$\mathbf{B} = \left(\begin{array}{c|cc} \cdot & \lambda & 0 \\ \hline \mu & \cdot & f \\ 0 & r & \cdot \end{array}\right), \mathbf{A}_0 = \begin{pmatrix} \lambda & 0 \\ 0 & 0 \end{pmatrix}, \mathbf{A}_1 = \begin{pmatrix} \cdot & r \\ f & \cdot \end{pmatrix}, \mathbf{A}_2 = \mathbf{B}_{2,1} = \begin{pmatrix} \mu & 0 \\ 0 & 0 \end{pmatrix}. \tag{5.18}$$

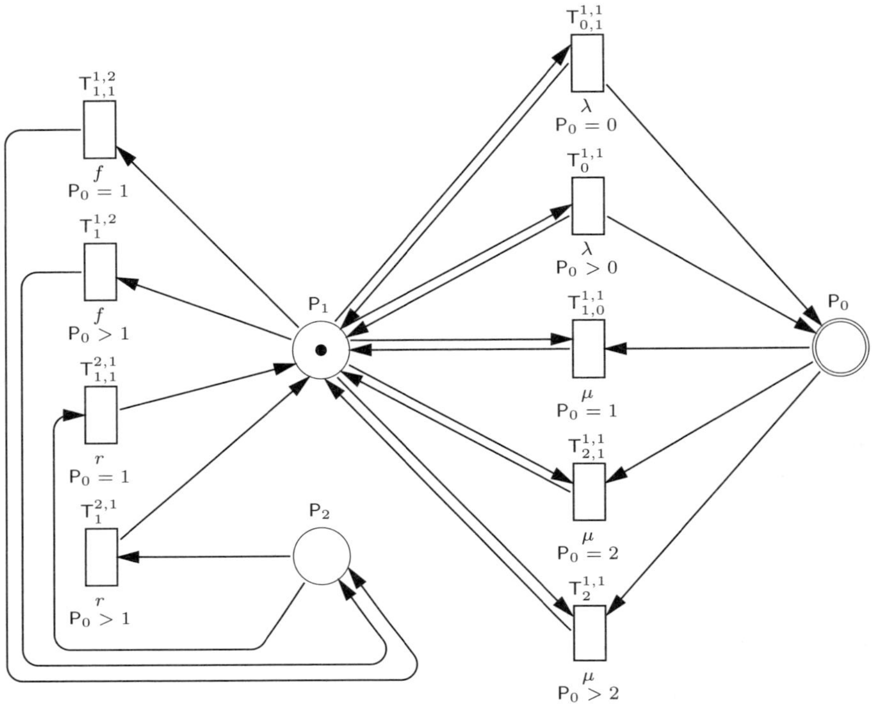


**Fig. 5.7.** The iSPN corresponding to the QBD process in Figure 5.6.

As we are dealing with a QBD of order $(1, 2)$, two bounded places $P_1$ and $P_2$ are needed in addition to the unbounded place $P_0$. Furthermore, since the transition rate matrices contain 9 positive entries, the iSPN construction process yields an iSPN with 9 transitions; it is depicted in Figure 5.7.

Note that this iSPN can be simplified to yield the iSPN in Figure 5.8, since several transitions with identical rate and input- and output arcs can be replaced by single transitions:

- $T^{1,1}_{0,1}$ and $T^{1,1}_{0}$ can be summarized as transition arr, since the enabling function of exactly one of these two transitions always yields 1.
- $T^{1,1}_{1,0}$, $T^{1,1}_{2,1}$ and $T^{1,1}_{2}$ can be replaced by a transition serve with enabling function $f_{P_0>0}$. As $P_0$ is an input place of serve, the enabling function can be omitted.
- Similarly, transitions $T^{1,2}_{1,1}$ and $T^{1,2}_{1}$ can be replaced by transition fail, and $T^{2,1}_{1,1}$ and $T^{2,1}_{1}$ can be replaced by rep. In both cases, the resulting enabling function is $P_0 > 0$.

From the simplified iSPN, it is easy to see that the underlying Markov chain of the iSPN is identical to the original QBD process.

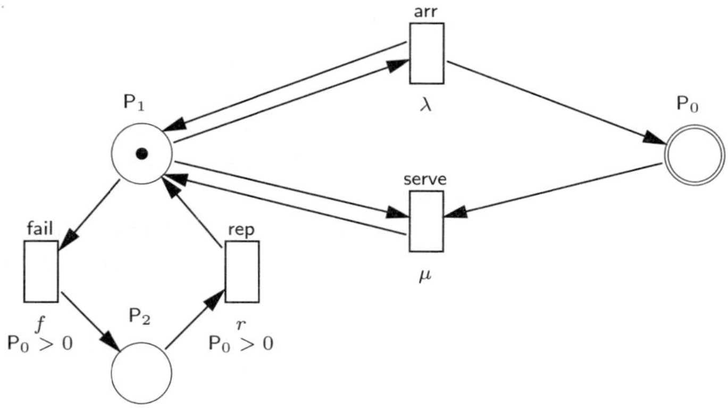


**Fig. 5.8.** Simplified iSPN derived from Figure 5.7 by aggregating transitions.

## 5.3 Implementation Issues

One major step in the development of an iSPN-based performance modeling and evaluation tool is the derivation of a QBD from an iSPN specification. The proofs conducted in Section 5.2 provide extensive guidance for the implementation of appropriate algorithms. This section is devoted to the discussion of several issues which have not been important from a formal point of view, but which have to be addressed if appropriate algorithms have to be developed and implemented in a time- and memory-efficient manner. In Section 5.3.1, we discuss how the number of boundary states of the QBD process underlying an iSPN can be minimized by choosing $j_{\min}$ as tight as possible. In Section 5.3.2, we comment on the derivation of the (infinite) reachability set of an iSPN, being a prerequisite for obtaining the QBD description. As we did not account for the special impact of immediate transitions in our considerations so far, we elaborate on this in Section 5.3.3. Finally, we conclude this section with an approach for modeling platooned customer arrivals and departures to and from the unbounded place of an iSPN in a way that does not require changes to the theory laid out in this chapter, and which can thus be applied without an increase in complexity of the underlying algorithms.

### 5.3.1 Tightly Choosing $j_{\min}$

The value $j_{\min}$ has a crucial influence on the efficiency of an iSPN-based performance evaluation environment. As it directly governs the number of states contained in the boundary level of the QBD underlying an iSPN, it determines the size of the boundary equation system, and thus impacts both computational speed and the memory required for its solution. In the first part of this section, we will improve the initial selection of $j_{\min}$ suggested in Proposition 2. In the second part, we sketch an entirely different approach for dynamically choosing $j_{\min}$ as small as possible.

**Improving the Current Choice of $j_{\min}$.** As already mentioned in Section 5.2.2, the QBD described by the generator matrix in Eq. (5.15) is not necessarily a QBD with minimum boundary. As the number of states in the boundary level and the border level determine the size of the set of linear equations to be solved for the QBD's boundary, as many states as possible should be attributed to the repeating levels.

The mapping illustrated in Eq. (5.15) represents a QBD with the following special properties, which are not required to assure the QBD structure as shown in Eq. (4.1):

- Transitions *within* the border level are identical to those in all subsequent repeating levels (with the rates contained in $\mathbf{A}_1$).
- The transition rates from the border level to the boundary level (denoted by $\mathbf{B}_{r,r-1}$) are identical to the rates to predecessor levels in the repeating levels. Therefore, all non-zero entries in $\mathbf{A}_2$ have corresponding entries in $\mathbf{B}_{r,r-1}$.[5]
- We have that the transitions from repeating level 1 to the border level are identical to $\mathbf{A}_2$.

In many systems, job arrivals do not depend on the number of elements in the buffer. At the iSPN level, this means that the enabling of transitions which increase the number of tokens in the unbounded place is always independent of the number of tokens in it. Thus, matrix $\mathbf{B}_{r-1,r}$ in Eq. (5.15) often contains the same transitions as $\mathbf{A}_0$. Given that the $j$-sets $S_{r-1}$ and $S_r$ contain the same submarkings, the number of states in the boundary of the QBD can be reduced by mapping the border level to $j$-set $S_{r-1}$.

In fact, the choice of $j_{\min}$ in Proposition 2 can lead to a QBD boundary which unnecessarily contains *two* repeating levels. As an example, consider the iSPN in Figure 5.9(a), which models an M|M|3 queueing system, where up to three servers are active as soon as enough jobs are waiting for service. The underlying Markov chain is shown in Figure 5.9(b). Due to the enabling function $f_{\mathsf{P}_0>2}$, we have $j_{\min} = 3$ according to Proposition 2. In this case, the boundary and border part of the QBD process is given as illustrated in the generator matrix on the left-hand side of 5.9(c). It can be observed that the requirement that *any* transition which originates from a marking in $j$-set $S_{j_{\min}}$ can occur for all $j \geq j_{\min}$ is not necessary to guarantee a QBD structure. Transitions within the border level and from repeating level 1 to the border level can be chosen freely. Considering this, a partitioning of the QBD states as illustrated by the right matrix in Figure 5.9(c) can be chosen.

More formally, a tighter selection of $j_{\min}$ is obtained by considering those transitions of the underlying Markov chain which depend on the number of tokens in the unbounded place, i.e., they correspond to the firing of an iSPN transition $t \in T$ with an enabling function that involves $\mathsf{P}_0$. Now, a better

[5] We thus have that $\mathbf{A}_2$ is "almost" a submatrix of $\mathbf{B}_{r,r-1}$ in the sense that $S_{r-1}$ may contain submarkings which do not appear in $S_r$, and vice versa.

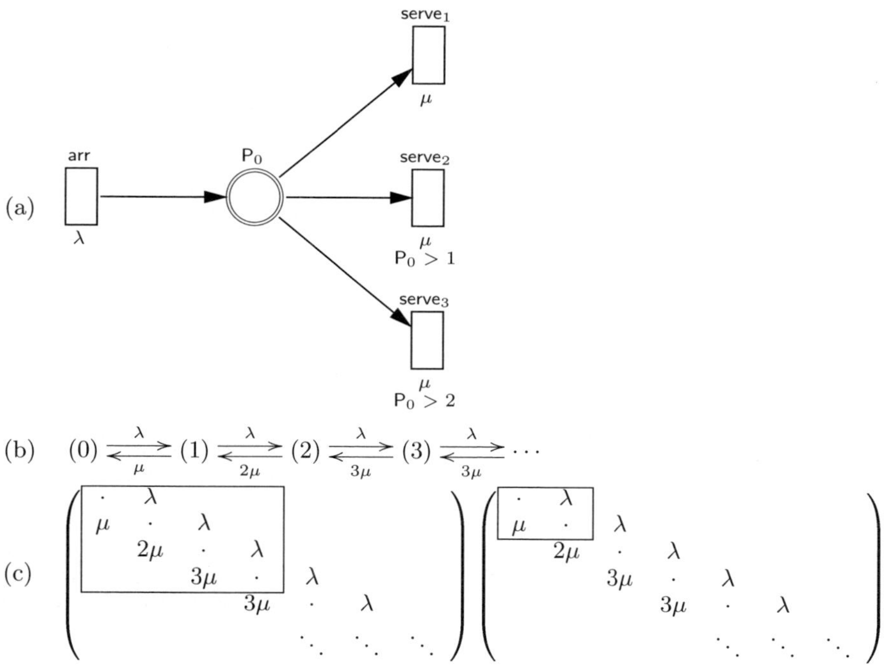


**Fig. 5.9.** (a) Example iSPN where $j_{\min}$ is larger than necessary; (b) shows the underlying Markov chain. In (c), the generator matrix is illustrated with a standard ($j_{\min} = 3$) and a minimal ($j_{\min} = 1$) boundary definition.

choice of $j_{\min}$ also takes into account whether a transition in the underlying Markov chain leads to a state in a higher $j$-set (i.e., a $j$-set with a larger value of $j$), a lower $j$-set, or within the same $j$-set:

- If $(j_{\min}, \bar{\mathbf{m}}) \xrightarrow{t} (j_{\min} + 1, \bar{\mathbf{m}}')$ then $(j, \bar{\mathbf{m}}) \xrightarrow{t} (j + 1, \bar{\mathbf{m}}')$ must hold for all $j \geq j_{\min}$.
- If $(j_{\min} + 2, \bar{\mathbf{m}}) \xrightarrow{t} (j_{\min} + 1, \bar{\mathbf{m}}')$ then $(j, \bar{\mathbf{m}}) \xrightarrow{t} (j - 1, \bar{\mathbf{m}}')$ must hold for all $j \geq j_{\min} + 2$.
- If $(j_{\min} + 1, \bar{\mathbf{m}}) \xrightarrow{t} (j_{\min} + 1, \bar{\mathbf{m}}')$ then $(j, \bar{\mathbf{m}}) \xrightarrow{t} (j, \bar{\mathbf{m}}')$ must hold for all $j \geq j_{\min} + 1$.

At the iSPN level, such a $j_{\min}$ is most easily characterized by partitioning the set of iSPN transitions in three sets $T_+, T_-$ and $T_0$, depending on whether their firing increases, decreases, or does not alter the number of tokens in $\mathsf{P}_0$. This is accomplished by investigating the input- and output-arcs of a transition. Then, Eq. (5.11) of Proposition 2 is refined to account for the type of a transition to yield

$$
\begin{aligned}
j_{\min} = \max(\{ & j \in \mathbb{N} | \exists t \in T_+ : \\
& \qquad f_{\mathsf{P}_0<j} \in E(t) \vee f_{\mathsf{P}_0=j-1} \in E(t) \vee f_{\mathsf{P}_0>j-1} \in E(t)\} \\
& \cup \{j \in \mathbb{N} | \exists t \in T_0 : \\
& \qquad f_{\mathsf{P}_0<j+1} \in E(t) \vee f_{\mathsf{P}_0=j} \in E(t) \vee f_{\mathsf{P}_0>j} \in E(t)\} \\
& \cup \{j \in \mathbb{N} | \exists t \in T_- : \\
& \qquad f_{\mathsf{P}_0<j+2} \in E(t) \vee f_{\mathsf{P}_0=j+1} \in E(t) \vee f_{\mathsf{P}_0>j+1} \in E(t)\} \\
& \cup \{1\}). \qquad (5.19)
\end{aligned}
$$

Note that this definition of $j_{\min}$ at the iSPN level might still lead to larger values than necessary. The reason for this is that there may be transitions in $T_+, T_0$ or $T_-$ which are never enabled, or whose enabling function never comes into effect due to some other condition which is never fulfilled. For an optimum choice of $j_{\min}$, the *reachability* of markings must be considered as well. Though this is feasible, such an approach requires a more elaborate implementation. In most (non-pathological) cases, the estimate given in (5.19) yields the correct, minimum value.

It is important to note that a choice of $j_{\min}$ according to Eq. (5.19) has no impact on the results of Section 5.2. The relevant proofs of Propositions 3 and 5 do not assume a more stringent repetitive transition structure than the one presented here. The reason for this is that in any considered sequence of transitions, the *first* transition is always an upward transition into a higher $j$-set, and all subsequent transitions appear between markings that belong to this or higher $j$-sets.

It should be noted that too large an estimate of $j_{\min}$ does not necessarily alter the definition of the boundary of the QBD process. According to Theorems 1 and 2, the boundary level is chosen such that it contains the first repetitive submarking set $\bar{\mathbf{S}}_r$ with $r \geq j_{\min}$. If the first appearance of such a repetitive structure occurs two levels after the optimum value of $j_{\min}$, then the possible error made in the original definition of $j_{\min}$ in Proposition 2 is irrelevant.

However, in many cases the repetitive structure occurs immediately at or after the optimum value of $j_{\min}$. In this case, the merits of a tighter selection should not be underestimated. Though a difference of merely two boundary states in the example presented above seems negligible, there can be a substantial gain for large repeating levels. For example, the study conducted in Section 6.2 leads to QBD processes with 765 states in the repeating levels. With similar boundary level sizes, a difference of two levels already leads to boundary equation sets of a size which approaches the limit of what can be dealt with by direct numerical solution methods (which, depending on the condition of the system, is in the range of a few thousand equations).

**Ineffective Enabling Functions.** Another reason why $j_{\min}$ might be chosen much larger than necessary is that it has been assumed that a change of the enabling functions associated to a transition always affects its enabled-ness. It is, however, very well possible that an enabling function like $f_{\mathsf{P}_0>1000}$ has been assigned to a transition, but that transition becomes never enabled because some other necessary condition is never fulfilled. Thus, a value of $\jmath_{\min} = 1000$ would be used, despite of the fact that much smaller values might be sufficient. Such situations can thus lead to a dramatic over-estimation of the QBD boundary size.

According to Section 5.1.3, the conditions which might "override" the effect of an enabling function concern

- the availability of tokens in the predecessor places;
- the value of other enabling functions associated to the same transition;
- implicit disabling;
- the existence of firable transitions with higher priority.

Unfortunately, for investigating whether any of these conditions applies, it is necessary to consider the dynamic behaviour of the iSPN, i.e., to explore the entire reachability set of the iSPN. Only then a decision can be taken whether one of these conditions permanently disables the considered transition above some limiting $\jmath$-set.

In Section 5.3.2, an algorithm for generating all reachable markings up to the beginning of the repetitive transition structure will be developed. The algorithm basically generates all $\jmath$-sets $S_0, S_1, \cdots$ until two submarking-equivalent $\jmath$-sets $S_r$ and $S_{r+k}$ *with* $r$ *larger or equal to* $\jmath_{\min}$ can be found. An approach to circumvent the problem of an optimal $\jmath_{\min}$ not being known *a priori* is to look for it dynamically while generating the state space. Such an interwoven approach would look for *any* submarking-equivalent $\jmath$-sets $S_r$ and $S_{r+k}$ regardless of $r$, and *afterwards* investigate whether the enabling functions which involve $\mathsf{P}_0$ are relevant to the transition structure between markings in $S_r, \cdots, S_{r+k}$. They are not relevant, if

- either $r$ is large enough so that these enabling functions will yield the same value in all subsequent $\jmath$-sets (this corresponds to the original definition of $\jmath_{\min}$), or
- all transitions which involve such enabling functions are disabled due to one of the other necessary enabling condition mentioned above.

If this applies to all markings in $S_r, \cdots, S_{r+k}$, then the repeating structure has been found. No additional transitions are possible in further $\jmath$-sets, since the firing conditions which do not involve $\mathsf{P}_0$ will not change. Similarly, none of the possible transitions will be disabled in further $\jmath$-sets, as all transitions which depend on $\mathsf{P}_0$ are already disabled.

To simplify our considerations for the state space generation algorithm, we will, however, not further elaborate on this approach. Furthermore, it may be argued that "proper" iSPN models do not contain enabling functions which never come into effect, and that the additional effort of keeping track *why* a transition is not enabled during the state space generation procedure is not worth the potential gain. The approach might, however, be an interesting

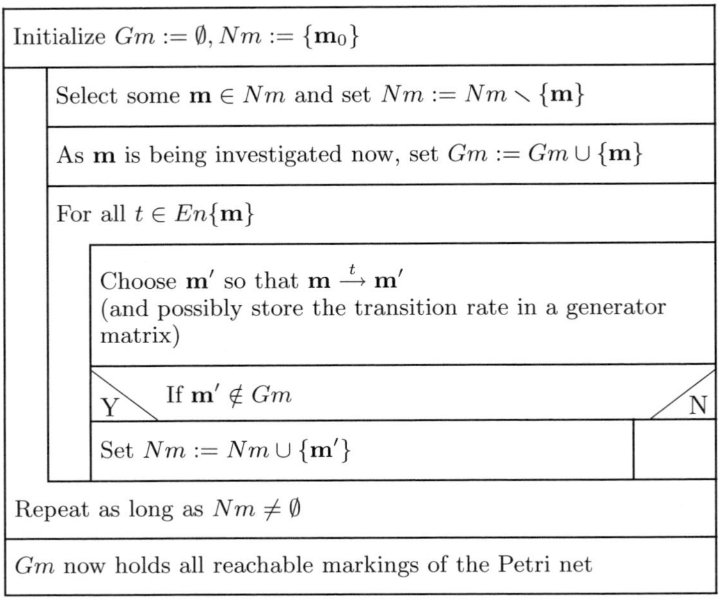


**Fig. 5.10.** A standard algorithm for generating the finite reachability set of a conventional, bounded Petri net.

option in future versions of the tool environment, as the discovery of ineffective enabling functions may be helpful to indicate potential design flaws during the development of large iSPN models.

### 5.3.2 State Space Generation

The focus of this section is the generation of the underlying Markov chain of an iSPN. From a theoretical point of view, it became clear in Section 5.2 that if the reachability set of an iSPN is available, then we can readily derive a QBD description of the underlying Markov chain. However, if this transformation is to be done by a practical tool implementation, it is not possible to generate the complete infinite reachability set. After quickly reviewing the conventional approach for SPNs with finite reachability sets, we present two key results which will then be used by an appropriate state space generation algorithm for iSPNs which is presented at the end of this section.

**Approach for Finite Reachability Sets.** In conventional Petri nets, it is assumed that the reachability set $Rs(\mathcal{P})$ of a Petri net $\mathcal{P}$ is *finite*. This is to guarantee that the underlying Markov chain of an SPN is finite as well, and thus standard numerical techniques can be applied for the numerical analysis. To obtain the Markov chain in these cases, a simple iterative procedure like the one shown in Figure 5.10 can be employed. The idea is that the state space generation starts by investigating all successor markings of the initial marking (which must be part of $Rs(\mathcal{P})$), and then continues by successively investigating those markings which have been generated so far for successor markings. The two sets $Gm$ and $Nm$ play a key role in the algorithm. At the end of each iteration, their semantics is as follows:

- A set $Gm \subseteq Rs(\mathcal{P})$ (for "generated" markings) holds all those reachable markings which have been investigated for possible successor markings.
- $Nm$ (for "new" markings) holds all those markings which still have to be examined.

Initially, $Nm$ only holds the initial marking $\mathbf{m}_0$ and $Gm$ is empty. After a finite set of iterations, the algorithm terminates as $Nm$ is empty. At that time, $Gm$ holds the entire reachability set of the Petri net.

**The Theory for Dealing with iSPNs.** Obviously, the algorithm presented above can not be applied to iSPNs, as it would not terminate due to the infinite reachability set of this SPN class. However, as we have seen in Section 5.2, the entire reachability set is not necessary to obtain a QBD description of the underlying Markov chain. After two submarking-equivalent $j$-sets $S_r$ and $S_{r+k}$ with $r \geq j_{\min}$ have been found, the state space generation algorithm can be stopped, since all further $j$-sets are identical to $S_r, \cdots, S_{r+k}$. The development of a proper state space generation algorithm is, however, still non-trivial, as the algorithm must be able to decide whether all states in a set $S_j$ have been generated yet. Only then a search for submarking-equivalent $j$-sets makes sense.

The algorithm that we propose employs two sets $Gm$ and $Nm$ in the same spirit as the original algorithm for bounded SPNs does. We look at these sets at time instants immediately following an iteration of the original algorithm, thus $\mathbf{m}_0 \in Gm$ is always valid. For a given $Gm$, the corresponding set $Nm$ holds all markings $\mathbf{m}'$ which are directly reachable from any marking $\mathbf{m} \in Gm$, and which are not contained in $Gm$:

$$Nm = \{\mathbf{m}' | \exists \mathbf{m} \in Gm : \mathbf{m} \rightarrow \mathbf{m}' \wedge \mathbf{m}' \notin Gm\}. \tag{5.20}$$

Analogous to the introduction of $j$-sets for partitioning $Rs(\mathcal{P})$, we will use a partitioning of $Gm$ into sets $G_j$ for notational convenience. All markings in $G_j$ contain the same number of tokens $j$ in the unbounded place. Naturally, $Gm = \cup_{j \geq 0} G_j$ holds.

We also employ a variable $j^*$ which denotes the minimum number of tokens in the unbounded place in any marking in $Nm$:

$$j^* = \min(\{j \in \mathbb{N}_0 | \exists \bar{\mathbf{m}} : (j, \bar{\mathbf{m}}) \in Nm\}). \tag{5.21}$$

Our algorithm is based on two key ideas:

- When selecting new markings from $Nm$ for investigation, we give priority to markings which belong to lower $j$-sets. This way, the generation of

markings which belong to lower $j$-sets is finished before new markings in the remaining part of the state space are investigated.
- We formulate a criterion that, depending on the markings in $Nm$ which remain to be investigated, guarantees that all reachable markings up to some limiting $j$-set have been generated. We can then search for submarking-equivalent $j$-sets below this bound.

The second point uses a key property of iSPNs which is given by the following theorem. For proving this property, the following intermediate result is needed, which states that any reachable marking that has not been generated yet can be reached from any set $G_j$, if $j$ is below a limiting value that depends on $j^*$:

**Proposition 6.** Let $Gm, Nm$ and $j^*$ be defined as mentioned above. For any reachable marking $\mathbf{m}'$ of an iSPN with $\mathbf{m}' \notin Gm$ and any value $j$ with $0 \leq j < \max\{j^*, 1\}$, there exists a marking $(j, \bar{\mathbf{m}}) \in Gm$ with $\mathbf{m} \xrightarrow{*} \mathbf{m}'$.

*Proof.* First note that if $j^* \leq 1$ holds, only $j = 0$ has to be considered, for which the proposition trivially holds as the initial marking $(0, \bar{\mathbf{m}}_0)$ is contained in $Gm$. From now on, we thus assume $j^* > 1$.

The subsequent considerations take advantage of the following property: any marking $(j', \bar{\mathbf{m}}')$ which is *directly* reachable from a marking $(j, \bar{\mathbf{m}}) \in Gm$ with $j < j^* - 1$ is also contained in $Gm$. This is due to the bounded jump size (Proposition 1), which limits the destination level $j'$ to $j' < j^*$, and the fact that $Nm$ (which always contains all directly reachable markings which have not been generated yet) only holds markings with *at least* $j^*$ tokens in the unbounded place.

Now, suppose that there exists a sequence of $n \geq 1$ transitions from the initial marking $(0, \bar{\mathbf{m}}_0)$ of the iSPN to some reachable marking $\mathbf{m}'$ which has not been generated yet (i.e., $(j', \bar{\mathbf{m}}') \notin Gm$):

$$(j_0, \bar{\mathbf{m}}_0) \xrightarrow{t_1} (j_1, \bar{\mathbf{m}}_1) \xrightarrow{t_2} \cdots \xrightarrow{t_{n-1}} (j_{n-1}, \bar{\mathbf{m}}_{n-1}) \xrightarrow{t_n} (j', \bar{\mathbf{m}}'), j_0 = 0.$$

As assumed in the proposition, select some arbitrary $j$ with $1 \leq j < j^*$ (for the case $j = 0$, the proposition holds, as already mentioned above). First, consider the case that all visited levels $j_i$ are smaller than the selected level number $j$, i.e., $\forall i : j_i < j$ holds. As $j < j^*$, this implies $\forall i : j_i < j^* - 1$, which, by successive application of the property mentioned above, yields that all $(j_i, \bar{\mathbf{m}}_i) \in Gm$ and also $(j', \bar{\mathbf{m}}') \in Gm$, contradicting the assumption $(j', \mathbf{m}') \notin Gm$.

Thus, there must exist some $j_i$ with $j_i = j$; let $i'$ denote the smallest index so that $j_{i'} = j$ holds (note that $i' \geq 1$ as $j \geq 1$ and $j_0 = 0$). Due to the minimality of $i'$, all preceding visited levels must be smaller than $j$, i.e., $\forall i < i' : j_i < j$. Again, as $j < j^*$, this implies $\forall i < i' : j_i < j^* - 1$. Similar to the previous case, the successive application of the above-mentioned result yields $(j_{i'}, \bar{\mathbf{m}}_{i'}) \in Gm$. As $j_{i'} = j$, $(j', \bar{\mathbf{m}}')$ is thus reachable from $(j, \bar{\mathbf{m}}_{i'}) \in Gm$, which proves the proposition. □

With this result, the following main theorem can be proved. It uses the repetitive transition structure to show that if submarking-equivalent sets $G_r$ and $G_{r+k}$ below some bound have been generated, then *all* reachable markings up to $j$-set $S_{r+k}$ have been generated.

**Theorem 4.** Let $Gm, Nm$ and $j^*$ be defined as mentioned above. Assume that there exist two submarking-equivalent sets $G_r$ and $G_{r+k}$ with $r \geq j_{\min}, k \geq 1$. Then, if $r+k < j^*$, the generated sets $G_0, \cdots, G_{r+k}$ are identical to the $j$-sets $S_0, \cdots, S_{r+k}$. Equivalently, there are no reachable markings $\mathbf{m}' = (j', \bar{\mathbf{m}}')$ with $j' \leq r+k$ and $\mathbf{m}' \notin Gm$.

*Proof.* Consider any reachable marking $(j', \bar{\mathbf{m}}')$ with $j' \leq r+k$. This marking is either contained in $Gm$, or, according to Proposition 6, as $r+k < j^*$ and thus $j' < j^*$, it is reachable by a finite number of transitions from a marking $(j, \bar{\mathbf{m}}) \in Gm$, for arbitrary values $j$ with $0 \leq j \leq r+k$. Select $j$ so that the number $n$ of transitions to $(j', \bar{\mathbf{m}}')$ is minimal. The theorem is proved by finite induction on $n$, and showing that $(j', \bar{\mathbf{m}}') \in Gm$ always holds.

(i) (*Basis*) For $n = 1$, $(j', \bar{\mathbf{m}}')$ is *directly* reachable from $(j, \bar{\mathbf{m}})$. All markings which are directly reachable from a marking in $Gm$ are either contained in $Gm$ or in $Nm$. The latter case does not apply, since $Nm$ only contains markings $(j', \bar{\mathbf{m}}')$ with $j' > r+k$ (as $j^* > r+k$), which contradicts the assumption $j' \leq r+k$. Thus, the proposition holds for $n = 1$.

(ii) (*Induction Step*) The induction hypothesis is that the proposition holds for all markings $(j', \bar{\mathbf{m}}'), 0 \leq j' \leq r+k$, which are reachable from some marking $(j, \bar{\mathbf{m}}) \in Gm, 0 \leq j \leq r+k$ within $n'$ transitions, where $n'$ is minimal and $1 \leq n' \leq n$ holds. For the case that $n+1$ transitions are needed, we have a sequence

$$(j, \bar{\mathbf{m}}) \xrightarrow{t_1} (j_1, \bar{\mathbf{m}}_1) \xrightarrow{t_2} \cdots \xrightarrow{t_n} (j_n, \bar{\mathbf{m}}_n') \xrightarrow{t_{n+1}} (j', \bar{\mathbf{m}}'). \tag{5.22}$$

First, it is important to note that $\forall i : j_i > r+k$ must hold, since

- if some $j_i \leq r+k$, then $(j_i, \bar{\mathbf{m}}_i) \in Gm$ holds due to the induction hypothesis, as this marking is reachable by $1 \leq i \leq n$ steps from $(j, \bar{\mathbf{m}})$;
- this, however, implies that $(j', \bar{\mathbf{m}}')$ is reachable in less then $n+1$ transitions from that marking $(j_i, \bar{\mathbf{m}}_i) \in Gm, 0 \leq j_i \leq r+k$, contradicting the assumption that $n+1$ is minimal.

Now assuming that *all* $j_i > r+k$, and using the fact that inter-level transitions are bound to a maximum jump size of 1, the only way to comply with $j \leq r+k$ and $j' \leq r+k$ is to select $j = r+k$ and $j' = r+k$. Thus, (5.22) becomes

$$(r+k, \bar{\mathbf{m}}) \xrightarrow{t_1} (r+k+1, \bar{\mathbf{m}}_1) \xrightarrow{t_2} \cdots \xrightarrow{t_n} (r+k+1, \bar{\mathbf{m}}_n) \xrightarrow{t_{n+1}} (r+k, \bar{\mathbf{m}}'). \tag{5.23}$$

As $r \geq j_{\min}$, the above transitions are independent of the level index and can be “shifted” $k$ levels to the left. Thus, the same sequence of transitions takes place if we start in marking $(r, \bar{\mathbf{m}})$:

$$(r, \bar{\mathbf{m}}) \xrightarrow{t_1} (r+1, \bar{\mathbf{m}}_1) \xrightarrow{t_2} \cdots \xrightarrow{t_n} (r+1, \bar{\mathbf{m}}_n) \xrightarrow{t_{n+1}} (r, \bar{\mathbf{m}}'). \tag{5.24}$$

We now consider this shifted sequence of transitions to analyze the original sequence (5.23). Due to the fact that $G_r$ and $G_{r+k}$ are submarking-equivalent by assumption, $(r+k, \bar{\mathbf{m}}) \in G_{r+k}$ implies that $(r, \bar{\mathbf{m}})$ is contained in $Gm$. Since $r+1 \leq r+k$, the induction hypothesis can be applied to the first transition in (5.24) to yield $(r+1, \bar{\mathbf{m}}_1) \in Gm$. Similarly, we have that $(r, \bar{\mathbf{m}}') \in G_r$, as it is reachable by $n$ transitions starting from $(r+1, \bar{\mathbf{m}}_1)$. Again using the submarking-equivalence of $G_r$ and $G_{r+k}$, we obtain $(r+k, \bar{\mathbf{m}}') \in G_{r+k}$, thus proving the proposition for $n+1$.

This concludes the induction proof. As we now have shown for all $n$ that any marking in $S_0, \cdots, S_{r+k}$ which is reachable by a (minimum) number of $n$ transitions from a marking in $G_0, \cdots, G_{r+k}$ (and thus, also from the initial marking $\mathbf{m}_0$) is already contained in $G_0, \cdots, G_{r+k}$, this proves the theorem. □

**A State Space Generation Algorithm for iSPNs.** With the aid of Theorem 4, it is possible to decide whether a state space generation algorithm like the one shown in Figure 5.10 has already generated all markings up to a certain $j$-set. This is accomplished by determining $j^*$ at the end of each iteration, and then looking for submarking-equivalent sets $G_r$ and $G_{r+k}$ with $r \geq j_{\min}, k \geq 1$ and $r+k < j^*$. If such sets exist, a repeating state space structure in accordance with Proposition 5 has been found, and the QBD description of the underlying Markov chain can be derived. Otherwise, further markings have to be generated until submarking-equivalent sets $G_r, G_{r+k}$ can be found.

It is important to note that such an algorithm will only terminate if we guarantee that $j^*$ increases. If, for example, there is some marking $(0, \bar{\mathbf{m}}) \in Nm$ which is never removed from $Nm$, then $j^*$ remains invariably equal to 0, and the state space generation will not end. It is, however, easy to assure that $j^*$ becomes arbitrarily large by always removing those markings from $Nm$ with the smallest number of tokens in the unbounded place (selecting a marking $(j^*, \bar{\mathbf{m}})$ is an obvious choice, as $j^*$ always denotes the lowest number of tokens in $\mathsf{P}_0$ for all markings in $Nm$). Since the number of reachable markings below any $j$-set $S_j$ is finite, $j^*$ must eventually increase. Then, according to Proposition 4, we must eventually encounter two submarking-equivalent $j$-sets, so that the termination of the algorithm is guaranteed. An extension of the original algorithm for finite state spaces, which takes all this into account, is shown in Figure 5.11.

It should be noted that if we assume that the underlying Markov chain of an iSPN is in compliance with the "simple case" as discussed in Section 5.2.2 (i.e., $k=1$), then the state space generation becomes particularly simple. In that case, it is only necessary to compare $G_{j^*-1}$ to $G_{j^*-2}$ at the end of each iteration. However, also for $k > 1$ an efficient implementation of the search

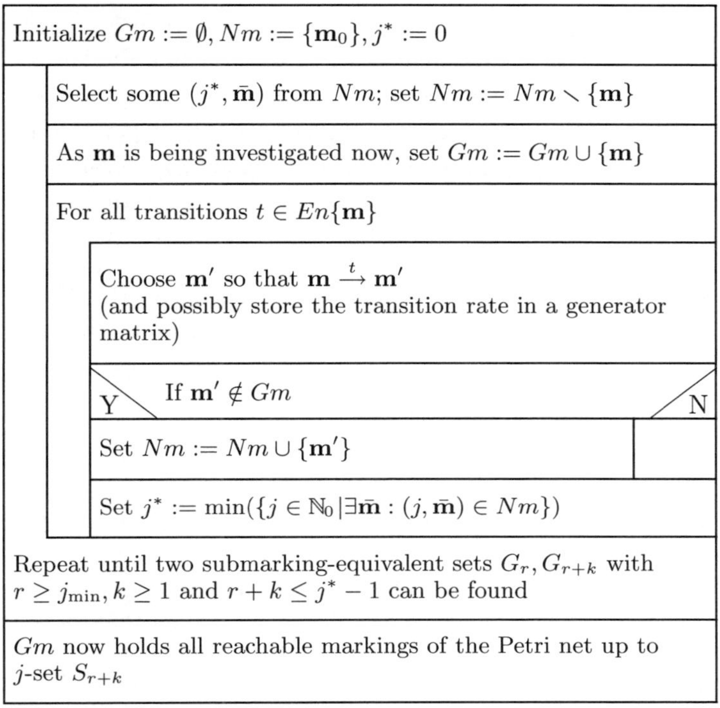


**Fig. 5.11.** Algorithm for generating the reachability set of an iSPN.

for submarking-equivalent $j$-sets is conceivable (for example, by defining an appropriate hashing function on $j$-sets).

### 5.3.3 Accounting for Immediate Transitions

Until now, we were not distinguishing between immediate transitions and timed transitions in this chapter. However, when generating the underlying Markov chain of an iSPN, we thus obtain infinitely large rate entries in the generator matrices, which have to be eliminated in order to obtain a proper QBD Markov chain which can then be analyzed.[6] In this section, we first discuss the influence of this state space reduction on the considerations which have been accomplished so far. As it is desirable to use immediate transitions

[6] Note that for finite Markov chains, an alternative solution approach which does not require the removal of infinite transition rates was suggested in [4, 46]. There, the stochastic process including infinite transition rates is considered as a semi-Markov process, whose steady-state distribution is given by considering the steady-state distribution of the embedded Markov chain at state transitions, and then weighting these probabilities with the sojourn times in the states of the original semi-Markov process. An extension of this approach to QBD processes with infinite rates is conceivable, but has not been performed yet.

in a more general way as it is allowed by the definition of iSPNs (Definition 4), we informally propose an extension to the original definition which does not violate any of the assumptions so far.

**Impact of Immediate Transitions in the Original iSPN Class.** The removal of infinite transition rates (or the corresponding associated *vanishing* markings at the iSPN level) can be performed either at the Markov chain level, or during the state space generation process discussed in the previous section.

At the iSPN level, this removal is reflected by the fact that the following sequence, where the firing of a timed transition $t$ is followed by $n$ immediate transitions $t_1, \cdots, t_n$,

$$(j, \bar{\mathbf{m}}) \xrightarrow{t} (j_1, \bar{\mathbf{m}}_1) \xrightarrow{t_1} \cdots \xrightarrow{t_n} (j', \bar{\mathbf{m}}'), \tag{5.25}$$

is replaced by a single transition $(j, \bar{\mathbf{m}}) \xrightarrow{t} (j', \bar{\mathbf{m}}')$. The transition rate $R(t)$ is then modified at the Markov chain level according to the weights $W(t_1), \cdots, W(t_n)$ (see e.g. [90] for details).

In the context of iSPNs, this represents a problem if the sequence (5.25) extends over more than two $j$-sets, i.e., if $|j - j'| > 1$. In that case, Proposition 1 no longer holds, and the underlying QBD Markov chain is no longer skip-free. However, for the iSPN class defined in Definition 4, this situation can not occur. Due to property (5.9), the firing of any immediate transition that involves the unbounded place will not change the number of tokens in it. Thus, the condition $|j - j'| \leq 1$ always holds. Furthermore, the reduction takes place in an identical manner in *all* $j$-sets above $j_{\min}$, so that all considerations which involve the submarking-equivalence of $j$-sets within and above $S_{j_{\min}}$ are still valid.

**Extension to a Useful Case for Practical Applications.** Often, it is desirable to use an immediate transition for removing or adding tokens to the unbounded place $\mathsf{P}_0$; this is, for example, useful for easily synchronizing job arrivals or services with the state of a complex job processing system. For this reason, we consider the special case where *one* immediate transition connected to the unbounded place $\mathsf{P}_0$ is allowed, without assuming the test arc property (5.9).

In that case, a skip-free QBD structure is still obtained if it can be guaranteed that prior to the firing of this immediate transition, a timed transition fires which does not change the number of tokens in $\mathsf{P}_0$ (i.e., the immediate transition is enabled just once between the firing of two of these timed transitions). It can even be allowed that the timed transition adds a token to the unbounded place if the immediate transition removes a token from $\mathsf{P}_0$ (or vice versa). Under the condition that such a timed transition always exists, the first transition in sequence (5.25) leads to some level $j_1 = j + d$ with $|d| \leq 1$. We now assume that subsequent transitions either do not alter the

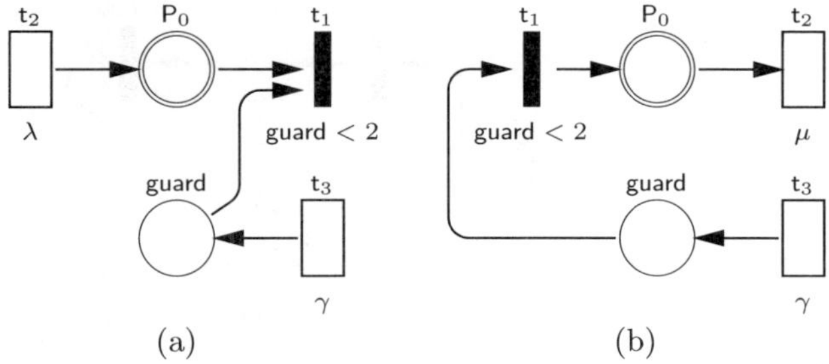


**Fig. 5.12.** Two examples where immediate arcs from (a) or to (b) the unbounded place are allowed.

level index, or contain at most one firing of the immediate transition connected to place $\mathsf{P}_0$, thereby adding $-d$ tokens to $\mathsf{P}_0$. Summarizing, we then obtain $|j - j'| \leq 1$.

As the above condition depends on the dynamic behaviour of iSPNs, we did not account for it in the original iSPN definition in order to simplify the formal considerations in this chapter. It is, however, possible to characterize two specific cases at the iSPN level which exactly lead to the special case mentioned above.

The corresponding iSPN constructions involve the introduction of a "guarding" place for asserting that a timed transition of the above-mentioned kind fires prior to each firing of the immediate transition that is connected to $\mathsf{P}_0$. Two example iSPN subnets for immediate job departures and arrivals are illustrated in Figures 5.12(a) and 5.12(b), respectively. In both cases, the following two restrictions are assumed:

- There is only one immediate transition (in this case, $\mathsf{t}_1$) connected to place $\mathsf{P}_0$. All other transitions connected to it (like $\mathsf{t}_2$) are timed.
- All transitions connected to the *guard* place (like $\mathsf{t}_3$) are timed. They either do not change the number of tokens in $\mathsf{P}_0$, or add (in the case of immediate job departures) or remove (for immediate job arrivals) a token from $\mathsf{P}_0$.

The enabling function of the immediate transition $\mathsf{t}_1$ ensures that it fires at most one time in a sequence of immediate transitions, even if the guard place contains more than one token. This way, it is avoided that more than one token enters (or leaves) the unbounded place at once, which would violate the limited jump size property in most cases. In both example iSPNs, it can easily be seen that the immediate transition $\mathsf{t}_1$ becomes firable only if some timed transition has put a token into the guard place before.

We will employ iSPN constructions like this several times in the application studies presented in Chapter 6. Note that it is often not necessary to explicitly introduce the enabling function $f_{\mathsf{guard}<2}$ for the immediate transition $\mathsf{t}_1$, as the number of tokens in guard is already limited to 1 due to the structure of the iSPN.

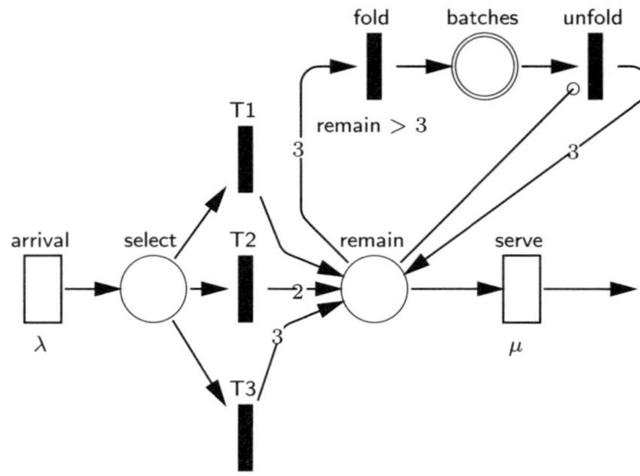


**Fig. 5.13.** Folding construction for obtaining a skip-free QBD in spite of batch arrivals.

### 5.3.4 Modeling Batch Arrivals and Departures

Throughout this chapter, is has been assumed that the underlying QBD of an iSPN is skip-free; this is guaranteed by limiting the multiplicity of arcs connected to the unbounded place to 1 in condition (5.7). In practice, it is, however, sometimes desirable to account for platooned customer arrivals (and departures) to the unbounded buffer place. From a theoretical point of view, this does not impose a problem, since non-skip-free QBDs can easily be transformed to an equivalent skip-free representation. Thus, all results which were derived in this chapter can be extended to the case where the multiplicity of transitions connected to the unbounded place is larger than 1.

It is, however, also possible to account for batch arrivals and departures solely at the iSPN-level, without violating condition (5.7). In this way, we circumvent a further increase in the complexity of the algorithms employed for deriving the underlying QBD process of an iSPN. Due to the fact that our current tool environment assumes that condition (5.7) holds, we employed such a "folding" procedure at the iSPN level to obtain a skip-free QBD for the comparison study accomplished in Section 4.5. The employed folding construction is similar to the one presented in [92]. We illustrate the procedure by considering batch arrivals for the example iSPN presented in Section 5.1.4, which was also used for the numerical investigation in Section 4.5.

The basic idea is to distribute the number of jobs $n$ which are waiting for service among *two* places. If the maximum batch arrival size is $S$, the first (unbounded) place holds $n_1 = \lfloor n/S \rfloor$ tokens. A second place contains $n_2 = n - n_1 \cdot S$ tokens, and from this place tokens are taken out to be served. Any time this place is empty and a token is available in the unbounded place, a token is removed from it and replaced by $S$ tokens in the service place; the actual number of jobs in the system is thus given by $n_1 \cdot S + n_2$. Figure 5.13

shows the appropriate modification of the arrival part of the original iSPN in Figure 5.1 for a maximum batch arrival size of 3. The two places mentioned before are represented by batches and remain, respectively. The select place is used to probabilistically choose a batch arrival size of 1, 2, or 3. As can be seen in the figure, arc multiplicities for arcs entering the unbounded place have been circumvented by only counting *multiples* of $S$ in the unbounded place. Note that an infinite folding/unfolding loop is avoided by the enabling function $f_{\mathsf{remain}>3}$.

It should be noted that this solution maps all states which belong to a buffer filling of $0, \cdots, S-1$ tokens to markings where the unbounded place is empty, instead of mapping just those states which correspond to an empty buffer. For larger values of $S$, this can lead to very large border levels in the underlying QBD process. This can be avoided if it is assured that *only* an empty buffer corresponds to 0 tokens in the unbounded place. One way to achieve this is to initialize place remain with $S$ tokens, and to introduce an additional enabling function $f_{\mathsf{batches}>0}$ for transition serve. Then, as soon as the first job enters the system, transition fold is enabled and place batches receives a token. Naturally, for obtaining the "real" number of jobs in the system, the initial $S$ tokens have to be deducted. This yields the formula $n_1 \cdot S + n_2 - S$ for the actual number of jobs in the buffer.

Concerning the immediate transitions connected to place batches, the constructions suggested in this section do not conform to the two special cases considered in Section 5.3.3. Nevertheless, it is readily verified that prior to each firing of fold and unfold either the timed transition arrival or serve must have fired. Thus, the underlying QBD process only contains transitions between markings in adjacent $j$-sets, and condition (5.7) holds.

## 5.4 Extensions for Buffer Resets and Quasi-Stationary Models

The iSPN framework can also be extended to easily account for the specification of systems which involve buffer resets or where the quasi-stationary distribution is of interest (see our comments in Sections 4.6.2 and 4.6.3).

One way to account for buffer resets is to introduce special input arcs originating at the unbounded iSPN place which implicitly have a multiplicity that equals the current number of tokens in the unbounded place. Optionally, a finite value may be associated to this arc that denotes the number of tokens that shall remain in the unbounded place when the associated transition fires. In this case, a firing of the associated transition will not lead to a marking in $S_0$, which must be taken into account when determining the value $j_{\min}$.

If quasi-stationary measures are of interest, then the iSPN framework needs to account for the fact that the underlying QBD must be transient. To this end, a new class of transitions could be introduced which is used without specifying an output arc. The firing of these transitions can then be reflected by transitions to the absorbing state in the underlying QBD.

The formalization and implementation of these extensions to the proposed iSPN framework seems straightforward; however, we do not further elaborate on these issues here.

## 5.5 Summary and Concluding Remarks

While QBD processes were identified in the last chapter as an attractive class of stochastic processes in the field of model-based performance evaluation, this chapter provides an extensive framework for taking them to the application level. By employing infinite-state SPNs as a high-level description method, the performance modeling engineer can take advantage of the merits of QBD processes without leaving the popular SPN paradigm, which he often already used before.

It should be emphasized that iSPNs represent an *extension* of a powerful class of SPNs rather than a restriction. The characteristic property of iSPNs, the unbounded place, provides an additional means to efficiently account for a modeling situation which frequently occurs in practice—system components which can be represented as large (or unbounded) buffers arise at any level in the design hierarchy of computer and communication systems, from task scheduling queues at the operating system level to segmentation and reassembly buffers in packetized communication services like TCP/IP and ATM.

While iSPNs have originally been suggested in [88, 89], the main contribution of this chapter is to provide a thorough formal basis for the approach. Founding on this formalization, we were able to derive the following two main results:

- It has been shown that the expressiveness of iSPNs is *equivalent* to the direct specification of QBD processes at the Markov chain level. Thus, from both a formal and a practical point of view, there is no reason to resort to the cumbersome and error-prone direct specification of QBD processes.
- We were able to obtain close bounds and efficient algorithms for deriving the QBD process underlying an iSPN, being the prerequisite to efficiently apply the numerical techniques for QBD processes.

A tool environment for the specification and numerical evaluation of iSPNs has been developed in [73, 118]. Recently, a user interface for the graphical specification of iSPNs has also been added. See Appendix B for an overview on this tool environment.

The formal foundation for iSPNs which has been laid in this chapter proved to be useful for showing important properties of this SPN class. In future work, this basis should be employed to extend the approach for dealing with batch arrivals in a "direct" manner (i.e., without using the auxiliary construction proposed in Section 5.3.4). It also seems worth-while to formalize and to implement the ideas for dynamically determining $j_{\min}$ during the state space generation procedure, as it can further increase the efficiency of the iSPN-based performance evaluation environment.

While we dealt with iSPNs in a thorough manner from a formal point of view, the justification of this work can only be given by its practical application. A small application example has already been given in Section 5.1.4; other small studies were presented in [88, 89, 95]. To prove that iSPNs can also be employed for the evaluation of much more complex systems, we focus on several larger studies in the next chapter. In these applications, real-world problems in the area of ATM connection management, World Wide Web and TCP/IP performance issues, and traffic modeling will be covered.

# 6. Application Examples: Node Analysis

The aim of this chapter is to illustrate the suitability of QBD models to model the network nodes of complex computer and communication systems, and to demonstrate how particular modeling requirements can be formulated in a way that fits the QBD-based framework.

The first two case studies presented here are rather system-oriented and investigate closely coupled pairs of traffic and server subsystems. In Section 6.1, connection management alternatives for delivering bursty video traffic are investigated. In Section 6.2, we focus on the impact of TCP/IP window flow control on the performance of World Wide Web applications.

It will be shown that correctly modeling the workload of these systems is of major importance for correct results. For that reason, the third study presented in Section 6.3 explicitly concentrates on the impact of so-called *self-similar* arrival traffic on queueing performance. The results of this chapter are summarized in Section 6.4, where we will also draw some important conclusions which motivate the work presented in the following chapters.

## 6.1 Connection Management for Video Traffic

In this case study a special connection management scheme for establishing connectionless services in connection-oriented communication networks is studied. Most broadband network infrastructures, like the ATM/B-ISDN framework [34, 85] which motivated this study, are connection-oriented to reduce the routing overhead. However, many services like remote procedure calls, bridging of connectionless local area networks, or IP over ATM, require connectionless services. One way to implement connectionless services on top of a connection-oriented bearer service has been described in [94, 95]. Since the benefits of this approach depend heavily on the shape of the traffic, and since video applications are a major application in broadband communication networks, we are particularly interested in the behaviour of the connection management mechanism for systems subject to video traffic. Note that the results presented here represent an extension and a more thorough description of the work presented in [73, 157].

In the following section the connection management scheme and the structure of video traffic are described in more detail. Then, an iSPN model accounting for connection management and video traffic is presented. We then present numerical results derived from the analysis of the model, and finally conclude with a summary of the results.

### 6.1.1 System Description

*Connection Management.* For the realization of connectionless services in a connection-oriented environment, two main approaches are conceivable: either separate connections are established for each individual packet, or permanent connections between potential peers are maintained. While the first approach implies a large amount of signaling traffic and large delays, the second approach suffers from a large amount of bandwidth which is reserved unnecessarily. To alleviate these drawbacks, the OCDR (on-demand connection with delayed release) connection management scheme has been proposed in [94, 95]; note that the need for such an algorithm has already been identified in [112] in the context of packet train models. In OCDR, connections are established on-demand for individual packets, however, they are not released immediately, but are maintained for a certain timeout period after a packet has been delivered. If other packets have to be delivered to the same destination within that period, they can use the same connection without additional overhead for a connection setup. Naturally, this implies that connections are sometimes kept unnecessarily if no successor packets to the same destination occur, but using the OCDR scheme it is possible to adapt the costs of reserved bandwidth to the tolerable packet delay. For timeout values of 0 and infinity, the OCDR algorithm coincides with the two extreme approaches presented first, i.e., with per-packet connection setups and permanent connections, respectively.

*Video Traffic.* There are basically two ways for submitting the traffic which originates from a video source to the transport network. One approach is to submit the data to the network as soon as it is available, at the maximum possible speed. In that case, there is a packet burst at the beginning of each time frame (which is usually 1/30th or 1/25th of a second). No packets are generated during the remaining time of a frame. Alternatively, some kind of traffic shaping takes place, where the video data is buffered and then distributed as uniformly in time as possible. The latter approach involves more hardware overhead and is less interesting with respect to its interaction with the connection management mechanism, as the traffic stream ideally converges to a constant packet interarrival time. Instead, we focus on the first (bursty) packet generation method to investigate the interactions between bursty arrival traffic and the OCDR approach.

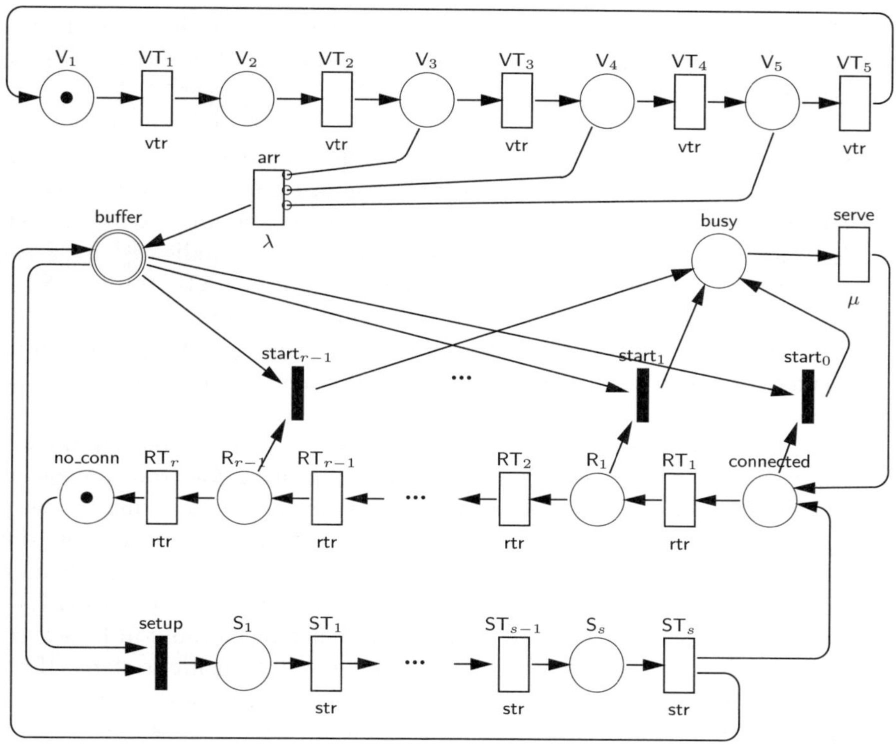


**Fig. 6.1.** Petri net model for connectionless services managed by OCDR.

### 6.1.2 Model Development

The overall iSPN model, including bursty video arrivals and the OCDR connection management policy is illustrated in Figure 6.1 (note that the corresponding SPN2MGM specification files can be found in Appendix C.3).

For generating the bursty video traffic, we resort to an approach proposed by Saito [175]. The idea is to model each timeframe by a sequence of Erlang stages, representing the deterministic nature of a time frame. Arrivals can occur only during the first, "active" stages of this distribution, i.e., only during the first part of a time frame. At the Petri net level, this can be modeled by a token which circulates in a sequence of places (see places $\mathsf{V}_i$ in the upper part of Figure 6.1). The number of places in this sequence corresponds to the number of stages of the Erlang distribution used to model the duration of a time frame (5 stages in Figure 6.1). The actual packet arrivals are generated by transition arr, however, this transition is disabled in the last three stages of the Erlang-5 distribution (as indicated by the inhibitor arcs from $\mathsf{V}_3$, $\mathsf{V}_4$ and $\mathsf{V}_5$. Note that the choice of an Erlang-5 distribution with two active arrival stages is a modeling assumption here; in practice, the number of stages and active phases depends heavily on the characteristics of the video traffic and the speed at which data is submitted to the network by transition arr.

In the model, both connection setup time and the OCDR connection release timeout can be modeled as Erlang distributions. Thus, we are in particular able to account for the more deterministic nature of the timeout-based connection release mechanism. If a cell enters the input buffer (place buffer) and no connection has been established yet (as indicated by a token in no_conn), it successively passes the $s$ transitions $\mathsf{ST}_1$ through $\mathsf{ST}_s$. Afterwards, the token is put back in the buffer place and a token in connected indicates the established connection. Via transition $\mathsf{start}_0$, a token in the buffer place can now immediately enter the busy place to be served by transition serve.

After each service completion, the connection indication token is put back in place connected. If there are already tokens waiting for transmission in the buffer place, the next one is immediately processed by entering the busy place via $\mathsf{start}_0$. Otherwise, if the buffer is empty after service completion, the token in place connected successively passes the $r$ transitions $\mathsf{RT}_i$, thereby modeling the connection release timeout. As long as the last transition $\mathsf{RT}_r$ has not been passed yet, cells arriving in the buffer place can still be served by entering the busy place via one of the immediate transitions $\mathsf{start}_i$. If no new cells arrive during the release timeout period, transition $\mathsf{RT}_r$ fires and a token in no_conn indicates that the connection has been released.

### 6.1.3 Parameterization

In our experiments, the burst cell arrival rate $\lambda$ (firing rate of transition arr) has been varied between 0 and 300. The duration of a frame (1/30th of a second) was modeled by an Erlang-5 distribution, as indicated by the upper row of places in Figure 6.1. Consequently, the firing rate $vtr$ in each stage was set to 150. Burst cell arrivals only occur during the first two stages of this distribution, leading to an effective arrival rate of $\lambda_{\text{eff}} = 2/5 \cdot \lambda$. If not stated otherwise, a value of 150 has been assumed for the service rate $\mu$.

For the connection setup time we selected an exponential distribution with mean 0.1, thus $s = 1$ and $str = 10$. Due to its deterministic character, the connection release timeout was modeled as Erlang-8 distribution, where we investigated two cases with mean 0.1 and 0.2 (thus $r = 8$ and $rtr = 80$ or $rtr = 40$, respectively) as well as a variation of this parameter between 0.02 and 0.4.

### 6.1.4 Numerical Results

In our investigations we were mainly interested in the costs and merits of the proposed connection management scheme. Since its behaviour depends heavily on the offered system load, this is investigated first. Afterwards, we focus on what can be gained at what price by employing OCDR. The investigations are then concluded by some comments on the numerical complexity of the model evaluation.

**Impact of System Load on OCDR.** The intensity of the video traffic entering the system heavily influences the performance of the OCDR algorithm, since for very small cell arrival rates the connection release timeout expires more often then for high arrival rates. The impact of this is illustrated by investigating the mean response time of the system for different burst cell arrival rates. Furthermore, we compare the *used* bandwidth to the *reserved* bandwidth, since the relation of these values represents a notion of costs. The used bandwidth equals the input rate at which cells arrive at the system, $\lambda_{\text{eff}}$ (usually expressed in bits per second). The reserved bandwidth equals the probability of an existing connection, multiplied by the bandwidth of that connection. The connection bandwidth is given by the parameter $\mu$; the probability $p_{\text{conn}}$ of an established connection can be formulated in a reward-based way at the Petri net level as follows:

$$\mathsf{conn} = \#\mathsf{busy} + \#\mathsf{connected} + \sum_{i=1}^{r-1} \#\mathsf{R}_i \tag{6.1}$$

$$p_{\text{conn}} = P(\mathsf{conn} > 0). \tag{6.2}$$

Note that all places $\mathsf{R}_i$ have to be considered for the computation of reward $\mathsf{conn}$, as a connection is not released before transition $\mathsf{RT}_r$ fires. Taking into account only places $\mathsf{busy}$ and $\mathsf{connected}$ is not sufficient.

The mean response time of the system is obtained by Little's law

$$\mathrm{E}[R] = \mathrm{E}[N] \cdot \frac{1}{\lambda_{\text{eff}}} = \frac{5}{2\lambda} \cdot \mathrm{E}[N], \tag{6.3}$$

where the mean number of cells in the system $\mathrm{E}[N]$ can be derived at the Petri net level using the following reward definition:

$$\mathsf{N} = \#\mathsf{buffer} + \#\mathsf{busy} + \sum_{i=1}^{s} \#\mathsf{S}_i. \tag{6.4}$$

This definition also accounts for the places $\mathsf{S}_i$, as tokens in $\mathsf{buffer}$ which lead to the establishment of a new connection temporarily leave the $\mathsf{buffer}$ place to pass all of the connection setup transitions $\mathsf{ST}_i$.

In Figure 6.2, the reserved bandwidth is illustrated for burst cell arrival rates between 0 and 300, for two different mean connection release timeout values. The used bandwidth is identical for both values of the release timeout. It can be observed that the reserved bandwidth is always larger for the larger mean timeout value than for the smaller one, due to the longer time an existing connection is maintained after a cell has been delivered. The increase

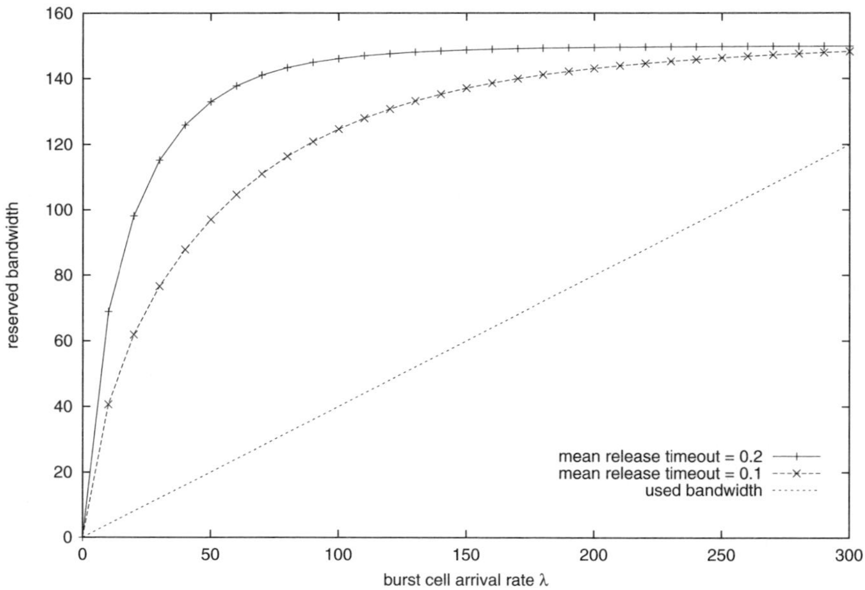


**Fig. 6.2.** Used bandwidth and reserved bandwidths for different burst arrival rates.

of the reserved bandwidth for larger burst arrival rates is due to the fact that the probability that the input buffer of the system (place buffer) remains empty for a period longer than the release timeout becomes smaller as the arrival rate increases, thus connections are released more seldom in higher arrival rate scenarios.

As another point, it can be observed that the cost of the OCDR scheme is excessive for very low cell arrival rates, since then the system almost always waits in vain for successor cells to arrive in time to use an existing connection. This is especially true for large connection timeout values.

Though the cost of the OCDR approach looks high, it should be noted that for the two extreme alternative connection management algorithms the situation is even worse. In case of persistent connections, the reserved bandwidth is equal to the connection bandwidth $\mu = 150$, thus constantly being larger than in the OCDR case. For per-cell connection establishments (corresponding to a connection timeout of 0), the cell delays will be excessive, as can be seen in the following investigation of the response time behaviour.

The mean response time of the system, for two connection release timeout values, is illustrated in Figure 6.3. First note that for the higher connection release timeout the mean response time is invariably better than for the smaller timeout value. This confirms that the OCDR method does reduce response times as compared to per-cell connection establishments, since per-cell connection setups correspond to a timeout value of 0 and imply mean

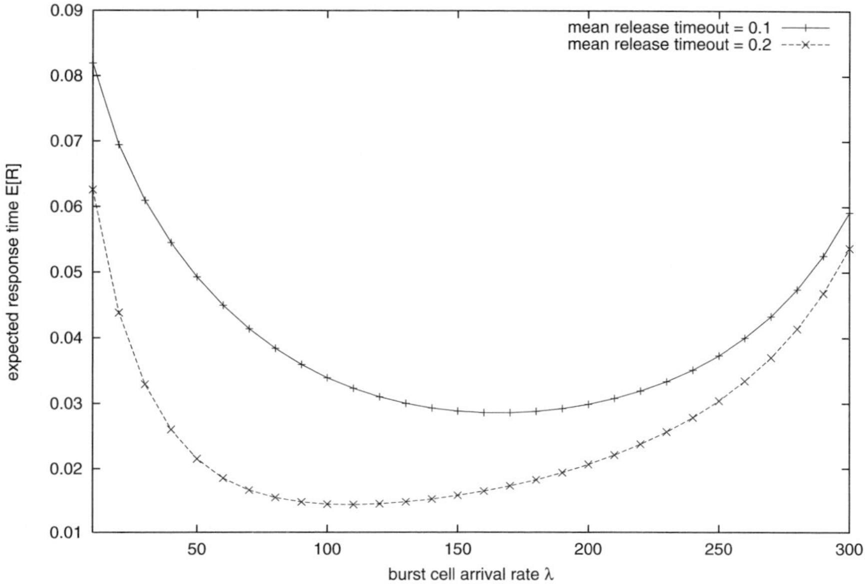


**Fig. 6.3.** Response time impact of system load for two different timeout values.

response times of at least $1/str = 0.1$. Without OCDR, this would render the system unstable even for moderate values of $\lambda$.

The development of the response time curves in Figure 6.3 is divided in two parts. First, for small traffic intensities, arriving cells only rarely encounter an existing connection since the cell interarrival time is often larger than the connection release timeout. The mean response time is rather high in this case, since a new connection has to be established then. As $\lambda$ grows, the mean response time becomes *better* due to the fact that the probability that an arriving cell encounters an existing connection increases. This development continues until some minimum mean response time is reached (at $\lambda \approx 110$ or $\lambda \approx 170$, depending on the release timeout). From then on, the mean response times increase since arriving cells almost always find an existing connection, but the traffic intensity is so large that the normal queueing behaviour comes into effect.

From a user perspective, one might be interested in adapting the burst cell arrival rate to obtain minimal mean response times (e.g., by changing the image quality in video conferencing applications). However, from a traffic engineering viewpoint it seems more natural to vary the timeout value for minimizing response times. As mentioned before, the minimum mean response time is obtained for persistent connections. This, however, might not be an option if one is also interested in keeping connection costs low. Thus,

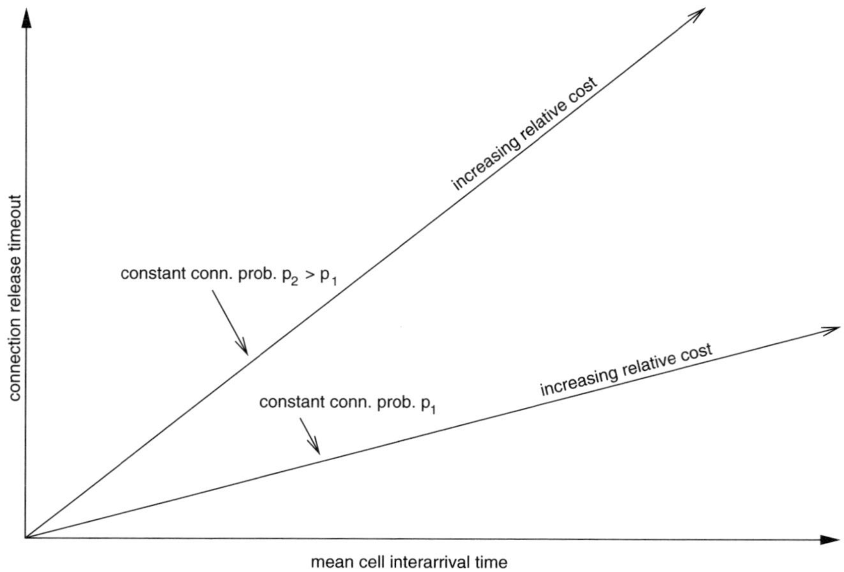


**Fig. 6.4.** Cost structures when employing OCDR.

the following investigation focuses on developing a notion of "cost" and relates it to the mean response time gain.

**Costs and Merits of the OCDR Method.** The idea of the OCDR approach is to reduce cell delays by keeping existing connections open for potential successor cells which can then use the same connection. The cost of this approach can be formulated by relating the reserved bandwidth to the actual input bandwidth. To illustrate the qualitative behavior of the OCDR costs, simple Poissonian arrivals under a deterministic timeout policy are considered in Figure 6.4.[1] There, the probability of an existing connection $p_{\mathrm{conn}}$ (which determines the reserved bandwidth) depends only on the ratio of the timeout value to the mean interarrival time; the two lines in Figure 6.4 correspond to two different values of $p_{\mathrm{conn}}$. Larger timeout values naturally imply higher connection probabilities, but also lead to higher costs as the reserved bandwidth increases. For $p_{\mathrm{conn}}$ fixed, it is interesting to note that the ratio of the reserved bandwidth to the input bandwidth increases as the mean interarrival time becomes larger, since the system load decreases, but the reserved bandwidth remains the same. Thus, for a given ratio of the connection timeout value to the mean interarrival time, OCDR is more expensive for lightly loaded connections.

[1] In this case, the connection probability is readily derived from the negative exponential probability distribution function as $1 - \mathrm{e}^{-x}$, where $x$ denotes the ratio of the release timeout to the mean interarrival time. The impact of queueing effects on connection management are neglected here.

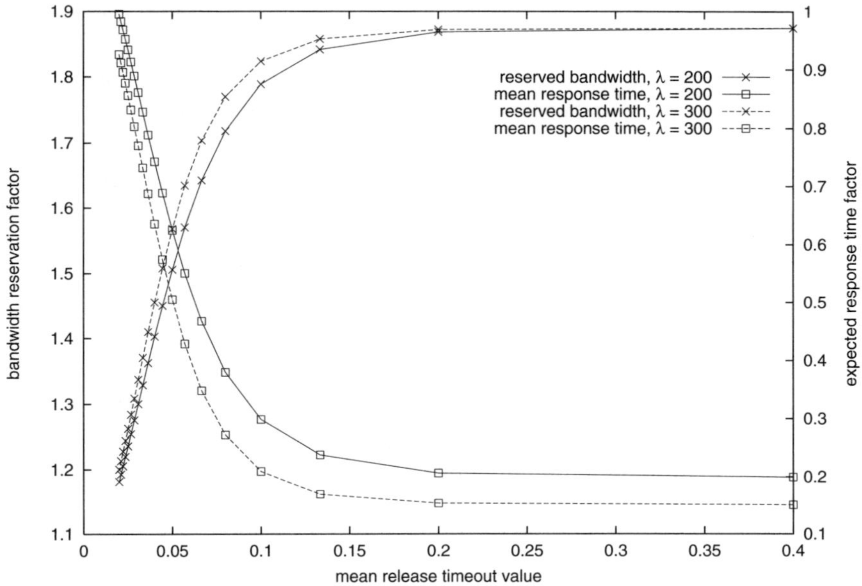


**Fig. 6.5.** Used bandwidth and reserved bandwidths for different burst arrival rates and connection release timeouts.

Focusing on our case study again, Figure 6.5 illustrates both the cost and the gain of the OCDR method with different connection release timeouts. The cost is reflected by a bandwidth reservation factor, defined as the ratio of the reserved bandwidth $p_{\mathrm{conn}} \cdot \mu$ and the used bandwidth $\lambda_{\mathrm{eff}}$. Naturally, the ratio is always larger or equal to 1. An upper bound for this cost measure is given by the persistent connection case, where it equals $\mu/\lambda_{\mathrm{eff}} = 150/(200 \cdot 2/5) = 1.875$. In the same diagram, the gain of employing the OCDR method is indicated by a response time factor. As the mean response time is not bounded due to the impact of queueing effects and connection timeouts on it, the mean connection setup time $s/str = 1/10$ has been taken as a reference point, as it represents the minimum mean response time for per-cell connection setups. The response time factor is then given by $\mathrm{E}[R]/(s/str)$, so that low values of it indicate improvement to per-cell connection setups.

In Figure 6.5 the impact of the connection release timeout is shown for two burst cell arrival rates, $\lambda = 200$ and $\lambda = 300$. Since the impact of the traffic intensity on OCDR has already been investigated in the previous section, it is kept constant in this experiment by setting $\mu = 150$ for $\lambda = 200$ and $\mu = 225$ for $\lambda = 300$. Thus, the role of plain queueing effects is minimized and only their impact in conjunction with connection management are considered.

Focusing on the case $\lambda = 200$ first (solid lines in Figure 6.5), one can observe that as the release timeout increases, the bandwidth reservation factor approaches its maximum value 1.875. Similarly, the corresponding values of the expected response time factor approach a limit of about 0.2. This factor corresponds to the queueing delay in a persistent connection system. By comparing the values of the bandwidth reservation factor and the response time factor for a particular timeout value it can easily be seen how large the response time gain is for a given increase in bandwidth. For example, the allocation of about 70% more than the input bandwidth (corresponding to a bandwidth reservation factor of 1.7) can be accomplished by selecting a release timeout of 0.08. This release timeout corresponds to a response time factor of 0.38. Thus, it is possible to reduce the mean response time of the system to just 38% of the minimum mean response time for per-cell connection establishments, being only about twice as high as the mean response time for persistent connections.

It is interesting to compare these results to the case where the traffic entering the system as well as the service rate $\mu$ is increased by 50% ($\lambda = 300$, $\mu = 225$). First note that the system now approaches the persistent connection case more quickly, as indicated by the bandwidth allocation factor which is constantly larger than for $\lambda = 200$. This is due to the fact that the connection release timeout expires less often when cells enter the system more frequently. While the qualitative development of the cost and gain curves is similar to the case with lower $\lambda$, it is remarkable that the mean response time of the system is always lower than in the case with $\lambda = 200$. The reason for this is that the expected number of customers in the system increases, but the higher service rate still provides better response times. Note that also the gain of employing OCDR increases for $\lambda = 300$; here, allocating 170% of the input bandwidth leads to a reduction of the mean response time to about 35% of the minimum mean response time for the per-cell connection setup policy, being better than the 38% reduction for $\lambda = 200$.

This fact, however, depends on the definition of the response time factor. If the persistent connection case is taken as reference point, then a 70% increase in bandwidth leads to a mean response time being twice as large for $\lambda = 200$ (as mentioned above), while for $\lambda = 300$ the response time factor equals $0.34/0.15 \approx 2.3$.

**Numerical Solution Issues.** The Petri net model proposed in Figure 6.1 led to an underlying QBD Markov chain with a boundary level size of 65 states. Each level in the repeating part of the state space contained 10 states, corresponding to 5 different states of the arrival process multiplied by 2 states indicating whether a connection has been established (note that the buffer can grow arbitrarily large before a connection establishment takes place). The generation of the underlying Markov chain, the derivation of the steady-state solution, and the computation of all reward-based measures with the tool SPN2MGM never took longer than one second on a SUN SparcStation

20 (clocked at 75MHz). Considering the current numerical limits of our tool, the model represents no computational challenge. However, by refining the structure of the arrival model to more stages or by employing Erlang distributions for modeling the connection setup time distribution the model's underlying Markov chain can grow very quickly.

### 6.1.5 Conclusion

The presented case study showed that the development of an iSPN model for the analysis of the OCDR connection management algorithm is straightforward under the taken assumptions on the video traffic structure and the nature of connection setup- and release time distributions. Important measures like mean response times and bandwidth allocation were easily specified at the Petri net level. Thanks to the usage of QBD processes as underlying stochastic model, the actual numerical analysis could be accomplished in a very efficient manner. For example, even for higher traffic intensities, the mean number of customers in the system could be derived very accurately without employing costly simulation studies.

Concerning the investigated application area, it can be said that the OCDR method proved to be useful for finding a compromise between the two extreme approaches of persistent connections and per-cell connection setups. Although OCDR can effectively reduce the mean response times to those experienced in the persistent connection case, the costs of employing OCDR are especially high for low cell arrival rates and large connection release timeout values. Thus, it is crucial to carefully select the OCDR parameters depending on traffic intensity, desired response time, and bandwidth costs. With the procedure described in the second part of Section 6.1.4, the basis for such a selection process has been laid.

## 6.2 World Wide Web Traffic and TCP/IP Congestion Control

While the study on connection management algorithms for video traffic in the previous section yielded interesting insights in the behaviour of the OCDR method, the aim of the following case study is to investigate a more widespread application area where extensive measurement data for the model instrumentation is available, and where the system's performance is very interesting from an economical point of view. Here, we focus on the World Wide Web (WWW) as a challenging application. The simple and easy-to-use WWW interface led to the enormous increase of the Internet's popularity in the last few years. Nowadays, a major portion of Internet traffic is due to HTTP (hypertext transfer protocol, [69]), the application layer protocol used between Web servers and client browsers.

Many factors are involved when analyzing the performance of such a system: the client issuing the request for Web pages, the server handling the request, the type of request (just text, text with embedded graphics, pictures, or even video), the Internet connecting the client and the server, as well as the transport protocols used by client and server (usually TCP/IP [56, 57, 181]). Taking all these aspects into account would render an analytical solution impossible, as also witnessed by the many measurement-based performance studies in this area ([12, 93]) and the high-level, queueing-network-based approaches presented in [140] . To the best of our knowledge, no detailed analytical performance studies for such systems have been reported yet.

The model presented in this study accounts for client characteristics (i.e., the request pattern and request types), Internet delays and packet losses, the server speed, and the influence of the underlying TCP/IP protocol. In particular, our model includes the explicit connection set-up (and release) phase of TCP/IP as well as its congestion avoidance windowing mechanism (known as *slow start* [111]). Using this model, we are able to show the impact of the interesting interactions between the TCP/IP and HTTP protocols on user-perceived performance. The results shown here have also been presented in [158] and [159].

As in the previous section, we propose an iSPN model after a more detailed description of the system under study. We comment on how parameters for the complex model can be derived, and then present the results of our numerical model analysis, including some remarks on necessary computational effort for the model solution. We finish with a concluding summary.

### 6.2.1 System Description

While HTTP is the application layer protocol used between peers in the WWW, it commonly uses TCP/IP as the underlying transport protocol suite (see Figure 6.6). TCP employs two window traffic control mechanisms to handle flow and congestion control. Therefore, each TCP connection is associated with two windows: a receiver window to synchronize the sender's speed with the speed a receiver is able to process incoming data, and a congestion window to avoid network overload. The amount of data which can be sent is given by the minimum of both windows. Assuming that the receiving station can handle incoming traffic sufficiently fast, the size of the congestion window is the limiting factor. It is one of the focal points of interest in this study.

The congestion window is used to avoid network congestion by limiting the number of packets traveling through the network simultaneously. It holds the remaining tolerable amount of bytes which can be fed into the network. Once a packet is acknowledged, the available congestion window space is increased by the number of bytes in that packet (which is assumed constant here). Since the tolerable number of packets which can be fed into the network prior to being acknowledged (i.e., the congestion window size) is not known *a priori*, TCP adopts the *slow start* mechanism suggested by Jacobson [111].

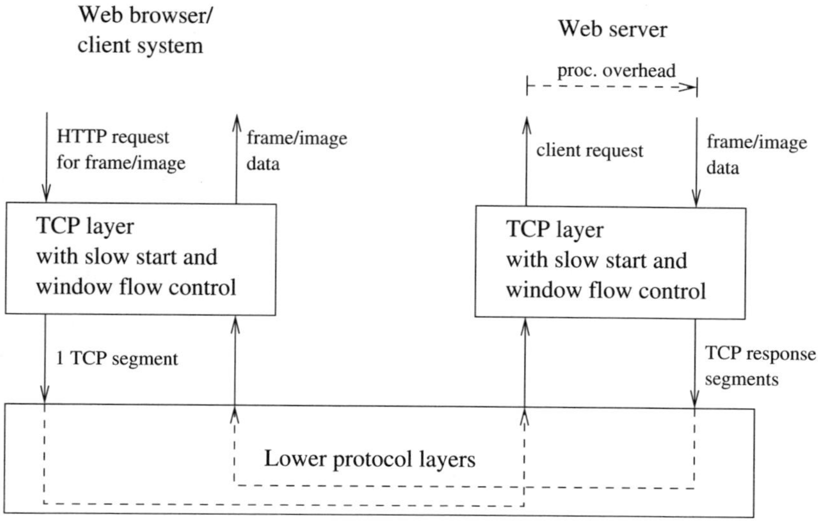


**Fig. 6.6.** Accessing WWW documents via TCP/IP.

The development of the congestion window size under slow start is illustrated in Figure 6.7. Slow start first initializes the congestion window size to one packet. Whenever a packet gets acknowledged, the window size is increased by one packet. Thus, the window size effectively grows exponentially. This exponential growth phase is bounded by a *maximum* window size parameter (denoted by max_cwin). Once it is exceeded, the congestion window size increases in a linear fashion. If a packet loss occurs, the maximum window size parameter is set to half the current window size, and the current window size is set to one packet.

While slow start provides a suitable algorithm for congestion avoidance, the initial phase of "probing for bandwidth" becomes a problem if a TCP connection is set up for transferring only a small amount of data. In particular, the HTTP protocol suffers from this fact, since getting a HTML document and the images referenced therein is accomplished by building separate TCP connections for each of them. Since the amount of data per item is typically small, the slow start mechanism will not reach reasonable congestion window sizes, thus leading to a situation where just a fraction of the available bandwidth is used. Furthermore, the establishment of each connection involves the usual TCP three-way handshake, increasing delay by one round trip time.

In version 1.1 of the HTTP protocol, the use of persistent connections for several requests (P-HTTP) is possible and recommended, thus minimizing the number of necessary slow start phases and connection setups. In this approach, connections are not released after satisfying single HTTP requests, but (depending on the Web server configuration) after a timeout has expired or a maximum number of requests has been answered. Thus, very similar to

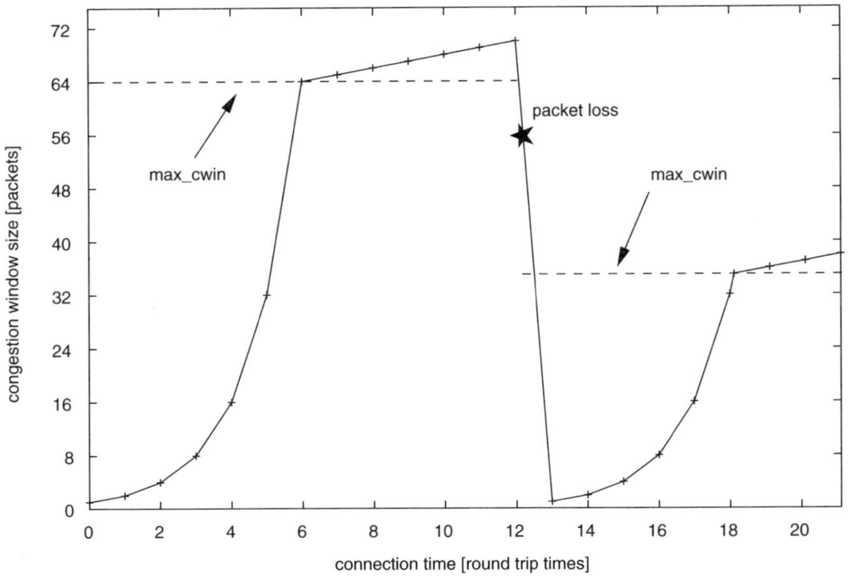


**Fig. 6.7.** Congestion window size development with the slow start algorithm.

the OCDR approach discussed in Section 6.1, one and the same connection is used for several successive data transfers in order to reduce the number of connection setups.

In the same spirit, other approaches like T/TCP [37, 38, 189] and UDP-based mechanisms like ARDP [149] were suggested and have been addressed in [93]. Note that modifications to the congestion control algorithm like fast retransmit and fast recovery [181, 182], which are commonly found in TCP/IP implementations, will also usually improve the situation; however, we will not consider them here.

### 6.2.2 Model Development

The model presented in this section has been developed having two main aims in mind. First, the model should account for the window flow control mechanism and capture the influence of the slow start procedure on transmission performance. Second, connection setups have to be considered, since we are interested in the gain obtained by using persistent protocol versions. As we eventually employ the iSPN-based numerical solution approach, complexity issues also played an important role during model design. Again, the complete SPN2MGM specification of the model developed in this section can be found in Appendix C.4.

**Server Model.** The model concentrates on the server-side of a client-server relationship. We assume that the main amount of traffic arises from server

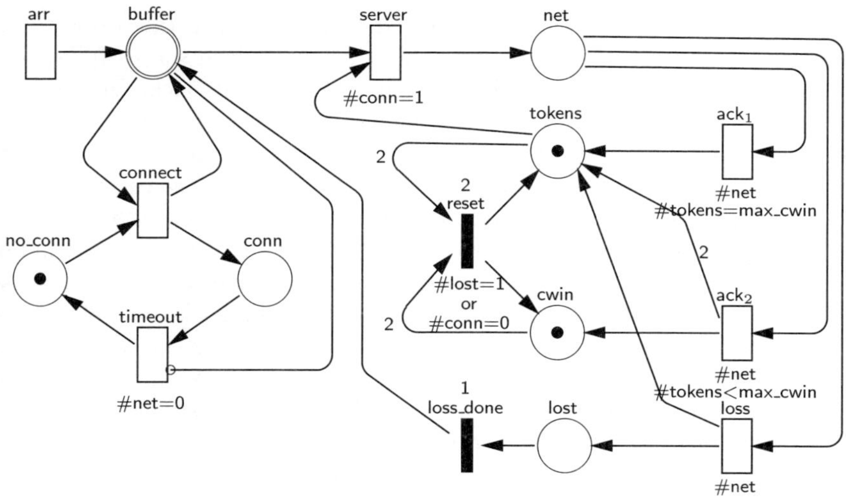


**Fig. 6.8.** Petri net for modeling TCP behaviour.

to client (as in the HTTP case), thus we neglect the impact of flow- and congestion control mechanisms on traffic which is sent to the server. Instead, different types of clients will lead to different patterns of generating segments (or packets) to be processed and transferred to the client. For the time being, we assume that the generation takes place according to a Poisson process.

The model is illustrated in Figure 6.8. The left-hand side of the model deals with connection setup, while the right-hand side represents windowing and the slow start mechanism. The segments to be delivered to the client are generated by the transition arr and put in the unbounded buffer place buffer. If there is no connection between server and client yet (as indicated by a token in no_conn), transition server is disabled due to its enabling function and a connection setup is performed by firing transition connect. The connection will not be released until all segments in buffer have been processed (inhibitor arc to timeout) and delivered (place net empty), and an additional delay (modeled by transition timeout) has passed.

Each segment in buffer is the result of some operation of the server. In the HTTP case, the server has at least to perform some kind of database (or disk) access to retrieve the desired document. This overhead is modeled by the transition server (note that this represents the overhead *per segment* of the server reply, not the overhead per client request) Since connection-setup and server overhead have been accounted for now, segments are ready for transmission. A prerequisite for submitting a segment to the network is that there is still enough space left in the congestion window. The total size of the congestion window is reflected by the place cwin. It initially holds one token, representing the size of one segment. The available size of the congestion window which currently remains for transmission is represented by the place tokens. Each segment submitted to the network takes one token from this place. After the segment has been acknowledged successfully (represented by transitions $\mathsf{ack}_1$ and $\mathsf{ack}_2$), the token is put back. Only one of these transitions is enabled at a time, depending on the current state of the slow-start process. Transition $\mathsf{ack}_2$ is enabled if the maximum allowed window size (max_cwin) has not been reached yet. In this case, *two* tokens are put back into place tokens, and one additional token is put into cwin to reflect the enlarged congestion window. Transition $\mathsf{ack}_1$ is only enabled if the maximum congestion window size has been reached. In this case, the congestion window is not enlarged any further, and just one token is put back into tokens (we do not account for a further linear increase as featured by the original slow start procedure). As a last possibility, a segment may be lost in the network, represented by transition loss. In that case, the segment first enters the place lost. This enables the immediate transition reset, which has a higher priority than loss_done and effectively resets the congestion window size to one by repeatedly taking two tokens out of cwin and putting back one token. Once the size of the congestion window has been reset, transition loss_done is enabled and the lost segment can re-enter the buffer for retransmission (this implies that the corresponding lost segment experiences the server delay induced by server for a second time; for excessive packet loss ratios, the server rate has to be adjusted appropriately). Transition reset is also enabled if no connection exists, thus initializing the congestion window for the next connection.

Note that after a packet loss occurred, the congestion window is actually only reduced by the size of tokens still available in the congestion window, since transition reset is also used to reset place tokens. Using a dedicated resetting transition token_r for that place, a construction like the one shown in Figure 6.9 could avoid this. Here, by transition net_r, unacknowledged packets are put back in the buffer place for retransmission. However, this leads to a non-skip-free QBD process, since the number of segments in buffer increases instantly by more than one segment if the immediate transition net_r fires several times in succession (i.e., more than one packet has not been acknowledged yet at the loss instant). The maximum jump size of this QBD process is quite large (up to the maximum congestion window size), thus, due to numerical constraints of our analysis tools, this alternative has not been realized. An alternative approach which distinguishes between "invalid" tokens (i.e., tokens which must not be used to increase the congestion window size since a packet loss occurred) and newly-issued "regular" tokens would also imply an unacceptable increase in the QBD's level size.

The exponentially distributed timing of transitions in the model surely is an approximation of the real-world system; it may be appropriate to model network delays, but timeout values usually have a less stochastic character. However, an approach to correctly model deterministic timeout values by Erlang distributions has already been presented in Section 6.1. To maintain a

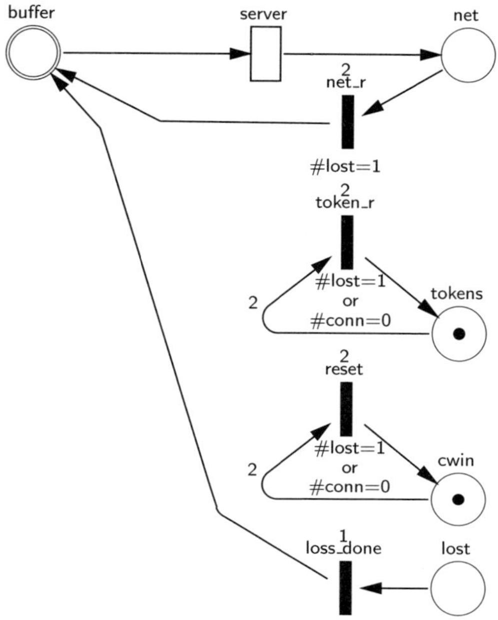


**Fig. 6.9.** Alternative submodel for correctly modeling window resets.

moderate size of the already complex model, we resort to exponential timeout modeling in the present study.

The firing rates of $\mathsf{ack}_1$, $\mathsf{ack}_2$ and loss are proportional to the number of segments submitted to the network (place net). This infinite-server behaviour for network delays accounts for the fact that the increase of network load by submitting single segments is negligible. It is conceivable to describe network delays more accurately, e.g. by using appropriate phase-type distributions as suggested in [68], albeit at the cost of an enlarged state space.

The most striking difference between this model and the real slow start mechanism is that the maximum congestion window is set to a constant size (max_cwin), instead of setting it to half its value each time a segment is lost. Introducing an additional place for holding the current maximum window size would drastically increase the state space and has thus been omitted. Since the maximum window size can not shrink in our model, the results obtained by analyzing the presented model provide an optimistic approximation.

**Workload Models.** As mentioned in Section 6.2.1, the slow start mechanism leads to poor bandwidth utilization when there is only a small amount of data to be transferred per TCP connection. In order to evaluate the merits of persistent connection approaches (like P-HTTP) under slow-start, an appropriate traffic model has to be developed. We focused on modeling HTTP workload using a characterization similar to the one presented in [93], considering the impact of one single client on one Web server, without modeling

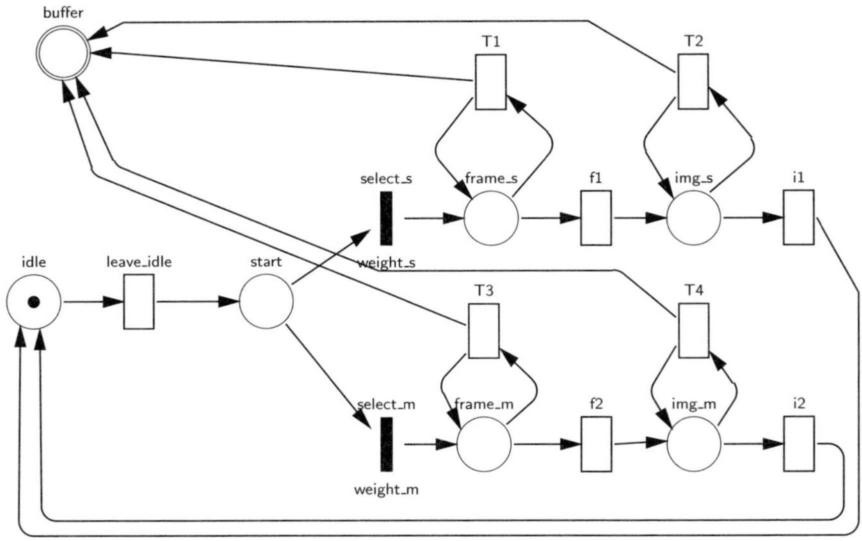


**Fig. 6.10.** Traffic generation model for HTTP.

any background traffic. The proposed workload models are coupled with the server model by the buffer place, and replace the Poissonian arrivals modeled by transition arr in Figure 6.8.

A user request for a web page is satisfied in two successive steps. First, the HTML document referenced by the URL is fetched from the server. Afterwards, all images referenced in this "frame" document are requested from the server. The traffic model must appropriately generate the segments which correspond to the server replies to these individual requests, and put them in the place buffer (Figure 6.8).

The proposed traffic model is shown in Figure 6.10. It is able to account for two different user request types (a small and a medium one, see next section for details). After an exponentially distributed user-idle time has passed (transition leave_idle), a probabilistic choice between small and medium request type takes place. Each request type consists of two phases, corresponding to generating segments to satisfy the frame request and segments belonging to the image requests. While the duration of each phase corresponds to the time needed to submit the corresponding request to the server, the rate at which segments are generated during that phase (transitions T1,T2,T3 and T4) corresponds to the number of segments the server will send in reply to the request.

As an alternative to the arrival model shown in Figure 6.10, we also investigate simplified versions of it, consisting of just one request type, simple IPP arrivals, and Poisson arrivals. Since all these models do not account for the fact that prior to submitting an image request the reply to the preceding

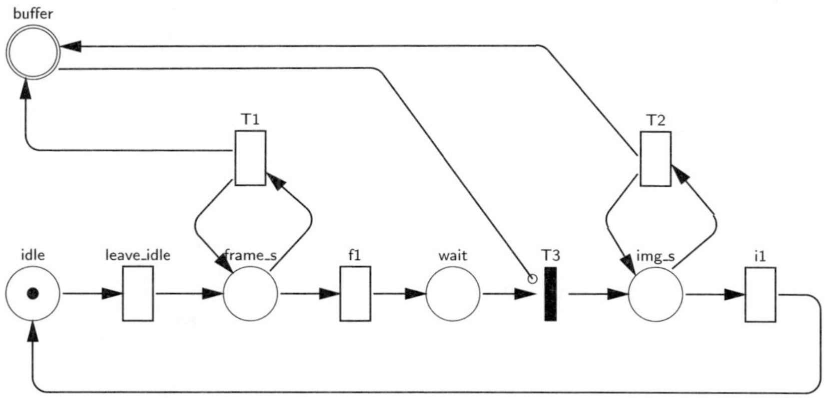


**Fig. 6.11.** Arrival model for one request type with blocking after submitting a frame request.

frame request has to be completed, we also investigate a blocking arrival model (see Figure 6.11). It deals with one request type only, where all frame segments have to be delivered before image segments are generated. This is realized by introducing an additional immediate transition which is disabled as long as there are any segments left in the server's outgoing buffer.

### 6.2.3 Parameterization

The parameters for the overall model can be split in three groups: server, network, and workload parameters (see Table 6.1). In the following paragraphs, we explain how transition rates for the iSPN model in Figure 6.8 and the workload models can be derived from these parameters (this is summarized in Table 6.2). Afterwards, we also comment on the choice of the actual parameters used in the numerical analysis.

**Server Performance.** Server-specific performance data is introduced in the model by the transition server. Its mean firing time corresponds to the average workload per segment of an answer to a client request. We assume that it is due to three parameters: computational effort per request, disk seek time per request and disk transfer time per segment. We neglect the potential parallel treatment of these parts by the Web server (e.g., by pipelining requests). Per-*segment* workloads are thus obtained by dividing the per-request overhead by the mean number of segments per request, denoted by $s_{\mathrm{mean}}$ (which is derived in the workload parameterization section). In conclusion, the firing rate of transition server is given by

$$\left(\frac{t_{\mathrm{seek}} + t_{\mathrm{comp}}}{s_{\mathrm{mean}}} + \frac{n_{\mathrm{MSS}}}{\lambda_{\mathrm{disk}}}\right)^{-1}. \tag{6.5}$$

| Server characteristics | |
|---|---|
| Computation time per request [s] | $t_{\text{comp}}$ |
| Disk seek time [s] | $t_{\text{seek}}$ |
| Disk transfer speed [B/s] | $\lambda_{\text{disk}}$ |
| Connection release timeout [s] | $t_{\text{release}}$ |

| Network characteristics | |
|---|---|
| Round trip time [s] | $t_{\text{rtt}}$ |
| Bandwidth [B/s] | $\lambda_{\text{bw}}$ |
| Loss ratio | $p_{\text{loss}}$ |
| Max. TCP segment size [B] | $n_{\text{MSS}}$ |
| Max. congestion window size [segs] | max_cwin |

| Workload characteristics | |
|---|---|
| User idle time [s] | $t_{\text{idle}}$ |
| Small request probability | $p_{\text{small}}$ |
| Small frame request size [B] | $b_{\text{sf}}$ |
| Number of small image requests | $n_{\text{si}}$ |
| Small image request sizes [B] | $b_{\text{si},1}, \cdots, b_{\text{si},n_{\text{si}}}$ |
| Medium frame request size [B] | $b_{\text{mf}}$ |
| Number of medium image requests | $n_{\text{mi}}$ |
| Medium image request sizes [B] | $b_{\text{mi},1}, \cdots, b_{\text{mi},n_{\text{mi}}}$ |

**Table 6.1.** Overview on all model parameters.

The last server-related parameter, the rate of the connection release transition timeout, is given by $t_{\text{release}}^{-1}$.

**Network parameters.** Apart from the usual packet transfer latencies, a connection setup requires an additional round trip time to account for the TCP three-way handshake, so the rate of connect is $t_{\text{rtt}}^{-1}$.

The time needed to acknowledge a segment sent from server to client consists of one round trip time plus the segment transfer time, which depends on segment size and bandwidth. Since we also consider packet losses, the rates of transitions $\mathsf{ack}_1$ and $\mathsf{ack}_2$ are given by $(t_{\text{rtt}} + n_{\text{MSS}}/\lambda_{\text{bw}})^{-1} \cdot (1 - p_{\text{loss}})$, and the rate of transition loss equals $(t_{\text{rtt}} + n_{\text{MSS}}/\lambda_{\text{bw}})^{-1} \cdot p_{\text{loss}}$.

**Workload parameters.** Due to the small size of client requests, we assume that the submission of a request takes on average one round trip time. Thus, the firing rates of transitions f1 and f2 equal $t_{\text{rtt}}^{-1}$. We also assume that all image requests are submitted simultaneously, thus again involving just one round trip time and yielding the rate $t_{\text{rtt}}^{-1}$ for transitions i1 and i2 as well.

The number of segments to be put into the server's buffer place depends on the size of the reply corresponding to a request. For small requests, it is given by $s_{\text{sf}} = \lceil b_{\text{sf}}/n_{\text{MSS}} \rceil$ for the initial frame request. The corresponding total number of segments to be transferred in reply to the $n_{\text{si}}$ image requests is $s_{\text{si}} = \sum_{k=1}^{n_{\text{si}}} \lceil b_{\text{si},k}/n_{\text{MSS}} \rceil$. These are the numbers of segments to be generated on average while a token is in places frame_s and img_s, so the rates of T1

| | Server iSPN (Figure 6.8) |
|---|---|
| server | $\left[(t_{\text{seek}} + t_{\text{comp}})/s_{\text{mean}} + n_{\text{MSS}}/\lambda_{\text{disk}}\right]^{-1}$ |
| ack1, ack2 | $\left(t_{\text{rtt}} + n_{\text{MSS}}/\lambda_{\text{bw}}\right)^{-1} \cdot (1 - p_{\text{loss}})$ |
| loss | $\left(t_{\text{rtt}} + n_{\text{MSS}}/\lambda_{\text{bw}}\right)^{-1} \cdot p_{\text{loss}}$ |
| connect | $t_{\text{rtt}}^{-1}$ |
| timeout | $t_{\text{release}}^{-1}$ |
| | **Workload iSPN (Figure 6.10)** |
| leave_idle | $t_{\text{idle}}^{-1}$ |
| f1, i1, f2, i2 | $t_{\text{rtt}}^{-1}$ |
| T1 | $s_{\text{sf}}/t_{\text{rtt}}$ |
| T2 | $s_{\text{si}}/t_{\text{rtt}}$ |
| T3 | $s_{\text{mf}}/t_{\text{rtt}}$ |
| T4 | $s_{\text{mi}}/t_{\text{rtt}}$ |

**Table 6.2.** Derivation of iSPN transition rates from the model parameters.

and T2 are $s_{\text{sf}}/t_{\text{rtt}}$ and $s_{\text{si}}/t_{\text{rtt}}$, respectively. The parameters for the medium size request type can be computed similarly. Also, using $n_{\text{si}}$, $n_{\text{mi}}$ and $p_{\text{small}}$, the mean number of segments per request can be computed as

$$s_{\text{mean}} = p_{\text{small}} \frac{s_{\text{sf}} + s_{\text{si}}}{1 + n_{\text{si}}} + (1 - p_{\text{small}}) \frac{s_{\text{mf}} + s_{\text{mi}}}{1 + n_{\text{mi}}}. \tag{6.6}$$

For completeness, the derivation of all iSPN transition rates from the server, network, and workload parameters is summarized in Table 6.2.

**Parameter selection.** As seen in the previous paragraphs, the firing rates of the transitions in the proposed model depend on a wealth of parameters. In order to assume sensible values, we decided to adopt results from existing measurement-based studies as often as possible.

*Server Characteristics.* The server characteristics heavily depend on the hardware and software environment under which the WWW server is operating. Since these properties usually vary in a very broad range, we are not trying to obtain parameters as exact as possible, but instead assume sensible (yet artificial) values. We suppose that the mean computation time per request is $t_{\text{comp}} = 0.01[\text{s}]$. The performance parameters of the disk are $t_{\text{seek}} = 0.01[\text{s}]$ and $\lambda_{\text{disk}} = 5 \cdot 10^6[\text{B/s}]$. If not mentioned otherwise, the connection release timeout $t_{\text{release}}$ has been set to 10 seconds.

*Network Characteristics.* Parameters concerning network performance have been taken from [68], where extensive Internet performance measurements have been accomplished. We selected two reference connections, a national one (RWTH Aachen to University of Karlsruhe, Germany) and an international connection (RWTH Aachen to Stanford University, U.S.A.). See Table 6.3 for the parameters of these connections.

| Parameter | National | International |
|---|---|---|
| $t_{\text{rtt}}$ | 0.020[s] | 0.270[s] |
| $p_{\text{loss}}$ | 0.008 | 0.125 |
| $\lambda_{\text{bw}}$ | $10^5$ [B/s] | |
| $n_{\text{MSS}}$ | 536 [B] | |

**Table 6.3.** Network parameters used in the experiments.

*Workload Characteristics.* The parameters for small and medium request types occurring in the model's workload characterization have been taken from [93], representing the structure of some popular Web pages. The small request type consists of an initial 6651 byte frame page, referencing two images of size 3883 and 1866 bytes. Medium requests are formed by a 3220 byte frame and three images of size 57613, 2344 and 14190 bytes. We assume that the probability for a small request is $p_{\text{small}} = 0.6$.

### 6.2.4 Numerical Results

The iSPN model proposed in Section 6.2.2 features a very large number of parameters, and a large number of performance measures can be obtained from its analysis. We will thus keep many parameters constant throughout all experiments, and focus on the analysis of a few aspects only.

Our investigations concentrate on three main areas. We first analyze the impact of different workload models on the connection probabilities, the mean number of elements in the server's input buffer, and the connection setup rate. We were also interested to which extent the maximum congestion window size (max_cwin) influences these performance characteristics. The connection release timeout plays an important role when dealing with protocols like persistent HTTP. In the next experiment, we thus show how changing this parameter affects the cost structure of the overall system, as represented by the connection probability and the connection setup rate. Finally, the influence of the segment loss probability on the mean buffer filling is investigated.

After presenting these numerical results, we will discuss the model's complexity and the computational solution effort.

**Workload Models and Congestion Window Size.** As a first point, we are interested in how far detailed workload models influence the results of our investigation. Since the workload model greatly enlarges the number of states of the Markov chain underlying the iSPN (e.g., the QBD level size of the arrival model shown in Figure 6.10 is five times as large as a model with simple Poisson arrivals), it is interesting to see whether this effort pays off.

Figure 6.12 shows the mean buffer size (number of tokens in place buffer) for different workload models. Clearly, increasing the maximum window size leads to a higher segment throughput, and thus reduces the buffer filling.

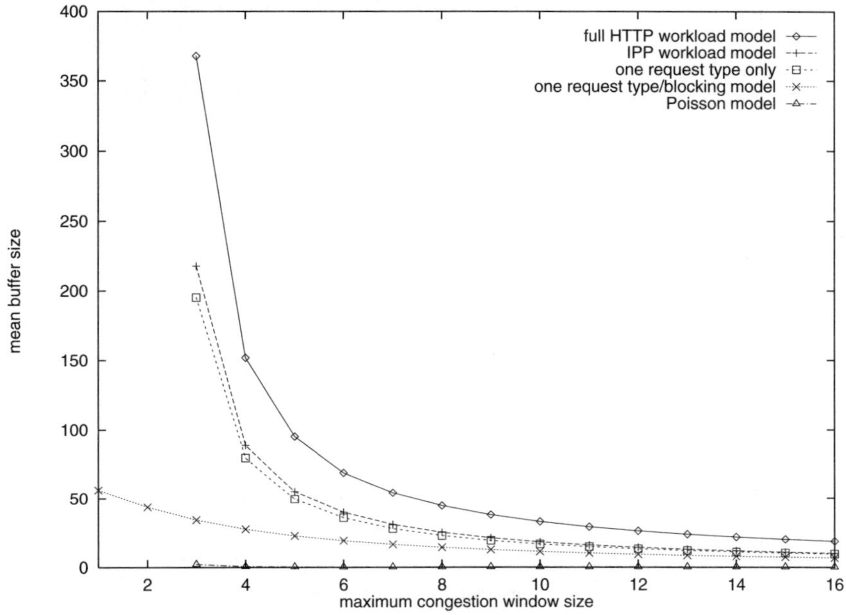


**Fig. 6.12.** Expected buffer size for the international connection and different workload models.

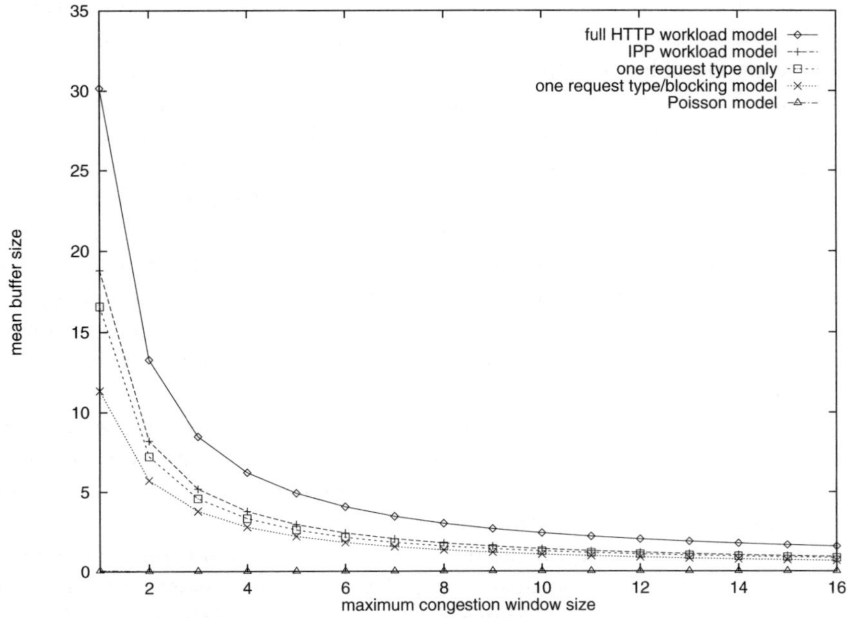


**Fig. 6.13.** Expected buffer size for the national connection and different workload models.

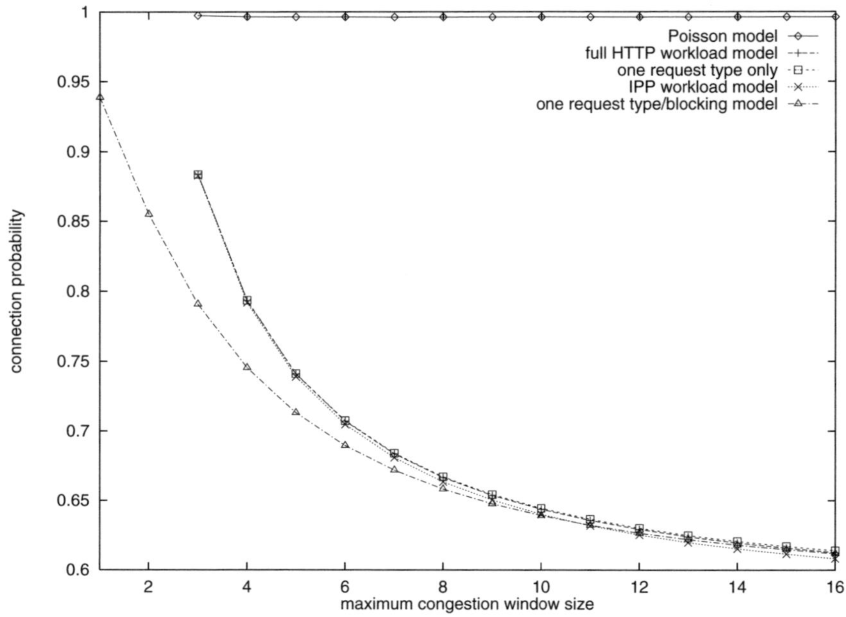


**Fig. 6.14.** Probability for an existing connection for different workload models.

Concerning the different workload models, the results for the full HTTP model (as shown in Figure 6.10) differ significantly from those of the simplified versions. The results of the one-request and IPP workload models (with identical mean segment arrival rate) are very similar and still capture the qualitative behaviour of the original model. However, due to ignoring the bursty arrival pattern accounted for by the other workload models, the approximation of the workload by a Poisson process leads to a dramatic underestimation of the expected buffer size. As an interesting point, the results for the blocking workload model (Figure 6.11) do not differ too much from the non-blocking one-request-type model, especially for larger maximum congestion window sizes. The absolute values are smaller, since the mean segment generation rate for this workload model is lower than for the other models (due to the additional waiting time in place wait).

Figure 6.13 illustrates the same performance measures for the national Internet connection. Due to the smaller round trip time, the average buffer filling is much smaller than in the international case. Again, the Poisson workload model yields much too low results.

From now on, we focus on the international connection type in all experiments. We investigate the steady-state probability for an existing connection (i.e., a token in place conn) in Figure 6.14. Small values of this probability are desirable, since they indicate a smaller bandwidth reservation and a smaller mean number of open connections for Web servers which serve several clients

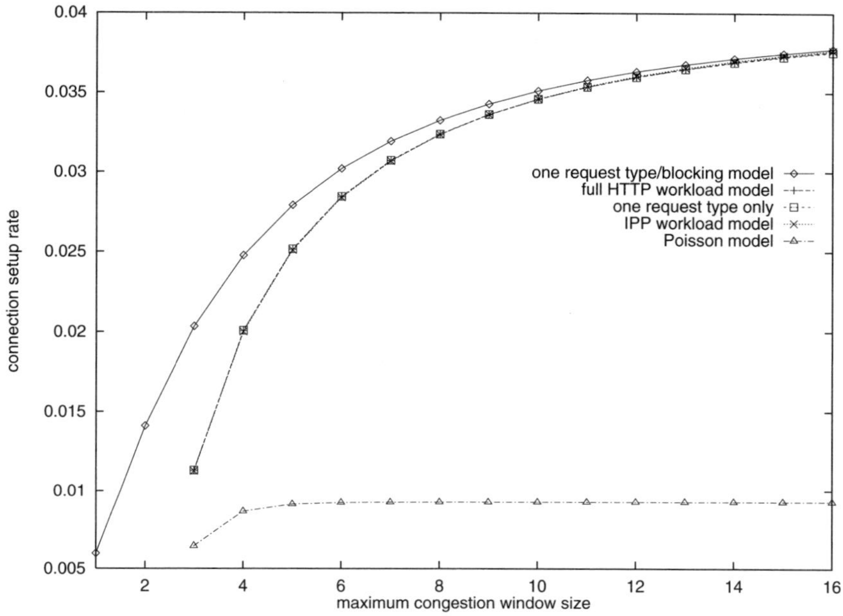


**Fig. 6.15.** Connection setup rates for different workload models.

simultaneously. While the results for all bursty workload models coincide, the Poisson workload distributes segments much more in time, leading to a situation where the connection timeout almost never expires due to the very high connection probabilities. Concerning the other arrival models, the segments in the server's buffer are delivered quicker to the client for larger maximum congestion windows due to the increased usable bandwidth. Thus, connections can be released quicker, resulting in smaller connection probabilities for larger values of max_cwin.

Figure 6.15 shows the connection setup rate for the different arrival models. This value can be considered as a cost factor, as more frequent connection setups increase both the user-perceived delay and the amount of signaling traffic. As can be observed, the setup rate increases for larger congestion window sizes, since requests are satisfied quicker and the release timeout expires more often in this case. Again, the Poisson model leads to significantly different results. The selection of a reasonable value for max_cwin depends on whether small bandwidth utilization or small connection setup rates are more important. Usually, large values of max_cwin are preferred, as the small connection setup rate for small values of max_cwin is usually outweighed by the increased response time due to queueing effects which increase for low bandwidth utilizations.

Summarizing, it can be said that ignoring the workload burstiness dramatically alters the performance measures obtained from the model, however,

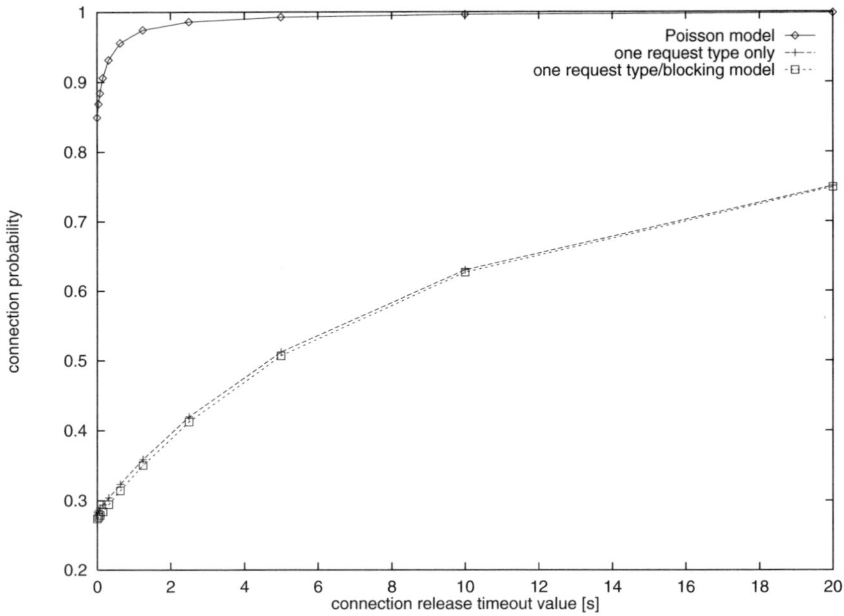


**Fig. 6.16.** Probability for an existing connection for different connection release timeouts.

thanks to our modeling environment, bursty arrival patterns can easily be accounted for. Furthermore, increasing the maximum congestion window above a minimum value of about 8-10 segments leads to much better bandwidth utilization, thus reducing the buffer size and the time connections are held, albeit at the cost of increased connection setup rates. Though increasing the bandwidth utilization by higher maximum congestion window sizes is generally desirable, this may also increase network congestion and packet losses. However, this aspect can not be considered with the single client-server model presented here.

**Influence of Connection Release Timeout.** The introduction of a connection release timeout is crucial for the reduction of connection-setups and delays when protocols like P-HTTP are employed. Clearly, when choosing this parameter it is important to compare the gain of less connection setups with the higher costs imposed by maintaining a (mainly unused) connection.

Figure 6.16 illustrates the probability of an existing connection for different values of $t_{\mathrm{release}}$ for a maximum congestion window size of 12. Obviously, the probability increases for larger timeout values. Since the amount of data to be transferred remains constant, an existing connection is often unused. On the other hand, the connection setup rate decreases for larger timeout values, as illustrated in Figure 6.17. This trade-off between costs of large release timeouts (in terms of high connection probabilities, leading to much reserved

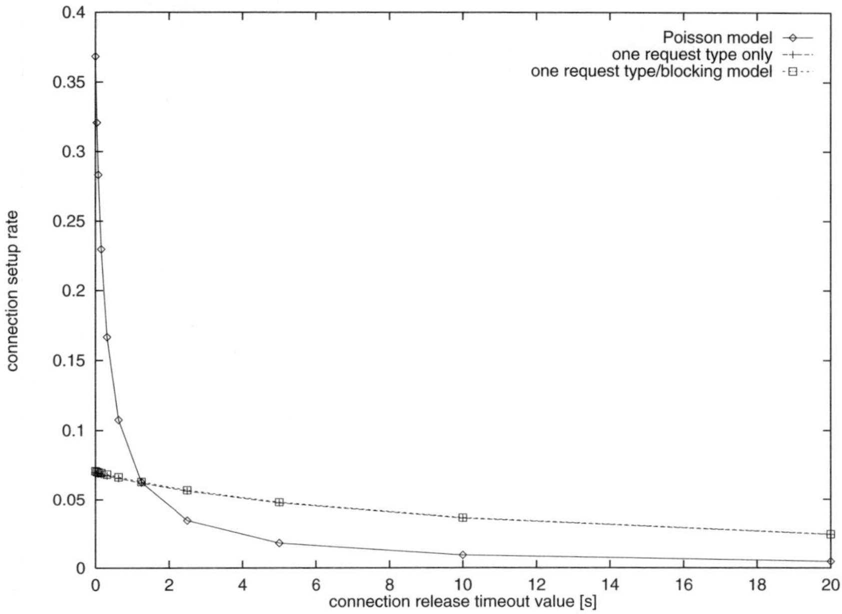


**Fig. 6.17.** Connection setup rates for different connection release timeouts.

bandwidth and many open connections) and their merits (i.e., a good response time behaviour due to low connection setup rates and less signaling traffic) is very similar to the situation we encountered in our analysis of the OCDR connection management scheme in Section 6.1.

**Influence of Segment Loss Ratio.** Apart from bandwidth and average round trip time, the packet loss probability heavily influences the performance of an Internet connection. The impact of losses on the windowing system is shown in Figure 6.18 for different values of max_cwin. It can be observed that the mean buffer size of the system increases dramatically for high loss ratios if max_cwin is chosen too small. The system's behaviour is much more robust concerning packet losses if the congestion window size is large enough. This behaviour is due to the fact that higher packet loss ratios effectively increase the amount of segments to be delivered by the network, since lost segments are re-submitted for transmission by transition loss_done. For a loss ratio of 0.5, every second packet has to be retransmitted. Since lost packets are again subject to loss in the next transmission try, the number of segments to be delivered effectively doubles. Since small values of max_cwin lead to small effective transmission performance, the buffer size is particularly sensitive to packet losses.

**Computational Effort.** The Markov chain underlying the investigated iSPN model can grow remarkably large. For example, for analyzing the server

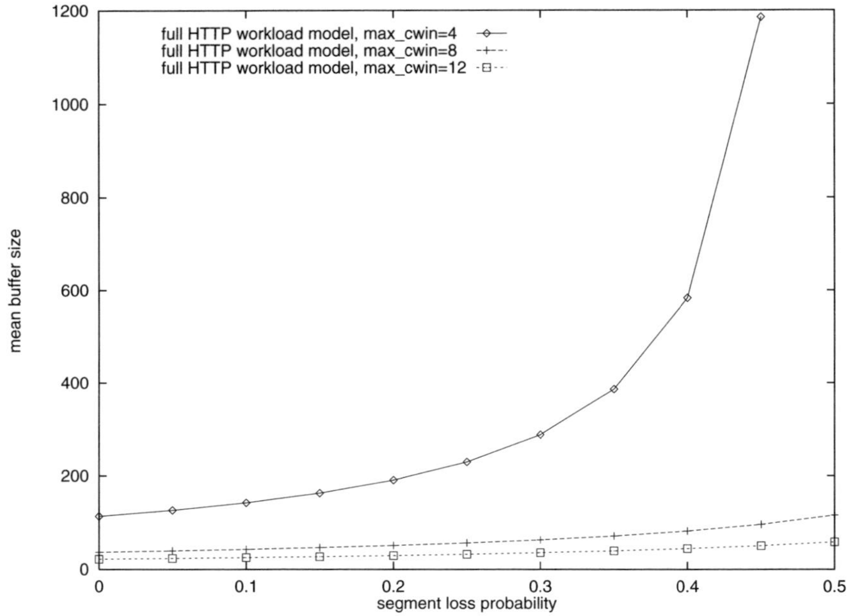


**Fig. 6.18.** Expected buffer size for different loss probabilities.

model with the full workload model as shown in Figure 6.10, an underlying QBD process with 765 states *per level* has to be investigated. The boundary level contained 1525 states. In some experiments (see e.g. Figure 6.18), we obtained *mean* buffer sizes around 1000 segments. Consequently, the evaluation of this system by using a large finite Markov chain would involve the investigation of several thousand levels, leading to a total number of several million states. While deriving steady-state measures of Markov chains of this size becomes a problem when using common numerical or simulation methods, the QBD-based solution approach leads to results in a quick and memory-efficient way.

We were able to solve the above-mentioned model in around 2.5 hours on a SUN SparcStation 20 clocked at 75MHz, including all steps from state space generation to computing the desired performance measures. However, it should be noted that models of this size currently represent the upper limit we are able to solve due to the numerical instabilities mentioned in Section 4.5.

### 6.2.5 Conclusion

It has been shown that the iSPN-based modeling framework is suitable for modeling even rather intricate flow control mechanisms like the slow-start procedure. At the same time, we were able to account for the structure of

HTTP traffic. We demonstrated that using persistent HTTP dramatically improves the user-perceived performance of Web-based applications; furthermore, reasonably large congestion windows are essential for a good bandwidth utilization, which is especially true for lossy connections.

Currently, the size of the presented model represents the upper limit of what is numerically tractable. In order to apply the modeling framework to the complex system under investigation, we had to perform several simplifications:

- The more intuitive approach of explicitly generating HTTP client had to be avoided.
- More detailed connection timeout distributions were not considered.
- The linear increase phase of the slow-start algorithm, which increases the window size after max_cwin has been reached, had to be omitted.
- Improvements to TCP/IP congestion control like fast retransmission and fast recovery could not be considered.

Furthermore, similar to conventional, non-QBD based modeling frameworks, the iSPN-based approach still suffers from the "curse of dimensionality", which quickly comes into play if we are, for example, interested in analyzing load-balancing algorithms, where several Web servers appear in one model. This problem represents one of the main points covered in this work, and we will further discuss it in Section 6.4.

## 6.3 Accounting for Self-Similar Traffic

In the previous sections, it has been shown that the iSPN-based modeling approach is suitable to model rather complex node behaviour like OCDR connection management or TCP's slow-start algorithm and Web services. In these studies, one main concern was to account for the properties of the traffic entering the nodes, and it has been shown that these properties seriously impact the results derived in the model analysis.

While the proposed traffic models seem reasonable, they do not account for properties which recently became well-known under the notions of "self-similar", "long-range-dependent", or "fractal" traffic. The presence of these properties in Ethernet traffic has initially been shown by Leland et al. in an extensive measurement study [132]; later studies also proved their existence in wide area traffic [52, 162] and signaling traffic [61], as well as in high-speed networks [44, 72] and in multimedia and video traffic [23, 45, 79].

In many of these and other studies (like [63] and [68]) it has been shown that self-similar traffic heavily impacts the queueing behaviour of a system, and that ignoring this aspect usually leads to a dramatic underestimation of e.g. mean buffer sizes and overflow probabilities. Even worse, common (finite) Markovian traffic models are not capable to account for self-similarity due to their finite memory. Current models for self-similar traffic are either computationally very expensive or analytically hardly tractable; furthermore, it is very undesirable to completely abandon the Markovian modeling approach. Therefore, the aim of this study is to investigate whether it is possible to sufficiently *approximate* the impact of self-similarity by an appropriate Markovian model.

We first briefly introduce the mathematics of self-similar stochastic processes in Section 6.3.1. Then, we review existing modeling approaches for self-similar traffic and develop a Markovian arrival model for approximating self-similarity in Section 6.3.2. The evaluation of the applicability of this traffic model is based on an measured, self-similar trace, and an appropriately fitted version of the arrival model. Both are presented in Section 6.3.3. Based on this, we investigate the behaviour of a simple queueing system subject to the original trace by trace-driven simulation, and compare these results to those derived numerically from the proposed Markovian model in Section 6.3.4. Finally, the results are summarized in Section 6.3.5.

### 6.3.1 Self-Similar Stochastic Processes

In this section, we introduce the mathematical foundations of self-similar stochastic processes. Based on this, a simple method for determining the degree of self-similarity in traffic measurements is presented.

**Self-Similarity and Long-Range Dependence.** While the two notions "long-range dependence" and "self-similarity" synonymously describe one and the same mathematical property of a stochastic process, they can be considered as two different ways of looking at this property (as will be shown later). Prior to going into detail, it is worth noting that there exists a considerable amount of non-equivalent definitions for self-similarity, as also mentioned in [188] and [198]. Here, we follow the work of Beran [22] and Cox [51].

Let us consider a discrete-time stochastic process $\mathcal{X} = \{X(t), t = 0, 1, \cdots\}$, representing e.g. the number of packet arrivals during the time interval $[t, t+1)$. It is assumed that $\mathcal{X}$ is covariance stationary, with an autocorrelation function $r(k)$ and the variance of $X(t)$ denoted by $\sigma^2$.

One way to characterize long-range dependence (also called the "Joseph Effect" [136] or the *persistence phenomenon* [102]) is in terms of its autocorrelation function. The stochastic process $\mathcal{X}$ is called long-range dependent, if

$$\lim_{k\to\infty} r(k) = c \cdot k^{-\beta}, c > 0, 0 < \beta < 1. \tag{6.7}$$

This means that the autocorrelation function asymptotically decreases in an hyperbolic fashion, and implies that $\sum_k r(k) = \infty$. The usual interpretation of this is that high-lag correlations become smaller and smaller, but their cumulative effect can not be neglected. This is in strong contrast to short-range dependent processes like those based on Markovian assumptions, whose autocorrelation function approaches $c \cdot \rho^k, 0 < \rho < 1$ and is thus summable over $k$. In the frequency domain, the non-summability of the autocorrelation function corresponds to the fact that the spectral density function $f(\lambda) = \sum_k r(k)e^{ik\lambda}$ is infinite for $\lambda = 0$.

Besides looking at the autocorrelation function, self-similarity can equivalently be defined in terms of the *aggregated* processes $\mathcal{X}^{(m)}$. Intuitively, these processes average the number of arrivals over non-overlapping intervals of size $m$; they are defined by

$$X^{(m)}(t) = \frac{1}{m} \sum_{u=0}^{m-1} X(mt + u), \; m > 1, t = 0, 1, \cdots \tag{6.8}$$

Since $\mathcal{X}$ is covariance-stationary, $\mathcal{X}^{(m)}$ also is. The corresponding autocorrelation function is denoted by $r^{(m)}$.

The process $\mathcal{X}$ is called *asymptotically second-order* self-similar,[2] if the variance of the aggregated processes $\mathcal{X}^{(m)}$ is asymptotically given by

$$\lim_{m \to \infty} \operatorname{Var} X^{(m)} = c \cdot m^{-\beta}, 0 < \beta < 1. \tag{6.9}$$

For short-range dependent processes, $\lim_{m \to \infty} \operatorname{Var} X^{(m)} = c \cdot m^{-1}$ holds, i.e., they correspond to $\beta = 1$. Thus, $\beta$ can be considered as a measure of self-similarity. Usually, the degree of self-similarity is given by the *Hurst* parameter $H = 1 - \beta/2$, which is always smaller than 1, and indicates short-range dependence for $H = 0.5$.

The slow decay of the variance of the aggregated processes can be used to optically illustrate the impact of self-similarity. Figure 6.19 shows the original Ethernet measurements presented in [132], depicting the number of arrivals per time unit at different aggregation levels. It can be observed that the traffic retains its bursty nature even for very high aggregation levels. Conventional Markovian arrivals quickly converge to an almost constant number of arrivals for larger time units.

Finally, it interesting to note that the autocorrelation function of the aggregated processes approaches a *fixed* function:

$$\lim_{m \to \infty} r^{(m)}(k) = \frac{1}{2} \left[ (k+1)^{2H} - 2k^{2H} + (k-1)^{2H} \right]. \tag{6.10}$$

For short-range dependent processes, $r^{(m)}(k)$ approaches 0 for all $k$. Also note that the right-hand side of Eq. (6.10) actually represents the autocorrelation function of fractional Gaussian noise (FGN) [137], which is an important class of self-similar stochastic processes due to its exact self-similarity.

[2] Note that other classes of self-similarity are also well-defined; however, in the context of traffic modeling, exact and non-asymptotic self-similarity are of no practical relevance.

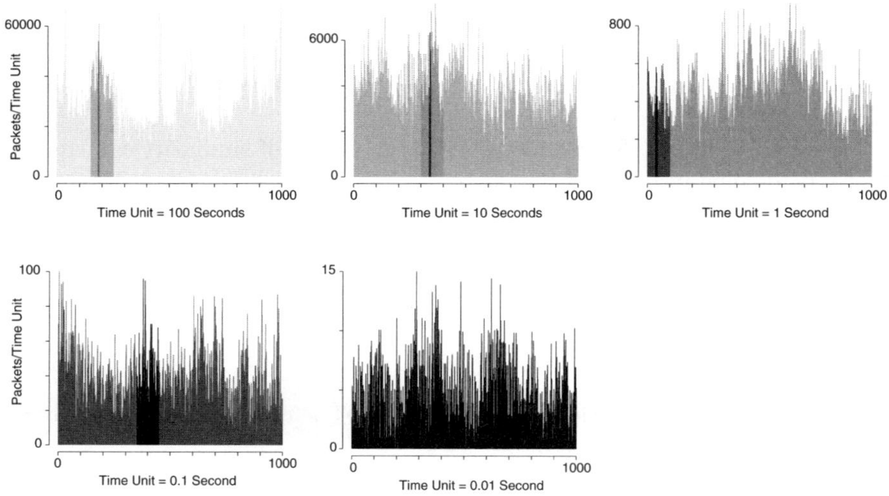


**Fig. 6.19.** Pictorial evidence for self-similar traffic. The traffic shape remains bursty over many time scales (figure after [132]).

**Testing for Self-Similarity.** Several approaches for testing the presence of self-similarity in measured traffic traces have been suggested (see e.g. [133, 185] for a survey). One of the most intuitive approaches is based on Eq. (6.9). Taking logarithms on both sides of this equation yields

$$\lim_{m \to \infty} \log \left( \operatorname{Var} X^{(m)} \right) = -\beta \log(m) + \log(c). \tag{6.11}$$

By considering the aggregated trace variances instead of $\operatorname{Var} X^{(m)}$ and plotting it in a log-log plot versus the aggregation level $m$, we asymptotically obtain a line with slope $-\beta$. For self-similar processes, $\beta$ is smaller than 1 and the Hurst parameter is given by $H = 1 - \beta/2$. Otherwise, the variance-time plot converges to a line with slope $-1$.

### 6.3.2 Self-Similar Traffic Models

The purpose of this section is two-fold. In its first part, we review the main approaches for modeling self-similarity. Based on one of these approaches, our own class of arrival processes for approximating self-similar traffic is presented in the second part.

**Existing Approaches.** Many of the measurement studies mentioned above conclude that the presence of self-similarity in network traffic heavily impacts the performance of systems subject to such traffic. Thus, considerable effort has been put in developing appropriate traffic models which can be employed in model-based performance evaluation studies.

Many approaches focus on the formal definition of self-similar stochastic processes. The most important approaches in this area are:

- *Increment processes* of exactly self-similar processes, like fractional Brownian motion (FBM) [22], which is derived from the differentiation of fractional Gaussian noise (FGN) [137].
- *Fractional autoregressive integrated moving average (F-ARIMA) processes* [101] are an extension of standard ARIMA [35] processes. Compared to e.g. FBM, they have a higher flexibility in accounting for short-range correlation structures as well.
- *Aggregation approaches* [184] are based on the superposition of a large number of identical renewal processes. The inter-renewal-time distribution has to be *heavy-tailed* (i.e., it has infinite variance, also referred to as the *Noah effect* by Mandelbrot [137]).

Approaches like these were applied successfully in several case studies (see e.g. [60, 63, 67, 154, 188]), however, the investigated systems had to be heavily simplified due the hard analytical tractability of these traffic models. Also, the generation of artificial self-similar traffic traces for simulation purposes based on these approaches is computationally very intensive. *Chaotic maps* [65] are a very efficient means for generating self-similar traffic in simulation studies, but this approach is also hardly tractable analytically.

The hard tractability of analytical approaches led to several approaches to develop Markovian approaches for *approximating* self-similar traffic, which can easily be included in existing models and which are amenable to powerful numerical algorithms. This approach became particularly interesting in view of the recently emerged discussion on the impact of self-similarity on performance evaluation (see e.g. [64, 83, 98, 171]). The proponents of short-range dependent models essentially indicate that the autocorrelation behaviour beyond some *critical time scale* [171] does not significantly affect the system performance.

Several Markovian models for approximating self-similar and highly-correlated traffic have been suggested. In [9], superposed MMPPs are suggested. Li et al. propose a certain class of MMPPs for matching measured traces in the frequency domain. In [170], an approach for matching measured traffic with MMPPs is presented as well. The approach presented by Robert et. al [166, 167] explicitly aims at approximating self-similar stochastic processes by so-called *pseudo self-similar* MMBPs, and offers an efficient matching algorithm. This approach drew our special attention, since the MMBP has a particularly simple structure, and the results presented in [166, 167] indicate a good approximation of self-similar traffic. We thus adopted this model for our modeling framework.

**Pseudo Self-Similar Phase-Type Distributions.** The approach suggested in [166, 167] was motivated by the fact that in practice self-similar conditions can not be met due to the finiteness of all traces. Therefore, the

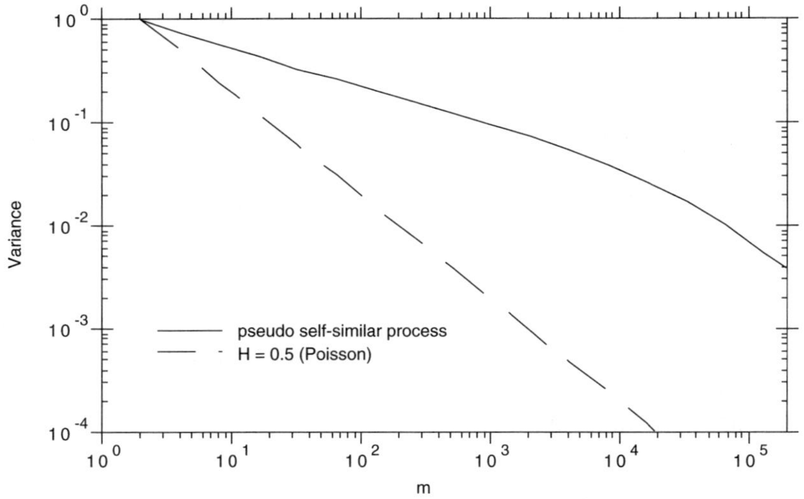


**Fig. 6.20.** Variance-time plot of pseudo self-similar traffic (figure after [166]).

limit for $m \to \infty$ occurring in Eq. (6.9) can never be taken in practice. The authors thus propose the notion of *pseudo* long-range dependence, with

$$\lim_{m \to m_{\max}} \operatorname{Var} X^{(m)} = c \cdot m^{-\beta}, 0 < \beta < 1, \tag{6.12}$$

where $m_{\max}$ is large. When checking for self-similarity by plotting the sample variance versus the aggregation factor in a log-log plot (as suggested above), pseudo self-similarity manifests (as shown in Figure 6.20) as a curve which first indicates self-similarity by a slope larger than $-1$, but eventually (as the *domain of validity* is exceeded) approaches slope $-1$ at an aggregation factor of about $10^4$.

The authors obtain this behaviour by employing a special MMBP, where the transition probability matrix is given by

$$\mathbf{P} = \begin{pmatrix} 1 - \frac{1}{a} - \frac{1}{a^2} - \cdots - \frac{1}{a^{n-1}} & \frac{1}{a} & \frac{1}{a^2} & \cdots & \frac{1}{a^{n-1}} \\ \frac{q}{a} & 1 - \frac{q}{a} & 0 & \cdots & 0 \\ (\frac{q}{a})^2 & 0 & 1 - (\frac{q}{a})^2 & \cdots & 0 \\ \vdots & \vdots & \vdots & \ddots & \vdots \\ (\frac{q}{a})^{n-1} & 0 & 0 & \cdots & 1 - (\frac{q}{a})^{n-1} \end{pmatrix},$$

with $q < a$ and $q, a > 0$

(6.13)

and arrivals occur upon entry to the first state. As an example, the transition structure for a model with $n = 4$ states, $a = 10$ and $q = 2$ is illustrated in Figure 6.21(a). Intuitively, the transition probabilities from the first state to

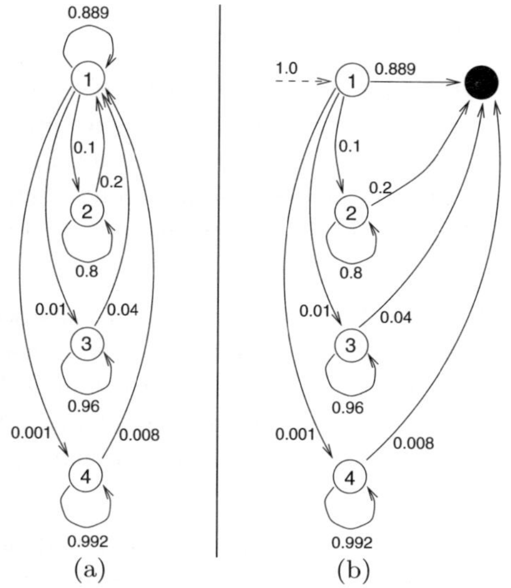


**Fig. 6.21.** Robert's initial MMBP representation (a) and the corresponding discrete-time PH analogue (b).

higher states become smaller and smaller, while the sojourn time in these states becomes larger and larger.

Note that since arrivals occur upon entering the first state only, the interarrival times form a renewal process. Furthermore, a phase-type distribution for the interarrival time can easily be obtained from $\mathbf{P}$ by replacing all transitions to the initial state by transitions to the phase-type distribution's absorbing state, as illustrated in Figure 6.21(b). Thus, an equivalent representation of the model proposed in [166, 167] is given by the discrete-time phase-type distribution $(\alpha, \mathbf{T}_{\mathrm{d}})$ with

$$\alpha = (1, 0, \cdots, 0), \; \mathbf{T}_{\mathrm{d}} = \begin{pmatrix} 0 & \frac{1}{a} & \frac{1}{a^2} & \cdots & \frac{1}{a^{n-1}} \\ 0 & 1 - \frac{q}{a} & 0 & \cdots & 0 \\ 0 & 0 & 1 - (\frac{q}{a})^2 & \cdots & 0 \\ \vdots & \vdots & \vdots & \ddots & \vdots \\ 0 & 0 & 0 & \cdots & 1 - (\frac{q}{a})^{n-1} \end{pmatrix}. \tag{6.14}$$

We use this representation for transforming the model in the continuous-time domain. This is accomplished by choosing a continuous-time phase-type distribution $(\alpha, \mathbf{T})$, with $\mathbf{T} = \mathbf{T}_{\mathrm{d}} - \mathbf{I}$. In [39] it has been shown that this model is equally suitable for modeling pseudo-self-similarity as the original approach proposed in [166, 167]. Subsequently, we will refer to this distribution as *pseudo self-similar phase-type distribution* (PSSP).

### 6.3.3 Parameterization

Since we are interested in whether self-similar traffic can be approximated by PSSP arrival models, the investigations in the next subsection are based on comparing the results for PSSP traffic with those obtained from analyzing an actual trace. Here, we shortly present the matching algorithm employed for deriving appropriate PSSP models for a given trace, and afterwards describe the trace which was used in our experiments

**Matching Algorithm.** In [166, 167], a procedure has been given for deriving the model parameters $a$ and $q$ to obtain a given mean interarrival time and Hurst parameter. The number of states of the underlying model has to be given; it basically influences the *domain of validity* where the model exhibits self-similarity. The authors state that employing 5 to 6 states is usually sufficient. The procedure involves two steps:

- First, the interarrival time of the model is matched by iteratively determining an appropriate value for $q$. Here, the authors use the fact that the mean interarrival time is independent of $a$ due to the special structure of the matrix $\mathbf{P}$. Its value is given by

$$\frac{1/(1-q)^n}{1-1/q}. \tag{6.15}$$

- In a second step, $a$ is chosen appropriately to obtain the desired Hurst parameter. This is again accomplished iteratively. In each iteration, the Hurst parameter is determined by investigating a variance-time plot. The plot can be generated easily since the authors derived an explicit expression for the variance of the aggregated processes $\operatorname{Var} X^{(m)}$.

The first step of the matching procedure can be adopted directly for PSSP models, since for given $q$ and $n$ the mean interarrival time is identical to the one in the original model. Concerning the second step, an explicit expression for $\operatorname{Var} X^{(m)}$ has not been derived in [39]. Instead, for $a$, $q$ and $n$ given, a trace is generated and then analyzed statistically (using the approach presented in Section 6.3.3) to obtain an estimate for the Hurst parameter. Though this is computationally more expensive than an explicit expression, the iteration scheme for obtaining $H$ converges rather quickly and only few traces have to be generated.[3] Usually, less than 20 iterations are needed to obtain a model that approximates a given Hurst parameter within a 1% error bound. On a moderately-sized workstation, this can be accomplished in less than 15 minutes even if very long intermediate traces are employed to estimate $H$.

**Trace Description.** A variety of traces have been investigated in [39]. Here, we focus on one particular trace, representing the World Wide Web requests

[3] Note that in [151] an explicit expression for $\operatorname{Var} X^{(m)}$ is given for phase-type distributed interarrival times. However, it has not been used in our studies.

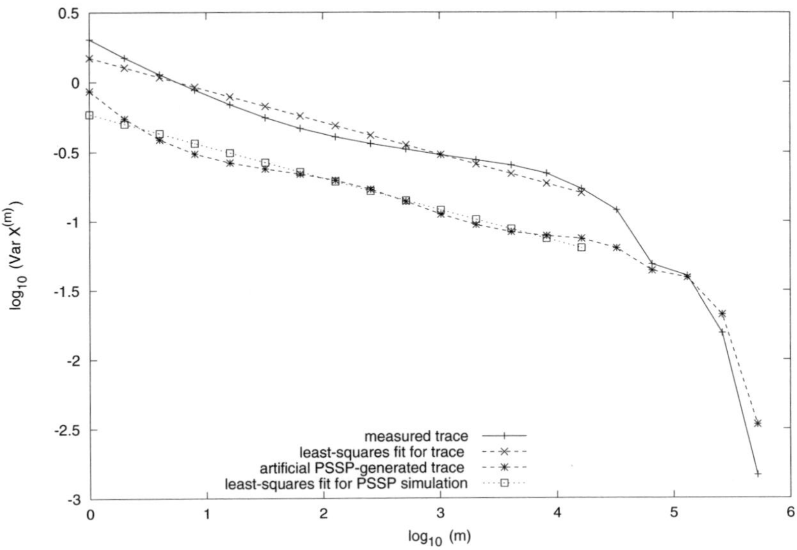


**Fig. 6.22.** Variance-time plots of the measured trace and of the corresponding PSSP model.

to the Sunsite Web Server at the RWTH Aachen during a 13-day period in May 1998. The trace contains 746638 events, corresponding to a mean interarrival time of 0.686 seconds. The variance-time plot for this trace is given in Figure 6.22; the fitted least-squares line has a slope of $-0.23$, indicating self-similarity with a Hurst parameter of $H = 0.885$.

This trace has been matched with a 6-state PSSP with identical $H$ and interarrival time and parameters $a = 1740$ and $q = 3.178$. As can be observed in Figure 6.22, it captures the self-similar behaviour of the original trace very well. Note, however, that in spite of the fact that the common slopes indicate a similar Hurst parameter, there is a difference between the absolute values of the variances of the trace and the PSSP.

### 6.3.4 Numerical Results

As we have seen in the previous section, the PSSP is able to correctly match the Hurst parameter of the measured trace. We are now interested in whether the PSSP is also an appropriate substitute for the original trace with respect to the performance measures obtained in a numerical evaluation study. This question is particularly interesting since such a modeling-oriented validation has not been accomplished for the original approach suggested in [166, 167].

We investigate this issue by focusing on a simple infinite-buffer queueing system which is subject to PSSP arrivals. We assume exponentially distributed service times to minimize the impact of the services on the results.

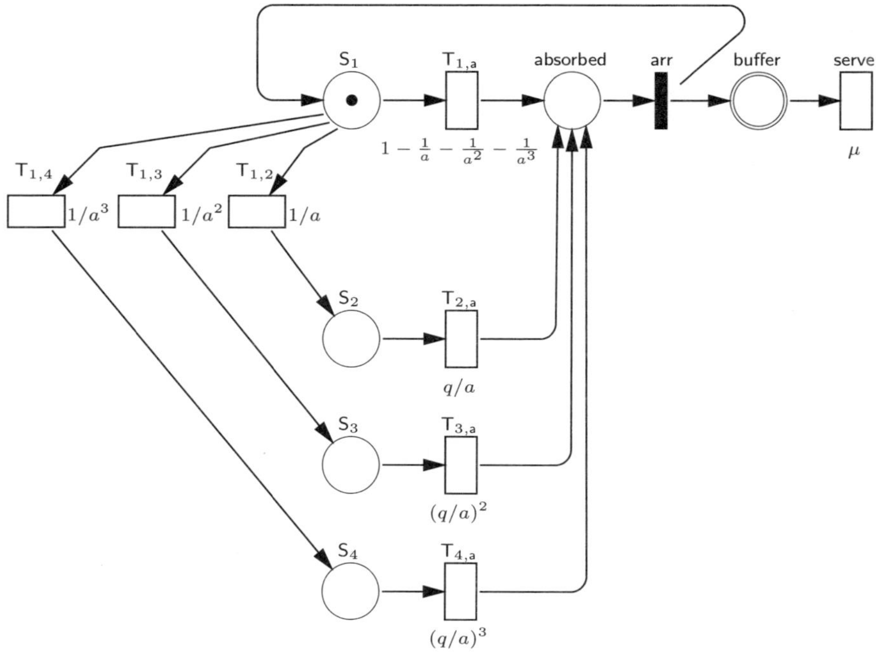


**Fig. 6.23.** iSPN for a PSSP|M|1 system with arrivals according to a 4-state PSSP.

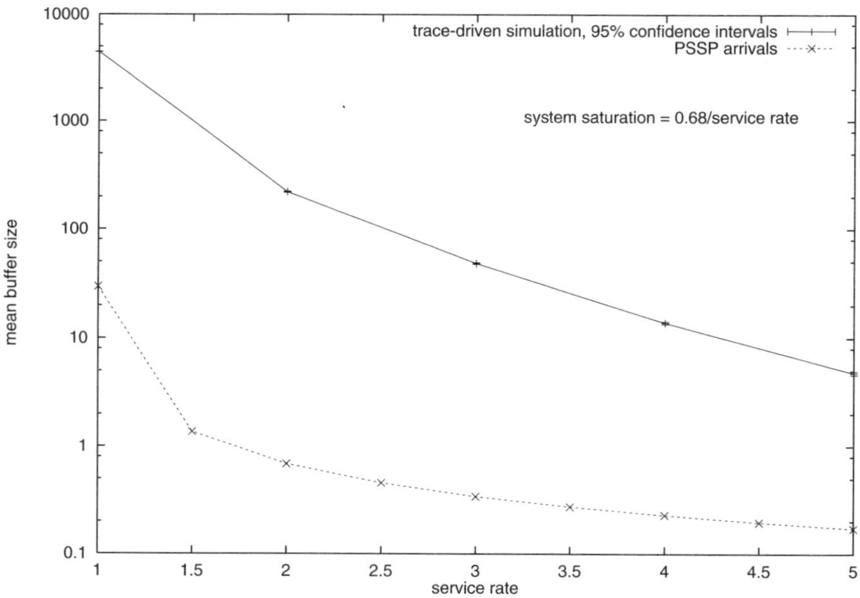


**Fig. 6.24.** The mean buffer length for different traffic intensities.

In our study, we use trace-driven simulation to evaluate the queueing performance of a server subject to the real trace, and then compare the results to those derived from a numerical analysis of the PSSP|M|1 system. The simulations were performed with the *Atlas* [55] simulation package. The numerical analysis was accomplished with the SPN2MGM tool by investigating an appropriate iSPN model. For a given PSSP, the iSPN model of a PSSP|M|1 system is derived in a straightforward fashion as illustrated by the example shown in Figure 6.23 (the specification files for this example can be found in Appendix C.5). In both the simulation study and the numerical analysis, we focus on the mean queue length and the queue length distribution as performance measures.

Figure 6.24 depicts the mean number of customers for different traffic intensities (obtained by varying the service rate). It can be observed that the PSSP model yields much too optimistic results, with mean buffer occupations being orders of magnitude smaller than those derived in the simulation case.

This dramatic discrepancy can be explained by looking at the queue-length distribution for both trace simulation and PSSP. For an example traffic intensity of 0.68, it can be observed in Figure 6.25 that the probability of small queue lengths is much larger in the PSSP case than for the real trace, except for queue lengths 0 and 1. In strong contrast to this, for queue lengths larger than about 200 (see Figure 6.26), the probabilities quickly approach zero in the PSSP case, but remain much larger for the original trace. Evidently, the heavy tail of the queue length distribution in the simulation case leads to much higher mean queue lengths than in the PSSP case. We have found the same behaviour for other traces as well [39].

The absence of this heavy-tailed property of the queue length distribution in the PSSP case may stem from the fact that we employ a finite-variance phase-type distribution in our model. In Section 6.3.2, however, it has been mentioned that self-similar processes can be constructed by superposing *infinite* variance renewal processes.

Another reason for the largely differing results might be that the PSSP approach (as well as the original MMBP model) is only capable to match the Hurst parameter and the first moment of the interarrival time distribution; however, higher-order statistics usually heavily impact the performance of a queueing system as well. In Figure 6.22, it can be observed that the *decay* of the variance of the aggregated processes is quite similar (and thus the approximation of the Hurst parameter is quite good), but that the *absolute* value of the variances differs by a factor of about $10^{0.4} \approx 2.5$ (note that for $\log(m) = 0$, Figure 6.22 depicts the interarrival time variances).

### 6.3.5 Conclusion

Though the model presented in [166, 167] and our related class of PSSP arrival processes is well suited for approximating self-similar traffic properties, PSSPs do not perform well if it comes to evaluating the performance of

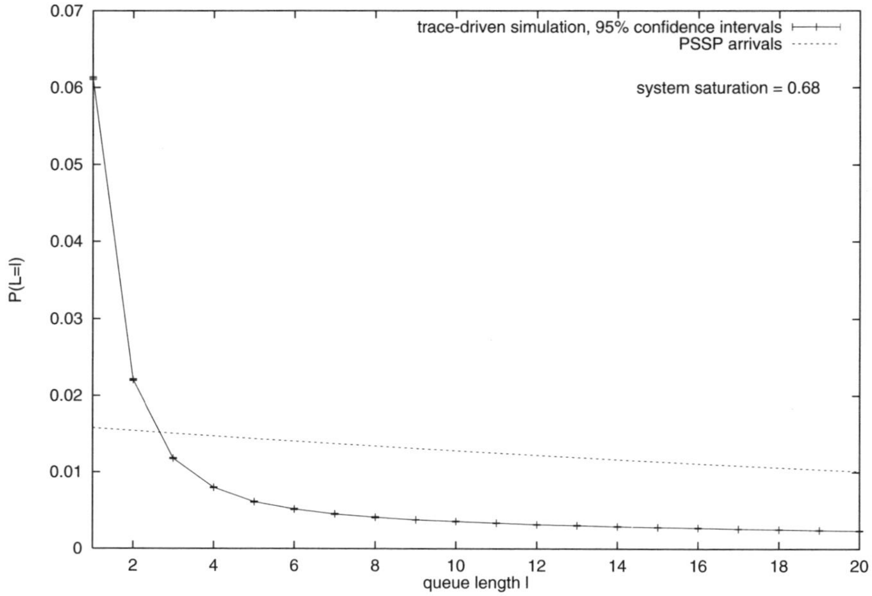


**Fig. 6.25.** Queue length distribution of the PSSP|M|1-system for small queue lengths (the buffer-empty probability of $1 - 0.68 = 0.32$ has been omitted for scaling reasons).

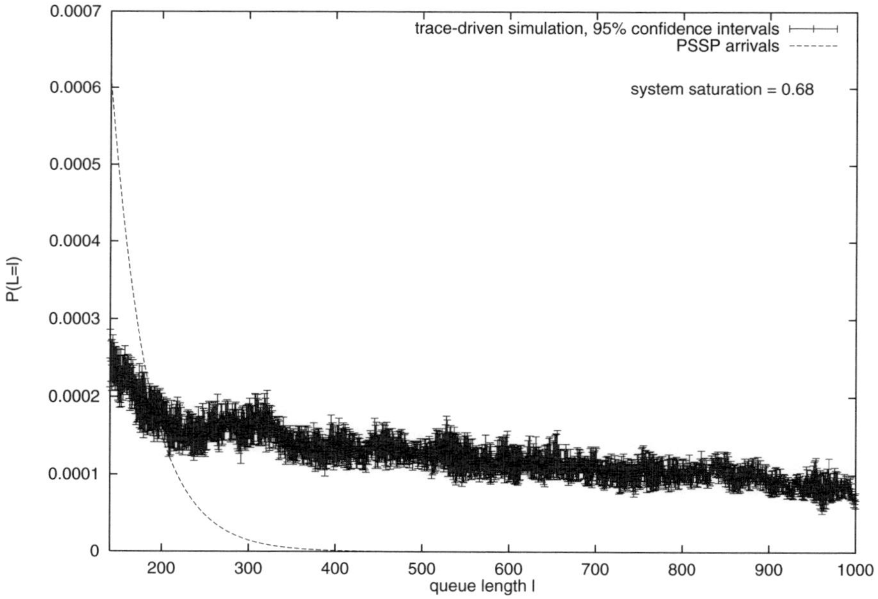


**Fig. 6.26.** Tail of the queue length distribution of the PSSP|M|1-system.

systems subject to such traffic. Since PSSPs share most of the properties of the original model, we believe that this is a weakness of the original model as well. Interestingly, except for a short note in [166, p. 124], no results have been published which investigate the suitability of this model for representing real-world arrival processes, and for obtaining realistic performance estimates from its analysis.

It should, however, be emphasized that the PSSP approach (as well as the original model) is suitable for approximating self-similar behaviour over several time scales. Especially in view of the recent discussion on the impact of self-similarity, a model which is able to account for correlations up to a large yet finite time-scale seems appropriate. Due to the small number of parameters and a convenient phase-type representation, PSSPs are a good starting point for developing improved models which approximate self-similar traffic.

Further developments should particularly focus on improving the ability of PSSPs to account for higher moments of the interarrival time distribution. One way to achieve this might be to introduce additional states to the PSSP for increasing the interarrival time variance. It also seems worthwhile to investigate the effect of changing the decay rate of the geometric series of the rates from the PSSP's starting state to higher numbered states, and from higher numbered states to the absorbing state. Currently, for states $k > 1$, these rates equal $(1/a)^{k \cdot f}$ and $(q/a)^{k \cdot f}$, respectively, where the decay rate $f$ is fixed to 1.

## 6.4 Summary and Concluding Remarks

In this chapter, we have shown the applicability of our iSPN/QBD-based modeling and evaluation framework towards the investigation of the nodes that appear in complex computer and communication systems. The approach is not only capable to account for the service behaviour of complex nodes, but also to account for the structure of the traffic that represents a node's workload.

Especially our study of ATM connection management for video traffic in Section 6.1 and the investigation of a WWW server subject to HTTP traffic in Section 6.2 proved that large QBDs are necessary to model these systems in sufficient detail. The specification of these QBDs can easily be performed at the iSPN level. Furthermore, high-level performance measures like response times and bandwidth allocation can easily be computed; the specification and derivation of these measures directly at the QBD level would represent a cumbersome task. The studies have also shown that the selection of system parameters like the connection release timeout or window sizes crucially affects the overall system performance. The correct selection of (near-)optimal values for these parameters (meeting the requirements of a particular workload) therefore represents an important task in practice, where our model-based performance evaluation environment provides the necessary tool for choosing among different alternatives.

In Section 6.3, a special class of renewal processes designed to account for the special properties of self-similar traffic has been integrated in our iSPN environment. The comparison of trace-driven simulation with measured self-similar WWW traffic to the corresponding iSPN model has shown that the approach does not provide satisfactory results in its current form. However, it could be observed that the correlation structure of traffic has a crucial impact on the node performance measures, confirming our finding that a correct description of arrival traffic patterns is essential.

All our investigations in this chapter focused on the analysis of isolated node models. For the concurrent analysis of several nodes, the modeling approach will quickly lead to QBDs of a size which is not tractable by existing solution methods. Also, the inclusion of several unbounded buffer leads to multi-dimensional QBDs, the solution of which still represents an open problem (see also Section 4.6.4). In the following chapter, we will therefore focus on a decomposition approach that allows the investigation of several interconnected iSPN nodes by analyzing the individual nodes in isolation, and by accounting for inter-node relations by appropriate traffic descriptions.

# Part III

# Network Analysis

# 7. Queueing Network Analysis Techniques

In Part II of this work, a powerful modeling and evaluation approach has been developed in order to resolve the node modeling limitations of our initial M|G|1-network based approach. By employing quasi-birth-and-death processes, an expressive stochastic framework has been identified, which is amenable to very efficient solution algorithms. By supplementing it with an intuitive high-level specification methodology based on stochastic Petri nets, individual nodes can be modeled in substantially more detail.

Considering the requirements which were identified for a modeling framework for modern telecommunication networks in Chapter 2, the results of Part II provide a major step towards fulfilling R1 (Expressiveness and Convenience) with tight orientation towards R3 (Fast Analysis). Requirement R4 (Service-Oriented Modeling) has already been tackled to a large extent by our first approach presented in Chapter 3. Thus, whereas the iSPN environment represents a powerful framework in its own right for the analysis of individual network nodes, the aim of this chapter is to develop approaches for embedding it in a network-oriented modeling and evaluation framework; this is reflected by the two requirements which still remain to be tackled:

- R2 (Non-Local Interactions) explicitly aims at providing a means for accounting for interactions between individual nodes and the traffic flows between them;
- R5 (Model Evaluation) is partly covered as the iSPN framework allows to derive detailed node properties; however, when extending the approach to a network of nodes, appropriate network-wide measures must be computed as well.

In Section 7.1, we summarize the main issues which need to be dealt with in view of these requirements. We also provide an overview on other approaches that have been developed in order to overcome the limitations of M|G|1-based approaches. One of them, the so-called queueing network analyzer (QNA) is of particular interest in our QBD/iSPN-based context, as several extensions of it are based on QBD processes; it will be discussed thoroughly in Section 7.2. Based on this, Sections 7.3 and 7.4 are then dedicated to the remaining issues which need to be tackled for integrating iSPNs in a network-based performance modeling and evaluation approach. We conclude the chapter in Section 7.5.

## 7.1 Main Problems and Existing Work

When queueing models were initially suggested to obtain performance measures of computer systems, the analysis was first based on the investigation of a single queueing station. Thus, there was no direct way to account for relationships among different nodes and for the traffic flowing between nodes. Clearly, this situation is very similar to the one we are facing with iSPNs right now.

When interactions between several nodes need to be accounted for, two "families" of approaches can be distinguished:

- *Product-form solutions* employ restrictions on the individual node models, so that the overall network-wide solution is given as a product of factors, each of them being determined by an individual node. Originating from open Jackson networks [110] and closed Gordon-Newell networks [81], one of the most prominent approaches in this area is the BCMP approach [19]. For an introduction to product form approaches, cf. [59].
- *Parametric decomposition approaches* concentrate on the analysis of individual nodes or of subnetworks of an overall network in order to reduce the complexity. The remaining part of the system (i.e., the part not considered in detail) is characterized by a few parameters only (e.g., by describing the traffic flowing into the considered subsystem). The initial approach presented in Chapter 3 can be considered as an example here, as nodes are analyzed in isolation, and the entire remaining system is abstracted by the rate at which jobs enter the considered node.

The first class of approaches usually imposes rather strict limitations on the node behaviour in order to guarantee a product-form solution. Developing a product-form approach for networks of nodes as flexible as those described by iSPNs therefore seems impossible, as also witnessed by our discussion of multi-dimensional QBD systems in Section 4.6.4. For this reason, we concentrate on developing a parametric approach in the remainder of this chapter. To start with, we summarize the main issues which need to be considered when developing a parametric decomposition framework in Section 7.1.1. In Section 7.1.2, we then provide a survey on existing approaches in this area, concluding with a discussion to which extent they are useful in the iSPN-based framework in Section 7.1.3.

### 7.1.1 Main Issues

Parametric decomposition approaches circumvent the generation and analysis of the overall state space of a network of node models. This is accomplished by focusing on individual nodes (or subsets of nodes) only and by accounting for the remaining system by a few parameters. In order to determine these parameters, a network-wide analysis of the interactions between nodes has to be performed. Then, after deriving node-specific performance results, these

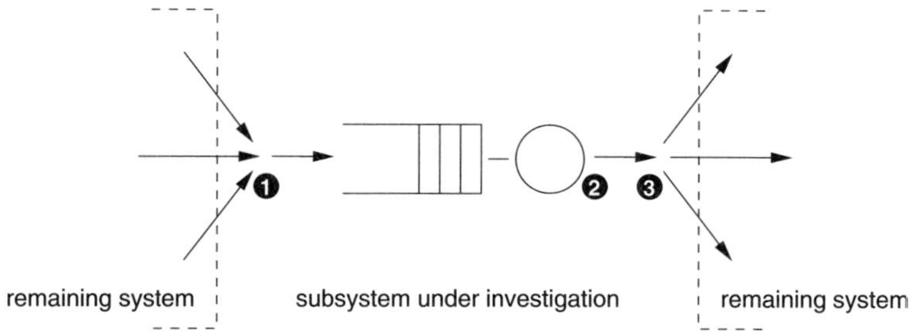


**Fig. 7.1.** The three main operations needed in a decomposition approach: merging of incoming traffic streams (1), determining the departure stream (2), splitting the departure stream (3).

need to be combined in order to derive network-wide results. The following main steps can thus be identified in a parametric decomposition approach:

1. Analysis of network-wide node interactions;
2. individual node (or subsystem) analysis, based on an abstraction of the behaviour of the remaining system;
3. integration of node-specific performance results to yield network-wide measures.

Clearly, the first step is of foremost importance in a network-oriented context, as it is only here where dependencies among different nodes are accounted for. In this context, the way in which jobs to be serviced by the individual nodes are flowing through the network is of major interest. The characterization of these *traffic flows* may comprise the mean interarrival time between successive jobs and higher moments of it; however, often even more detailed descriptions may be necessary, possibly explicitly describing the interarrival time distribution or the correlation between arrivals (this is particularly interesting in view of the impact of self-similar traffic properties on the node performance as discussed in Section 6.3).

In order to determine these traffic flows, one has to account for the fact that several flows can be directed towards one and the same processing station. In order to determine the overall traffic that enters a node, it is thus necessary to *merge* these traffic streams. Conversely, the traffic which originates from a processing node often needs to be *split* into the individual streams that are routed to different successor nodes. Finally, to complete the considerations of the network-wide traffic flows, a formalism is needed to yield the output traffic at a node that results from a given input traffic. Summarizing, the following steps need to be performed (see also Figure 7.1 for an illustration):

1. Merge all traffic flows directed to a node in order to determine the aggregate arrival stream which needs to be processed by the considered node;
2. given the aggregate arrival stream and a description of the node to be analyzed (e.g., in terms of a service time distribution or an iSPN), determine the output stream of the node;
3. split the output stream according to the routing policy; the split streams represent arrival streams to other nodes.

In the following section, we provide an overview on existing parametric decomposition approaches for queueing networks in order to discuss to what extent they can be used in an iSPN-based modeling environment.

### 7.1.2 Parametric Decomposition Approaches

**Positioning the M|G|1-Based Approach.** In the M|G|1-based approach developed in Chapter 3, the analysis of network-wide dependencies is accounted for by considering the mean job interarrival times between nodes. This step can be performed efficiently, as a network of $N$ nodes directly leads to a set of $N$ linear equations which can be solved at complexity $\mathcal{O}\left(N^3\right)$. As all occurring traffic streams are assumed to be Poissonian, streams can easily be merged by simply adding the individual rates (as reflected by Eq. (3.2) in Section 3.2.2). Similarly, streams can be split by simply multiplying the overall arrival/departure rates with appropriate routing probabilities (see Eq. (3.1)).

Clearly, the assumption that all traffic streams exhibit exponentially distributed interarrival times is an approximation; for streams departing from M|G|1 queueing stations, this is in general not the case. Furthermore, as also witnessed by the examples considered in Chapter 6, jobs do not enter the network in terms of Poisson streams, but according to much more complicated traffic patterns. Thus, while iSPN could be embedded in this framework in order to model complex service policies, this seems not to be appropriate in view of the fact that as soon as traffic leaves a node, it needs to be approximated by a simple Poisson stream.

**QNET and ΠNET.** Two approaches for accounting for more complex traffic structures are the QNET and ΠNET methods introduced in [86]. They are suitable for the analysis of open queueing networks and allow to model networks with batch arrivals and deterministic or probabilistic routing. The occurring traffic streams (and all service time distributions) are described by the first *and* second moment of the interarrival time distributions. The QNET method approximates the queueing network by a reflected Brownian motion (RBM) model; to obtain its solution, a partial differtial equation needs to be solved. However, the authors report that the computational effort to accomplish this task is excessive for networks with more than two queueing stations (according to [53], the complexity of the solution algorithm is

$\mathcal{O}(N^{10})$, where $N$ denotes the number of nodes in the network). The authors therefore suggest to employ a simplified approximate solution technique for the RBM model, leading to the ΠNET method which is very similar to the QNA method that is discussed in detail below; this also holds for the solution complexity $\mathcal{O}(N^3)$ of QNA. The main differences to the QNA approach are as follows:

- When merging traffic streams, the second moment of the overall stream is determined as a weighted average of the input streams. Here, QNA employs a more elaborate approach for better accuracy.
- The second moment of the interdeparture time distribution at a node is taken to be identical to that of the service time distribution; at high traffic intensities, this approaches the QNA solution. However, at low traffic intensities, the QNA approximations may provide better results.

Even though the authors report more accurate results when comparing the QNET approach to ΠNET and QNA, its application is prohibitive in our context, where large networks of nodes need to be analyzed. The quality of the ΠNET solution is, in turn, very similar to what is obtained by QNA. Whereas ΠNET is based on the RBM formalism, QNA is particularly promising in a matrix-geometric analysis environment, as will be shown subsequently. It should be emphasized that both in QNET and in ΠNET, all interarrival and service time distributions are described in terms of their first two moments. Although this is a substantial improvement over considering mean values only, this will not be sufficient to account for the complex traffic structures arising from large iSPN node models.

**Sequential Bottleneck Decomposition.** A hybrid approach to overcome the excessive computational effort required by the QNET method is the sequential bottleneck decomposition (SBD) approach presented in [53]. The key idea of SBD is that the analysis of the entire network by QNET is avoided. Instead, smaller subnets are investigated in isolation. Thereby, the remainder of the network is divided into a set of *underloaded* nodes (with lower node utilizations than those in the considered subset), and a set of *overloaded* nodes (where the traffic intensity is larger). For the analysis, arguments from heavy traffic theory are employed: the set of underloaded nodes is approximated by immediate services, and the overloaded nodes are represented as traffic sources and sinks. Comparing the results of SBD with those derived by simulation, the authors report better results than those obtained by QNA in many situations. As SBD is based on heavy-traffic theory, it often provides substantially better results than QNA for nodes close to saturation. However, at lower traffic intensities, the results of QNET are often worse; furthermore, the advantage of SBD is only marginal in many of the considered examples. Also note that the complexity of SBD is still $\mathcal{O}(N^{10})$ due to the underlying QNET approach, however, the size of the considered sub-networks is smaller than in QNET, leading to a substantial reduction of the absolute solution effort.

Compared to QNET, SBD provides an efficient framework when it comes to nodes which involve high traffic intensities; in general, the accuracy of the approach (compared to other approaches like QNA) is not necessarily better. Also, as SBD is based on QNET, all distributions are also only accounted for in terms of their first two moments. Similar to all approaches considered so far, SBD is limited to the analysis of open queueing networks with infinite buffers.

**Maximum Relative Entropy.** In contrast to the approaches mentioned above, the Maximum Relative Entropy (MRE) approach introduced in [187] deals with the analysis of *closed* queueing networks by means of decomposition. The decomposition is based on classifying the overall state space of a queueing network in different sets of states that are equivalent in terms of the distribution of probability among the contained states. The approach uses the concept of entropy between different distribution functions in order to find distributions that fulfill a set of given constraints (which are expressed in terms of mean values) best and that is maximally unbiased towards any other assumptions. In [187], the computational complexity of MRE is given as $\mathcal{O}\left(N^3 + K^2\right)$, where $K$ denotes the number of customers in the network.

Although the authors report encouraging results when comparing their results to exact solutions in a set of example networks, it must be said that the proposed decomposition scheme is limited to finite state spaces only; in its current form, the approach thus does not allow to account for an unbounded number of customers. Furthermore, the problem of merging and splitting traffic streams and determining the departure process of a node are not tackled within the MRE approach, as it solely aims at decomposing and aggregating the results from the analysis of individual subsystems. Thus, MRE remains an interesting approach to be followed once these basic issues have been resolved; here, we do not further consider this approach, but focus on these basic issues first.

**Solutions for the Departure Process of G|G|1 Queues.** The decomposition approach presented in [96] aims at the analysis of open queueing networks with finite buffers, thereby accounting for the first two moments of all occurring traffic streams. The method focuses on determining the departure stream of a G|G|1|$k$ queueing station. For merging and splitting operations, the author refers to the same methods as employed in the QNA approach. In the proposed method, a G|G|1|$k$ node is analyzed by approximating it by a M|G|1|$k$ or G|M|1|$k$ system, depending on the variance of the interarrival and service time distributions. Interestingly, the author employs a SPN-based description technique for creating larger network models. For the solution of network models, an iterative approach is suggested that is very similar to the one described in Section 7.2.2; in each iteration step, a set of linear equations needs to be solved.

For a simple feed-forward queueing network, the author reports promising results. In view of our aim to employ iSPNs for the specification and analysis of individual nodes, however, this approach is less suitable since replacing complex iSPN models by a simple service time distribution in terms of the first two moments does not seem to be appropriate.

In this context, the approach presented in [25] is especially interesting. Although the method is not explicitly designed towards the analysis of queueing network models, an approach is presented to derive exact performance measures of individual G|G|$s$ nodes. Interarrival and service time distributions can be formulated in terms of Coxian distributions, thereby allowing to account for rather complex systems. Using a generating function approach, the author derives (among other measures) the queue length distributions at various time instants, including those instants where job departures take place. Therefore, it seems viable to extend this approach towards deriving the departure process of a node, and to embed the methodology in a queueing network analysis approach.

**The Queueing Network Analyzer.** As an efficient approximate method for the analysis of large open queueing networks, the Queueing Network Analyzer (QNA) has been proposed by Whitt [196] as an extension of the initial work by Kühn [123]. It allows for service and interarrival time distributions which are characterized by their first two moments. As mentioned above, the approximations QNA introduces for merging and splitting traffic streams are also employed by several other approaches; they can thus be considered as a safe starting point for further improvements. Computationally, QNA involves the solution of sets of $N$ equations, having complexity $\mathcal{O}\left(N^3\right)$.

While the accuracy of QNA is good when compared to related methods like ΠNET, the approach has recently been extended substantially in several ways in order to extend the node model class and to increase the solution accuracy. Interestingly, matrix-geometric methods were employed to accomplish this task. In view of this fact and due to the promising results which have so far been obtained, QNA represents a promising starting point for developing an iSPN-based network modeling and analysis environment.

### 7.1.3 Conclusion

A survey on existing parametric decomposition approaches has been presented (see Table 7.1 for an overview). It has shown that all of them are limited to considering the first two moments of all traffic streams only. In addition, the existing approaches do not account for the correlation of traffic and service times, even though this property represents an issue which can severely affect network performance (as shown in Section 6.3). This is also confirmed by the problems most approaches experience when nodes are considered where departing jobs are immediately fed back into the system, which may lead to highly correlated arrivals. In most approaches, immediate feedbacks need to be removed by reconfiguring the network.

While the analysis of *open* queueing networks is appropriate for the analysis of telecommunication systems in many cases, it is often desirable to

<table>
<tr><th></th><th colspan="2">Network class</th><th colspan="4">Solution</th></tr>
<tr><th></th><th>open/ closed</th><th>finite/ infinite</th><th>stream description</th><th>departure derivation</th><th>merging/ splitting</th><th>com- plex.</th></tr>
<tr><td>M|G|1</td><td>open</td><td>infinite</td><td>Poisson</td><td>input=output</td><td>rate-based</td><td>$N^3$</td></tr>
<tr><td>QNET</td><td rowspan="2">open</td><td rowspan="2">infinite</td><td rowspan="2">first 2 moments</td><td colspan="2">RBM</td><td>$N^{10}$</td></tr>
<tr><td>ΠNET</td><td colspan="2">similar to QNA</td><td>$N^3$</td></tr>
<tr><td>SBD</td><td colspan="6">see QNET</td></tr>
<tr><td>MRE</td><td>closed</td><td>infinite</td><td colspan="2">analytically, based on entropy</td><td>not dealt, must be provided</td><td>$N^3$ $+K^2$</td></tr>
<tr><td>G|G|1 ([96])</td><td>open</td><td>finite</td><td>first 2 moments</td><td>M|G|1 or G|M|1 approximation</td><td>see QNA</td><td>$N^3$</td></tr>
<tr><td>QNA</td><td>open</td><td>infinite</td><td>first 2 moments</td><td>Marshall/KLB (see Sec. 4.2)</td><td>stationary interval/ asymptotic method</td><td>$N^3$</td></tr>
</table>

**Table 7.1.** Overview on the presented decomposition approaches. The solution complexity is given in terms of the number of nodes ($N$) and (for MRE) the number of customers in the network ($K$).

account for nodes with limited buffers as well. From the methods discussed above, this is only possible with the MRE approach which, however, is limited to closed queueing networks, and the methods presented in [96]. Also, the QNA approach has recently been extended to account for this requirement.

As none of the discussed approaches performs significantly better than QNA (in fact, results from QNA are often adopted), and due to the fact that recent extensions of QNA are based on matrix-geometric methods, we decided to take QNA as a starting point for developing an iSPN-based network analysis methodology.

## 7.2 The Queueing Network Analyzer

As mentioned in Section 7.1.2, the Queueing Network Analyzer (QNA) represents an approach for the approximate analysis of open queueing networks. All occurring traffic streams and service time distributions are accounted for by their first two moments. The purpose of this section is to provide a brief introduction to QNA, and to present existing and new improvements to the QNA method.

In Section 7.2.1, we summarize the basic QNA method, as originally suggested by Whitt in the mid-1980s [196]. Section 7.2.2 is then dedicated to the description of a recent QNA extension towards the inclusion of finite buffers in the QNA context; the proposed iterative solution approach will prove useful in our iSPN-based environment as well. In Section 7.2.3, we then develop a different view for looking at QNA node models, thereby establishing a connection to QBD systems. How this new viewpoint can be used to further improve

QNA by employing matrix-geometric solution methods will then be described in Section 7.2.4. Finally, we conclude with a summary in Section 7.2.5.

### 7.2.1 Basic QNA

In the QNA approach, a network consisting of $N$ queues with $m_i$ $(1 \leq i \leq N)$ servers each is assumed. The expected service time for server $i$ is denoted by $E[S_i] = 1/\mu_i$, and the corresponding squared coefficients of variation are $c_{si}^2$. External jobs can enter the network at any node as a renewal process with interarrival rate $\lambda_{0,i}$ and squared coefficient of variation $c_{0,i}^2$. Jobs are routed between the individual nodes queues according to the probabilities $p_{i,j}, i, j \in \{1, \cdots, N\}$, and they leave the network at node $i$ with probability $p_{i,0}$. In addition, single jobs may either be multiplied upon departure from a node or multiple jobs may be combined to single jobs. This is accounted for by a factor $\gamma_i$ which equals 1 if no multiplication or combination takes place at node $i$.

The analysis of this queueing network by QNA then involves four main steps:

- Elimination of immediate feedbacks, i.e., routing probabilities $p_{i,i} > 0$.
- Solving a linear set of linear equations, the so-called *traffic rate* equations. Here, the arrival rates at all networks nodes are determined.
- The solution of a second set of linear equations—the *traffic variability* equations—yields the squared coefficients of variation of the involved traffic streams.
- As all traffic characteristics are now described in terms of their first two moments, results for the individual nodes can be obtained. These can be combined to yield network-wide results (e.g. delay values).

We briefly summarize the tasks that need to be performed in these four main steps below.

**Elimination of Immediate Feedbacks.** Prior to starting the main calculations, immediate feedbacks in the network need to be eliminated. This step has already been proposed by Kühn [123]; the motivation for it is that in the original system with feedbacks, input and output processes can be strongly correlated. In such situations, the assumption of independent traffic streams which is needed in the subsequent solution steps is violated, leading to inaccurate results.

By eliminating feedbacks, jobs receive their total service time at once. Without going into details, this involves increasing the service time of a queue and changing the routing probability to the original conditional probability that a job is not fed back into the queue; for details, cf. [196].

Note that the elimination step neither changes a node's queue length distribution nor its waiting time distribution. However, other measures may need to be recalculated, in particular those based on visit ratios.

**Traffic Rate Equations.** In this step, the first moments of all traffic streams which occur in the network are determined. As no losses can occur, the first moment of a node's departure stream equals the sum of the first moments of its input streams. As splitting is performed probabilistically, the total rate $\lambda_i$ at which jobs arrive at a node is given by the sum of external arrivals and those jobs that are routed from other nodes towards it:

$$\lambda_i = \lambda_{0,i} + \sum_{j=1}^{N} \lambda_j \gamma_j p_{j,i}, \quad i = 0, 1, \cdots, N. \tag{7.1}$$

Introducing row vectors $\boldsymbol{\lambda} = (\lambda_1, \cdots, \lambda_N)$ and $\boldsymbol{\lambda}_0 = (\lambda_{0,1}, \cdots, \lambda_{0,N})$, this yields a set of linear equations that can be written in matrix form as

$$\boldsymbol{\lambda}(\mathbf{I} - \operatorname{diag}(\gamma_1, \cdots, \gamma_n)\mathbf{P}) = \boldsymbol{\lambda}_0. \tag{7.2}$$

By solving this $N$-dimensional system, the arrival rates $\lambda_i$ for all nodes are derived. With these values, the following results are readily available:

- $\lambda_{i,j} = \lambda_i \gamma_i p_{i,j}$, the rate at which jobs flow from node $i$ to node $j$;
- $r_{i,j} = \frac{\lambda_{i,j}}{\lambda_j \gamma_j}$, the share of all arrivals to node $j$ that originate from node $i$;
- $d_i = \lambda_i \gamma_i (1 - \sum_i p_{i,j})$, the rate at which jobs leave the network at node $i$;
- $d = \sum_i d_i$, the total rate at which jobs leave the network, and
- $\rho_i = \lambda_i/(m_i \mu_i)$ the utilization of node $i$.

If all node utilization values $\rho_i$ are less than 1, then the network is stable and the second moment of the traffic streams can be computed in the next step of QNA.

**Traffic Variability Equations.** Similar to the traffic rate equations, this step of the QNA method derives a set of linear equations in order to determine the squared coefficient of variation of all traffic streams. However, while the merging and splitting operations and departure stream derivation are simple when considering mean values only, dealing with the second moment is more difficult.

Figure 7.2 summarizes the traffic streams and variation coefficients that need to be considered in this step, along with and the involved probabilities. The (given) service time variability of node $i$ is denoted by $c^2_{\mathrm{s}i}$, the (unknown) variation coefficient of the departure stream is denoted by $c^2_{\mathrm{d}i}$, and $c^2_{\mathrm{a}i}$ refers to the (unknown) variability of the interarrival time distribution at node $i$. Furthermore, the variation coefficient of the traffic between nodes $i$ and $j$ is denoted by $c^2_{i,j}$.

The establishment of linear relationships between these values is the key contribution of the QNA approach; the results are largely based on approximate methods for the characterization of point processes, namely the *asymptotic method* and the *stationary interval method* (see [195] for details). In the

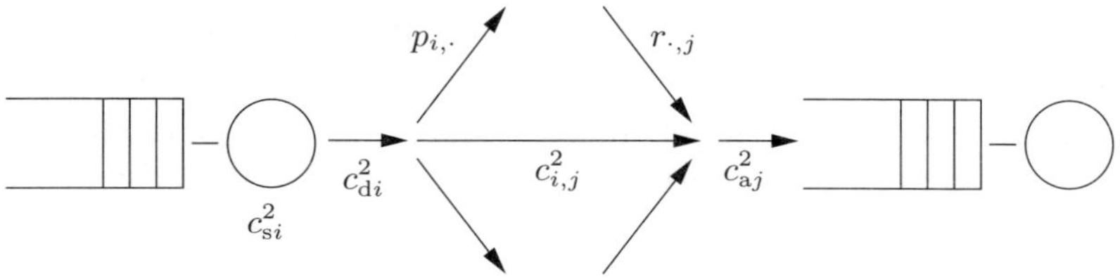


**Fig. 7.2.** Variation coefficients and probabilities considered in QNA's traffic variability analysis.

following, we briefly summarize the steps that are performed when merging traffic streams, splitting them, and how to derive the coefficient of variation of a node's departure stream.

*1. Merging Traffic Streams.* The squared coefficient of variation $c^2_{\mathrm{a}j}$ of the arrival process at node $j$ clearly depends on the variation coefficients $c^2_{i,j}$ of those traffic flows which are routed from any node $i$ to the considered node $j$. The asymptotic method yields a simple approximate linear combination, where the parameters are the probabilities that a job originates at node $i$:

$$c^2_{\mathrm{a}j} = \sum_i r_{i,j} c^2_{i,j}. \tag{7.3}$$

In [196], it is reported that a combination of this result and the result for the stationary interval method leads to more accurate results; however, the latter method yields a non-linear relationship which cannot be used here. Still, it was shown to be advantageous to set the correlation coefficient which would be obtained from the stationary interval method to 1, and then to use a weighted sum of (7.3) and the value 1:

$$c^2_{\mathrm{a}j} = w_j \sum_i r_{i,j} c^2_{i,j} + (1 - w_j). \tag{7.4}$$

According to experimental results by Whitt and others, the weighting parameter $w_j$ should be chosen according to

$$w_j = \left(1 + 4(1-\rho_j)^2(\nu_j - 1)\right)^{-1}, \quad \text{where} \quad \nu_j = \sum_i r^2_{i,j}. \tag{7.5}$$

Note that if all incoming traffic to node $j$ originates from one node $i$ (i.e., some $r_{i,j} = 1$ and all others equal 0), then Eq. (7.4) correctly yields the exact relationship $c^2_{\mathrm{a}j} = c^2_{i,j}$.

*2. Splitting Traffic Streams.* Under the assumption that the traffic flow leaving service station $i$ is a renewal process, the resulting partial flow to node $j$ is again a renewal process. Then, its variation coefficient $c^2_{i,j}$ is given by

$$c^2_{i,j} = p_{i,j} c^2_{\mathrm{d}i} + (1 - p_{i,j}). \tag{7.6}$$

Note that this equation actually represents an approximation, since the stream leaving the service station is in general not a renewal process.

*3. Departure Process Derivation.* In a last step, the unknown coefficients $c_{\mathrm{d}i}^2$ of the departure process (occurring in Eq. (7.6)) have to be determined. For single-server nodes, the stationary interval method coincides with Marshall's formula [138] for GI|G|1-queues, which yields

$$c_{\mathrm{d}i}^2 = c_{\mathrm{a}i}^2 + 2\rho_i^2 c_{\mathrm{s}i}^2 - 2\rho_i(1-\rho_i)\mu_i \, \mathrm{E}[W_i]. \tag{7.7}$$

Here, $\mathrm{E}[W_i]$ denotes the expected waiting time of a job queued for service at node $i$. Whitt suggests the approximation

$$\mathrm{E}[W_i] = \frac{\rho_i(c_{\mathrm{a}i}^2 + c_{\mathrm{s}i}^2)}{\mu_i \cdot 2(1-\rho_i)}, \tag{7.8}$$

which is based on a simplification of the Krämer and Langenbach-Belz (KLB) approximation [122]. Using this in Eq. (7.7), the simple formula

$$c_{\mathrm{d}i}^2 = \rho_i^2 c_{\mathrm{s}i}^2 + (1-\rho_i^2)c_{\mathrm{a}i}^2 \tag{7.9}$$

is obtained. QNA employs an extension of this formula for multi-server queues:

$$c_{\mathrm{d}i}^2 = 1 + (1-\rho_i^2)(c_{\mathrm{a}i}^2 - 1) + \frac{\rho_i^2}{\sqrt{m_i}}(c_{\mathrm{s}i}^2 - 1). \tag{7.10}$$

According to [196], the influence of the service time variability on that of the interdeparture times is in practice less than what this equation suggests. Therefore, it is suggested to replace $c_{\mathrm{s}i}^2$ in Eq. (7.10) by $\max(0.2, c_{\mathrm{s}i}^2)$.

Using Eqs. (7.4), (7.6) and (7.10), linear relationships between the variation coefficients of all traffic streams in the network have been established. Similar to the approach taken for determining the first moments, a set of linear equations can now be solved in order to derive $c_{\mathrm{a}i}^2$ for all nodes in the network.

**Individual Node Analysis and Network-Wide Results.** Once the job arrival rate and variability for each node in the network have been determined, it is possible to derive congestion measures of the individual nodes.

One of the most important characteristics of a node is the average waiting time for a job that enters a queue, $\mathrm{E}[W_i]$. In QNA, this value is already determined in the departure process derivation (see Eq. (7.8)) by applying a simplified version of the Krämer and Langenbach-Belz approximation for GI|G|1-queues. As it is now no longer necessary to obtain a linear relationship, the normal Krämer and Langenbach-Belz approximation can be applied, leading to more accurate results. Then, the expected waiting time is given by

$$\mathrm{E}[W_i] = (c_{\mathrm{a}i}^2 + c_{\mathrm{s}i}^2)\frac{\rho_i \, \mathrm{E}[S_i]}{2(1-\rho_i)} \cdot g(\rho_i, c_{\mathrm{a}i}^2, c_{\mathrm{s}i}^2), \tag{7.11}$$

where

$$g(\rho_i, c^2_{ai}, c^2_{si}) = \begin{cases} e^{-\frac{2(1-\rho_i)(1-c^2_{ai})^2}{3\rho_i(c^2_{ai}+c^2_{si})}} & \text{if } c^2_{ai} < 1 \\ 1 & \text{if } c^2_{ai} \geq 1 \end{cases} . \tag{7.12}$$

Note that in Eq. (7.8), the factor $g(\rho_i, c^2_{ai}, c^2_{si})$ was simply assumed to be 1.

Using this result for the average waiting time, Little's law can be used to obtain the mean queue length. Furthermore, QNA provides other measures like higher moments of the waiting time distribution at a node and buffer-empty probabilities.

Also, network-wide performance measures can be derived from these figures, e.g., by adding the waiting times of all nodes involved in a network-wide service, similar to the approach taken in Section 3.2.

**Summary.** The original QNA method represents an efficient method for analyzing open queueing networks; the solution trajectory basically involves the solution of two systems of $N$ linear equations, which scales well even for reasonably large networks. The first two moments of all traffic streams and service times are taken into account. QNA involves the following approximations:

- All interarrival times are modeled by renewal processes.
- When merging incoming streams to a network node, an approximate formula with empirically determined coefficients is used to derive the overall traffic variability.
- The squared coefficients of variation of the interdeparture times of jobs leaving a node are determined using a simplification of the approximate Krämer and Langenbach-Belz formula.
- The same approximate formula is used to derive per-node congestion measures.

Clearly, the last two points are related to the analysis of individual nodes, which becomes more difficult as the node models become more complex. Therefore, it also difficult two account, for example, for nodes with finite waiting room. As this is a situation which arises often in practice, an extension to QNA has been presented in [87] to enable this. As it also provides the basis for the integration of more detailed node models (and, eventually, iSPN models), the approach will be summarized briefly in the following section.

The first two issues, however, are related to the description and the analysis of the occurring traffic streams. In addition to the consideration of the first two moments only, especially the renewal assumption is a limitation of QNA (and many similar approaches) which can lead to very inaccurate results. As the introduction of new approaches in this area, together with a new node description approach, completely replaces the key concepts of QNA, we defer the discussion of this topic to Section 7.3.

### 7.2.2 Finite Buffers

The approximations for determining the node departure processes in original QNA are based on the fact that all nodes comprise *unbounded* queues. However, in many studies, unbounded queues are used as an approximation for very large, yet bounded buffers. In order to allow for smaller, finite buffers as well, the approach presented in [87] introduces appropriate extensions to QNA. As a key concept, the author presents a solution approach based on fixed-point iteration in order to remove the requirement of explicit and linear relationships among model parameters.

**Traffic Rate Equations.** The traffic rate equations for the finite buffer case are similar to the original QNA equations (7.1). However, due to limited buffers, a job that tries to enter a node might be discarded, and thus the rate at which jobs leave a node is smaller than the arrival rate. This fact is accounted for by distinguishing between the node arrival rates $\lambda_{\mathrm{a}i}$ and the node departure rates $\lambda_{\mathrm{d}i}$. The latter is given by $\lambda_{\mathrm{d}i} = \lambda_{\mathrm{a}i}(1-\beta_i)$, where $\beta_i$ denotes the steady-state probability of a full buffer at node $i$. Note that this approach represents an approximation for arrival processes other than Poisson; to be exact, the buffer-full probabilities *at job arrival instants* need to be considered. For Poisson processes, this is identical to the steady-state probability due to the PASTA (Poisson arrivals see time averages) property [200].

These considerations lead to a slightly modified version of the original traffic rate equations (7.1):

$$\lambda_{\mathrm{a}i} = \lambda_{0,i} + \sum_{j=1}^{N} \lambda_{\mathrm{d}j} \gamma_j p_{j,i}. \tag{7.13}$$

Clearly, the departure rates $\lambda_{\mathrm{d}i}$ (or, respectively, the buffer-full probabilities $\beta_i$) represent a set of $N$ additional variables; however, there are still only $N$ traffic rate equations available. Unfortunately, the buffer-full probabilities depend (in general) not only on the mean interarrival and service times at a node, but also on higher moments—for example, a high service time variability usually leads to a larger mean buffer filling, and thus to larger buffer overflow probabilities and smaller departure rates.

Therefore, in contrast to the original QNA approach, the traffic rate equations can no longer be considered in isolation, but they must be investigated together with the traffic variability equations, which are discussed below.

**Traffic Variability Equations.** The approach in [87] still employs a description of all occurring traffic streams in terms of their first two moments. Thus, the merging and splitting approximations of the original QNA method can be maintained. The derivation of the departure process, however, has to be changed for finite-buffer queues, as the introduction of finite buffers strongly affects the traffic flow leaving a service station.

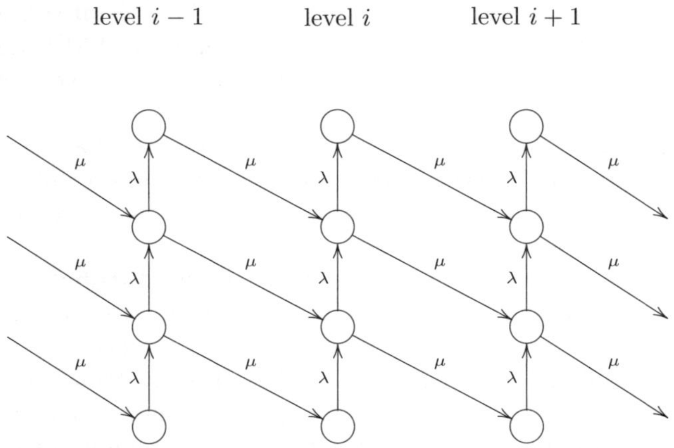


**Fig. 7.3.** State space of the Markov chain for obtaining the interarrival time distribution.

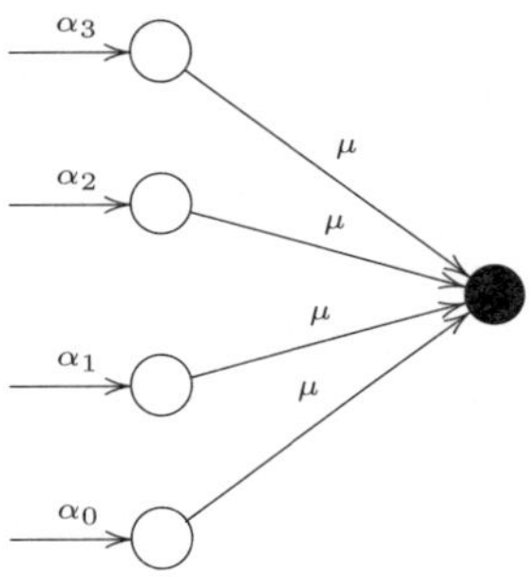


**Fig. 7.4.** Phase-type distribution approximating the interdeparture times.

The methodology proposed in [87] involves matching the interarrival and service time distributions by appropriate phase-type distributions, and using them to derive a phase-type representation of the interdeparture process. The method is best illustrated by a simple example where exponentially distributed interarrival and service times with rates $\lambda$ and $\mu$ are assumed. In the example, the queue is limited to three jobs.

First, an infinite, two-dimensional Markov chain is constructed by counting the number of jobs in the queue in the vertical (finite) direction, and counting job departures horizontally. The state space is illustrated in Figure 7.3.

Obviously, the time between two departures from this queue can be expressed as the time between entering a level and leaving it in order to enter the next level. This time is approximated by using a phase-type distribution as illustrated in Figure 7.4.

The initial probability vector $(\alpha_0, \alpha_1, \alpha_2, \alpha_3)$ for this distribution is obtained by combining the steady-state distribution of a M|M|1|3 queue with the conditional probabilities that a transition in Figure 7.3 corresponds to a departure, i.e., to a transition to the next level.

Once the phase-type distribution has been obtained, the first and second moments of it are easily derived using known expressions for phase-type distributions, thus yielding the departure rates $\lambda_{\mathrm{d}i}$ and the squared coefficients of variation of the departure process, $c^2_{\mathrm{d}i}$.

The extension of this scheme from exponentially distributed interarrival and service times to phase-type distributions is straightforward. In this case, the first two moments of these distributions are matched by appropriate hyper-exponential and hypo-exponential phase-type distributions; note that this leads to a larger number of states in the bounded dimension of the process shown in Figure 7.3. This also increases the number of states which form the interdeparture phase-type distribution, and to determine its initial probability vector, the steady-state probabilities of a PH|PH|1|$k$ queue have to be determined.

It is important to note that so far, it has been assumed that the interarrival time distributions at the nodes are already known. However, in order to determine their first two moments, the first two moments of the departure streams must be available for the solution of the traffic rate equations (7.13). The departure processes, however, are in turn only available after analyzing the traffic variability using the phase-type approach presented here. As this approach yields no explicit linear relationship to obtain the values of $\lambda_{\mathrm{d}i}$, it is not possible to build an overall system of equations to solve for all needed unknowns simultaneously. Therefore, it is suggested to employ an approach based on fixed-point iteration.

**The Iteration Scheme.** To derive the queueing network solution from the recursive set of nonlinear relationships between the system unknowns, an iterative procedure is suggested that circumvents the concurrent consideration of the traffic flow and the traffic variability.

Starting with an initial value for all arrival rates $\lambda_{\mathrm{a}i}$, and setting the variation coefficient of the node arrival processes to those of the external arrival streams at that node, the departure rates are first determined by applying the phase-type approximation. The results are then used to obtain new approximations for $\lambda_{\mathrm{a}i}$ by solving the traffic flow equations (7.13). After performing the splitting and merging operations, this procedure is repeated until the change between successive solutions is sufficiently small. The algorithm is summarized in Figure 7.5. Of course, it is still possible to account for unbounded queues in this approach as well by employing the original QNA traffic equations for these nodes. For brevity, this has been omitted in the presented algorithm.

**Discussion.** From a computational point of view, the iterative procedure is clearly more expensive than the original QNA method, as a set of equations

Initialize all $\lambda_{ai} := 0.1, c_{ai}^2 := c_{0,i}^2$

Compute $\lambda_{di}$ and $c_{di}^2$ using the phase-type approach.

According to the traffic flow equations:
$\lambda_{ai} := \lambda_{0,i} + \sum_{j=1}^{N} \lambda_{di}\gamma_i p_{i,j}$.

Arrival rates of jobs coming from other nodes:
$\lambda_{i,j} := \lambda_{di}\gamma_i p_{i,j}$.

Proportion of arrivals from predecessors: $r_{i,j} := \lambda_{i,j}/(\lambda_{dj}\gamma_j)$.

Apply Whitt's approximation for splitting traffic streams (Eq. (7.6)):

$$c_{i,j}^2 := p_{i,j}c_{di}^2 + (1 - p_{i,j}).$$

Apply Whitt's approximation for merging traffic streams (Eq. (7.4)):

$$c_{ai}^2 := (1 - w_i) + w_i \sum_i r_{i,j}c_{i,j}^2.$$

Repeat until the change of all $\lambda_{ai}$ and $c_{ai}^2$ is less than $\epsilon$.

**Fig. 7.5.** The iterative solution algorithm for finite queues in QNA.

has to be solved in each iteration step. In addition, each iteration involves the new phase-type approach to determine the departure processes of the nodes, which requires substantially more effort than applying the original approximation formulas. Nevertheless, even a larger number of iterations can be performed within reasonable time on modern computers.

Concerning the fixed-point approach, no formal proofs are provided in [87] concerning the convergence of the algorithm and regarding the existence of a unique solution; however, it is reported that the approach always converged to the correct solution during the investigation of over one hundred test cases.

From a qualitative point of view, the results reported by Haverkort in [87] are quite encouraging. For the investigated examples, the relative error compared to simulation results never exceeded 10%; however, errors become larger for increasing buffer sizes. Large buffer sizes, however, are more common than small ones when analyzing communication networks.

Large buffers also represent another problem in this approach when solving for the traffic variabilities: even in the case of simple M|M|1|$k$ nodes, the phase-type representation of the interdeparture time distribution holds $k+1$ states. To derive the initial probability vector $\boldsymbol{\alpha}$ it is necessary to obtain the stationary distribution of this M|M|1|$k$-queue. Unless matrix-geometric methods are employed to accomplish this (see the following section), this can quickly lead to problems concerning computational effort and numerical stability.

Besides the fact that the algorithm for finite queues in QNA still uses Whitt's approximate results for merging and splitting traffic streams, another inaccuracy is introduced by the way the interdeparture time distribution is characterized. When a departure takes place, this corresponds to entering the absorbing state in Figure 7.4. The current phase of the *arrival* process is lost, and it is only accounted for in a static manner by the initial distribution vector $\boldsymbol{\alpha}$.

In the following section, we will propose the introduction of matrix-geometric methods in order to alleviate these problems and to improve the overall accuracy of QNA even further.

### 7.2.3 From QNA Nodes to QBD Nodes

In Section 7.2.2, it was suggested to match the first two moments of the interarrival and service time distributions by appropriate phase-type distributions in order to derive a phase-type representation of the interdeparture time distribution. The method can therefore be considered as an approach to determine the interdeparture time distribution of a PH|PH|1|$k$ system.

Interestingly, the underlying Markov chain of such a queueing system is a quasi-birth-and-death process, so that matrix-geometric methods can be employed for its analysis. In this section, we briefly summarize the structure of the QBD Markov chain that results from a given phase-type representation of the interarrival and service time distributions; in Section 7.2.4, we then elaborate on the merits of adopting this QBD-oriented viewpoint.

**The Underlying Markov Chain of PH|PH|1 and PH|PH|1|$k$ Queues.** The underlying Markov chain of a PH|PH|1-queue can be considered as a special case of a quasi-birth-and-death process (see Chapter 4); each level of the QBD state space corresponds to a specific number of customers in the queueing system, and transitions between levels correspond to arrivals and departures.

As some kind of memory must be provided in the QBD for remembering the phase of the arrival process and the phase of the departure process, the number of states in one level corresponds to the possible phases of both these distributions. If we do not consider the embedded Markov chain at either arrival or departure instants, the number of states needed within each level equals the product of the order of the involved phase-type distributions.

In the following, we assume a queueing system where job interarrival times are distributed according to the phase-type distribution $(\boldsymbol{\alpha}, \mathbf{A})$, and service times are distributed according to $(\boldsymbol{\beta}, \mathbf{B})$. The orders of these distributions are denoted by $l$ and $m$, respectively. An example for $l = 3$ and $m = 2$ is given in Figure 7.6.

As departures can only take place if at least one job is in the buffer, only the phase of the arrival process needs to be accounted for at the boundary

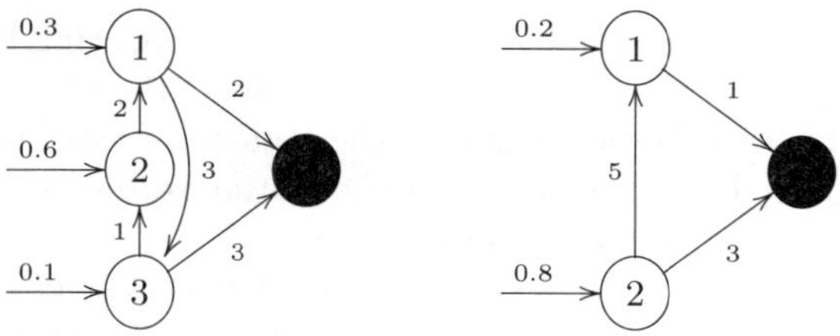


**Fig. 7.6.** Example phase-type distributions for interarrival and service times.

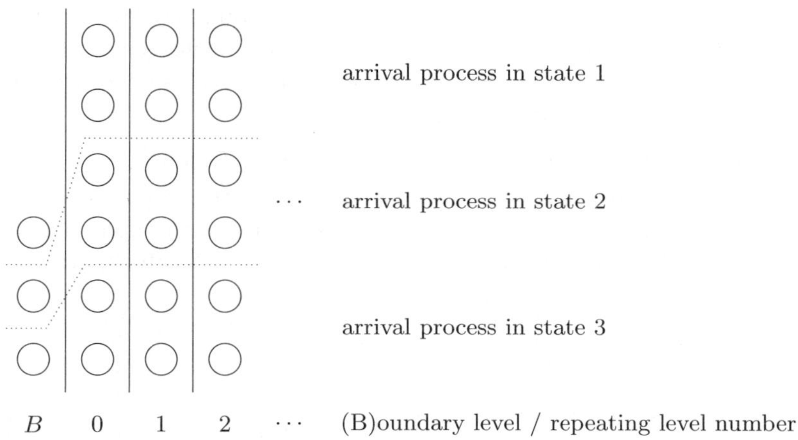


**Fig. 7.7.** State space structure of a QBD process for the example distributions. The state of the departure process is indicated by the two states between the dotted lines.

level of the state space, so the number of states in this level equals $l$. In the repeating levels, the phases of both processes must be accounted for, and thus the number of states in these levels is $l \cdot m$. The structure of the state space resulting from the two example distributions is illustrated in Figure 7.7.

The generator of this QBD process can be described conveniently by using Kronecker products:

$$\mathbf{Q} = \begin{pmatrix} \mathbf{A} & \mathbf{A}^0\boldsymbol{\alpha} \otimes \boldsymbol{\beta} & & & \\ \mathbf{I} \otimes \mathbf{B}^0 & \mathbf{A} \otimes \mathbf{I} + \mathbf{I} \otimes \mathbf{B} & \mathbf{A}^0\boldsymbol{\alpha} \otimes \mathbf{I} & & \\ \mathbf{0} & \mathbf{I} \otimes \mathbf{B}^0\boldsymbol{\beta} & \mathbf{A} \otimes \mathbf{I} + \mathbf{I} \otimes \mathbf{B} & \mathbf{A}^0\boldsymbol{\alpha} \otimes \mathbf{I} & \\ & & \ddots & \ddots & \ddots \end{pmatrix}. \tag{7.14}$$

The submatrices in this matrix are the result of the following considerations:

– Within the boundary level, only transitions which belong to the arrival process can occur, since no departures are possible. Transitions within this level which do not lead to an increased buffer size are contained in matrix **A**.

- Transitions from the boundary level to the first repeating level correspond to absorption in the Markov chain that represents the arrival time distribution. Since this distribution restarts, the transition rates to the absorbing state are replaced by transitions to the initial states, weighted with the initial probability vector $\boldsymbol{\alpha}$, yielding $\mathbf{A}^0\boldsymbol{\alpha}$.
  Furthermore, servicing starts since the buffer is no longer empty. The current states of the service time distribution Markov chain are thus distributed according to $\boldsymbol{\beta}$. Summing up, the overall transition matrix from level 0 to 1 is given by $\mathbf{A}^0\boldsymbol{\alpha} \otimes \boldsymbol{\beta}$.
- Transitions from the first repeating level to the boundary level correspond to absorption in the Markov chain of the service phase-type distribution. The phase of the arrival process remains unchanged, so the overall rate matrix is given by $\mathbf{I} \otimes \mathbf{B}^0$.

The remaining entries are derived by similar considerations.

**The Finite-Buffer Case.** When considering PH|PH|1|$k$-queues with finite waiting room where customers that encounter a full queue upon arrival are discarded, similar considerations lead to the (finite) generator matrix

$$\mathbf{Q} = \begin{pmatrix} \mathbf{A} & \mathbf{A}^0\boldsymbol{\alpha} \otimes \boldsymbol{\beta} & & & \\ \mathbf{I} \otimes \mathbf{B}^0 & \begin{matrix} \mathbf{A} \otimes \mathbf{I} \\ +\mathbf{I} \otimes \mathbf{B} \end{matrix} & \mathbf{A}^0\boldsymbol{\alpha} \otimes \mathbf{I} & & \\ \mathbf{0} & \mathbf{I} \otimes \mathbf{B}^0\boldsymbol{\beta} & \begin{matrix} \mathbf{A} \otimes \mathbf{I} \\ +\mathbf{I} \otimes \mathbf{B} \end{matrix} & \ddots & \\ & & \ddots & \ddots & \mathbf{A}^0\boldsymbol{\alpha} \otimes \mathbf{I} \\ & & & \mathbf{I} \otimes \mathbf{B}^0\boldsymbol{\beta} & \begin{matrix} (\mathbf{A} + \mathbf{A}^0\boldsymbol{\alpha}) \otimes \mathbf{I} \\ +\mathbf{I} \otimes \mathbf{B} \end{matrix} \end{pmatrix}. \tag{7.15}$$

Clearly, matrix-geometric solution algorithms for finite and infinite QBD processes can be employed here to derive the steady-state distributions associated to the generators in (7.14) and (7.15).

Note that in the iteration approach presented in Section 7.2.2, the steady-state probabilities of a PH|PH|1|$k$ needs to be determined; this step can be performed efficiently by employing matrix-geometric methods to solve the corresponding QBD. In the following section, several ways will be presented how the QBD viewpoint can be used to further improve QNA.

### 7.2.4 Using QBDs to Improve QNA

By considering QNA nodes as (bounded or unbounded) PH|PH|1 queueing stations, the underlying QBD structure can be used to improve QNA in several ways. After presenting a simple approach to obtain more detailed node results in QNA, a method to determine the exact moments of the departure process will be presented. Then, we introduce an approach to determine the loss process of queues with finite waiting room. Finally, an alternative way to determine the steady-state distribution of a PH|PH|1|$k$ queue is discussed.

**Node Analysis.** In [87, 196], it has been suggested to employ matrix-geometric methods in QNA's node analysis step in order to circumvent the application of the approximate Krämer and Langenbach-Belz formula and to obtain more detailed node results. This is achieved by matching the interarrival and service time distributions of the node under consideration by appropriate phase-type distributions (similar to the approach taken in Section 7.2.2) and then to derive the measures of interest by analyzing the QBD process that results from this description.

In the context of the QNA version which allows for finite queues (see Section 7.2.2), using this approach is obligatory to derive node-specific performance measures, as the approximation formula employed in standard QNA is only valid for bounded queues.

The advantage of investigating the underlying QBD instead of applying the approximation formulas is that more detailed performance measures can be obtained. Furthermore, it is now possible to employ arbitrary complex service time distributions. This also applies for nodes which are subject to external job arrivals only—then, complex job arrival patterns can be described as well.

The benefits of this approach should, however, not be overestimated. All of QNA's internal flow calculations are still based on the first two moments only, so deriving too detailed results from data based on this fact may be dangerous. Nevertheless, the results reported in [87] indicate that the matrix-geometric approach is more accurate than original QNA in those cases where both the interarrival times and the service times are subject to extremely low or extremely high variability. Otherwise, the approach performs about equally good as the original QNA approximation.

**Exact Derivation of the Moments of the Departure Process.** In addition to deriving more detailed and more accurate node performance measures, the QBD constructed from a PH|PH|1 or PH|PH|1|$k$ queue can also be analyzed in order to derive the *exact* first two moments of the interdeparture time distribution of these queues.

This is accomplished by applying a result on PH|PH|1|$k$ queues published in [31]. The calculations are based on the steady-state distribution of the underlying finite QBD process; due to limitations of the presentation in [31], we briefly summarize the steps that need to be performed.

In the following, the steady-state distribution of the underlying QBD process of the PH|PH|1|$k$ queue is as usual denoted by $(\mathbf{b}, \mathbf{v}_0, \cdots, \mathbf{v}_{k-1})$; the vector $\mathbf{b}$ thus corresponds to an empty queue, and the vector $\mathbf{v}_{k-1}$ to a full queue with $k$ elements in it. Furthermore, the following variables are used:

- $(\boldsymbol{\alpha}, \mathbf{A})$ and $(\boldsymbol{\beta}, \mathbf{B})$ correspond to the phase-type representations of the interarrival and service time distributions, respectively;
- $\mathbf{v}_{\mathrm{a}k-1} \in \mathbb{R}^m$ denotes the probability distribution among the states of the phase-type distribution of the service process *immediately preceding the arrival instants* where a full queue is encountered; and

– $\mathbf{b}_\mathrm{d} \in \mathbb{R}^l$ denotes the probability distribution among the states of the service time distribution *immediately after a job departure* occurred that leaves the queue empty.

The mean interarrival time of jobs equals the first moment of the interarrival time distribution. It is denoted by $1/\lambda_\mathrm{a}$ and as usual derived by

$$1/\lambda_\mathrm{a} = -\boldsymbol{\alpha}\mathbf{A}^{-1}\mathbf{1}. \tag{7.16}$$

According to [31], the state of the service phase-type distribution at time instants immediately preceding an arrival to a full queue is derived as

$$\mathbf{v}_{\mathrm{a}k-1} = \frac{1}{\lambda_\mathrm{a}}\mathbf{v}_{k-1}(\mathbf{A}^0 \otimes \mathbf{I}). \tag{7.17}$$

Clearly, by adding the elements of this vector, we obtain the probability $p_\mathrm{loss}$ that an arriving job encounters a full buffer and is discarded:

$$p_\mathrm{loss} = \mathbf{v}_{\mathrm{a}k-1}\mathbf{1}. \tag{7.18}$$

Using this probability, the rate $\lambda_\mathrm{d}$ at which customers leave the node is readily obtained as

$$\lambda_\mathrm{d} = \lambda_\mathrm{a}(1 - p_\mathrm{loss}), \tag{7.19}$$

which provides the first moment $1/\lambda_\mathrm{d}$ of the interdeparture time distribution.

In order to determine the variation coefficient of the departure process, the state probabilities of the arrival phase-type distribution at departure instants where the queue is left empty is derived in a way similar to Eq. (7.17):

$$\mathbf{b}_\mathrm{d} = \frac{1}{\lambda_\mathrm{d}}\mathbf{v}_0(\mathbf{I} \otimes \mathbf{B}^0). \tag{7.20}$$

In [31], the second moment of the interdeparture time distribution is obtained by considering the queue at departure instants. If the queue is not empty after a departure took place, then the time to the next departure is distributed according to the service time distribution. Otherwise, this time has to be added to the time until the next job has arrived. Using this fact, the variance of the interdeparture time distribution can be derived as the sum of the variance $\sigma_\mathrm{s}^s$ of the service time distribution and the variance $\sigma_0^2$ of the phase-type distribution $(\mathbf{b}_\mathrm{d}, \mathbf{A})$:

$$\begin{aligned}\sigma_\mathrm{d}^2 &= \sigma_\mathrm{s}^2 + \sigma_0^2 \\ &= 2\boldsymbol{\beta}\mathbf{B}^{-2}\mathbf{1} - (\boldsymbol{\beta}\mathbf{B}^{-1}\mathbf{1})^2 + 2\mathbf{b}_\mathrm{d}\mathbf{A}^{-2}\mathbf{1} - (\mathbf{b}_\mathrm{d}\mathbf{A}^{-1}\mathbf{1})^2.\end{aligned} \tag{7.21}$$

Using Eq. (7.19) for the mean interdeparture time, the squared coefficient of variation of the interdeparture time distribution is then given by

$$c_\mathrm{d}^2 = \lambda_\mathrm{d}^2\sigma_\mathrm{d}^2. \tag{7.22}$$

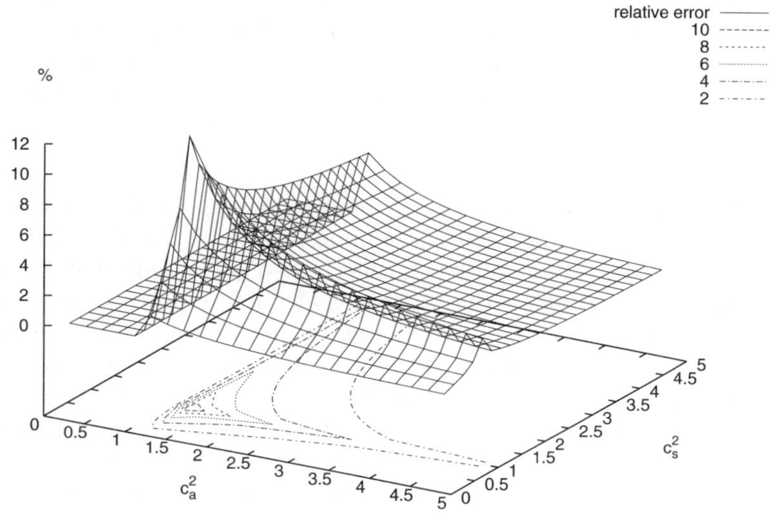


**Fig. 7.8.** Relative error between approximate and exact determination of the departure process variation of a PH|PH|1|20 queue.

While the method as described in [31] has been designed for finite PH|PH|1|$k$ queueing systems, it can easily be extended to unbounded queues by taking the limit $k \to \infty$. In this case, we have $\lambda_{\mathrm{d}} = \lambda_{\mathrm{a}}$, and Eqs. (7.21) and (7.22) remain valid.

In contrast to the approximate derivation of the moments of the interdeparture time distribution that was used in Section 7.2.2, the solution presented here is exact and does not suffer from neglecting the phase of the arrival process at departure instants.

In a first experiment, we compared the relative error of the approximate solution to the exact one in a PH|PH|1|20 system at utilization 1/3, depending on the variation coefficients of the arrival and departure processes; the relative errors of the approximate approach are depicted in Figure 7.8. It can be observed that the error grows as large as about 12%, with the largest errors occurring when both interarrival and service time variations are slightly larger than 1.2. Clearly, these figures are not representative—the approximation results can be better or worse, depending on the queue utilization and the buffer size. However, it must be expected that errors accumulate when analyzing networks of nodes, so that the improvement can be much larger in practice.

The above results were also confirmed by the studies in [173], where the exact solution method was used to replace the approximations made in the finite buffer version of QNA which was presented in Section 7.2.2. There, the same 9-node network has been considered that was investigated in [87]. Replacing the approximate formulae by the QBD-based approach, the relative error in the computation of the expected waiting time never exceeded 0.43% (at any node) when compared to exact simulation results. This represents a substantial improvement to the results in [87], where errors as large as 8% are reported.

**Calculating the Loss Process.** In many modeling situations, it is desirable that jobs that encounter a full queue upon arrival are not discarded, but are instead diverted to other nodes (e.g., specialized nodes dedicated to deal with overload situations). Also, the properties of the loss process itself are often of interest. To account for this, another improvement to QNA with finite buffers has been suggested in [173]; it is also based on the QBD Markov chain underlying the PH|PH|1|$k$ queue.

The key idea is that the entire finite QBD Markov chain is used to represent a phase-type distribution $(\boldsymbol{\delta}, \mathbf{D})$ which models the loss process. The transition matrix of this phase-type distribution is obtained by slightly modifying the bottom right submatrix of the QBD generator (7.15) to account for the fact that upon an arrival at a full buffer, the arrival process is no longer restarted; instead, absorption takes place. This leads to

$$\mathbf{D} = \begin{pmatrix} \mathbf{A} & \mathbf{A}^0\boldsymbol{\alpha} \otimes \boldsymbol{\beta} & & & \\ \mathbf{I} \otimes \mathbf{B}^0 & \begin{matrix}\mathbf{A} \otimes \mathbf{I} \\ +\mathbf{I} \otimes \mathbf{B}\end{matrix} & \mathbf{A}^0\boldsymbol{\alpha} \otimes \mathbf{I} & & \\ \mathbf{0} & \mathbf{I} \otimes \mathbf{B}^0\boldsymbol{\beta} & \begin{matrix}\mathbf{A} \otimes \mathbf{I} \\ +\mathbf{I} \otimes \mathbf{B}\end{matrix} & \ddots & \\ & & \ddots & \ddots & \mathbf{A}^0\boldsymbol{\alpha} \otimes \mathbf{I} \\ & & \mathbf{I} \otimes \mathbf{B}^0\boldsymbol{\beta} & & \begin{matrix}\mathbf{A} \otimes \mathbf{I} \\ +\mathbf{I} \otimes \mathbf{B}\end{matrix} \end{pmatrix}. \tag{7.23}$$

In order to determine the subvectors of the initial distribution vector $\boldsymbol{\delta} = (\boldsymbol{\delta}_b, \boldsymbol{\delta}_0, \cdots, \boldsymbol{\delta}_{k-1})$, the system must be considered at time instants where an arriving customer encounters a full buffer. This means that

- the queue contains $k$ customers, thus $\boldsymbol{\delta}_b, \boldsymbol{\delta}_0, \cdots, \boldsymbol{\delta}_{k-2} = \mathbf{0}$.
- The state probabilities of the states forming the phase-type presentation of the service process are distributed according to $\mathbf{v}_{\mathrm{a}k-1}$ (see Eq. (7.17)), and
- the arrival process restarts with initial probability vector $\boldsymbol{\alpha}$.

The last two points lead to $\boldsymbol{\delta}_{k-1} = \boldsymbol{\alpha} \otimes \mathbf{v}_{\mathrm{a}k-1}$. After renormalization, this yields the final $\boldsymbol{\delta}$:

$$\boldsymbol{\delta} = \frac{1}{(\boldsymbol{\alpha} \otimes \mathbf{v}_{\mathrm{a}k-1})\mathbf{1}}(\mathbf{0}, \cdots, \mathbf{0}, \boldsymbol{\alpha} \otimes \mathbf{v}_{\mathrm{a}k-1}). \tag{7.24}$$

Clearly, for large queues, the order of the phase-type representation of the interdeparture time distribution becomes exceedingly large. Nevertheless, the approach is feasible up to reasonably large buffer sizes, and the experimental investigations in [173] show that the approach provides a very accurate estimation of the moments of the loss process.

**Determining the Steady-State Distribution.** The above methods for obtaining the moments of the node departure process and for deriving the loss process require the steady-state distribution of the QBD process associated to the PH|PH|1|$k$ queue. Naturally, this can be achieved by applying any of the methods that were discussed in Chapter 4. While these approaches aim towards the solution of general QBDs, a solution that is specialized on those QBDs arising from PH|PH|1|$k$ queues has been proposed in [32]. As the description in [32] is unclear in several points, the algorithm is briefly reviewed here.

In order to determine the QBD's steady state solution $(\mathbf{b}, \mathbf{v}_0, \cdots \mathbf{v}_{k-1})$, the approach first introduces two matrices $\tilde{\mathbf{A}}, \tilde{\mathbf{B}} \in \mathbb{R}^{l\times m}$ with

$$\tilde{\mathbf{A}} = (\mathbf{1}_l\boldsymbol{\alpha} - \mathbf{I}_l) \otimes \mathbf{B} - \mathbf{A} \otimes \mathbf{I}_m \quad \text{and} \quad \tilde{\mathbf{B}} = \mathbf{A} \otimes (\mathbf{1}_m\boldsymbol{\beta} - \mathbf{I}_m) - \mathbf{I}_l \otimes \mathbf{B}. \tag{7.25}$$

These are used to define the matrices $\mathbf{W}_0 \in \mathbb{R}^{l\times lm}$ and $\mathbf{W}, \mathbf{W}_{k-1} \in \mathbb{R}^{lm\times lm}$:

$$\mathbf{W}_0 = -(\mathbf{A} \otimes \boldsymbol{\beta})\tilde{\mathbf{B}}^{-1} \tag{7.26}$$

$$\mathbf{W} = \tilde{\mathbf{A}}\tilde{\mathbf{B}}^{-1} \tag{7.27}$$

$$\mathbf{W}_{k-1} = -(\mathbf{A}^0\boldsymbol{\alpha} \otimes \mathbf{I}_m)[(\mathbf{A} + \mathbf{A}^0\boldsymbol{\alpha}) \otimes \mathbf{I}_m + \mathbf{I}_l \otimes \mathbf{B}]^{-1}. \tag{7.28}$$

Now, let the vector $\mathbf{u} \in \mathbb{R}^l$ represent the steady-state solution of a continuous-time Markov chain with generator $\mathbf{A} + \mathbf{A}^0\boldsymbol{\alpha}$, so that

$$\mathbf{u}(\mathbf{A} + \mathbf{A}^0\boldsymbol{\alpha}) = \mathbf{0} \quad \text{and} \quad \mathbf{u}\mathbf{1}_l = 1 \tag{7.29}$$

holds, and define the matrix $\mathbf{V} \in \mathbb{R}^{l\times l}$ by

$$\mathbf{V} = \mathbf{I} + \left(\sum_{i=0}^{k-2} \mathbf{W}_0\mathbf{W}^i + \mathbf{W}_0\mathbf{W}^{k-2}\mathbf{W}_{k-1}\right)(\mathbf{I}_l \otimes \mathbf{1}_m) \tag{7.30}$$

$$= \mathbf{W}_0\left[\left(\sum_{i=0}^{k-2} \mathbf{W}^i\right) + \mathbf{W}^{k-2}\mathbf{W}_{k-1}\right](\mathbf{I}_l \otimes \mathbf{1}_m). \tag{7.31}$$

According to [32], the boundary solution vector $\mathbf{b}$ is then derived by solving the following set of equations:

$$\mathbf{b}\mathbf{V} = \mathbf{u}. \tag{7.32}$$

The remaining steady-state probability vectors $\mathbf{v}_i$ are then given by

$$\mathbf{v}_i = \begin{cases} \mathbf{b}\mathbf{W}_0\mathbf{W}^i, & i = 0, \cdots, k-2 \\ \mathbf{b}\mathbf{W}_0\mathbf{W}^{k-2}\mathbf{W}_{k-1}, & i = k-1. \end{cases} \tag{7.33}$$

Note that this solution also exhibits a matrix-geometric form, except for the last vector $\mathbf{v}_{k-1}$.

Unfortunately, this solution approach exhibited severe numerical instabilities in our experiments [172], so that the application of matrix-geometric methods for the analysis of general QBD processes seems preferable.

### 7.2.5 Conclusion

In this section, we first introduced to the queueing network analyzer, an efficient framework for the approximate analysis of open queuing networks which accounts for the first two moments of all involved distributions. Based on an approach for accounting for finite buffers, a QBD description for the underlying stochastic process of the network nodes has been derived. Subsequently, it has been shown how this representation can be used to further increase the modeling expressiveness and the accuracy of the QNA approach.

The QBD-oriented view has been adopted for deriving node-specific performance measures and to determine the moments of the interdeparture time distribution. Individual nodes are now described in terms of a service time distribution which is specified as an arbitrary phase-type distribution. While this is much more flexible than the moment-based approach used in standard QNA, the question arises how appropriate detailed phase-type descriptions of the service time distribution can be derived. As the iSPN framework presented in Chapter 5 represents a convenient way to model complex servers as QBD processes, this problem can be resolved by describing nodes in terms of iSPNs; this will be discussed in Section 7.3.

Another limitation of QNA, which has not been addressed in this section, is the fact that all traffic streams are still only described in terms of their first two moments, thereby employing the original QNA approximations for splitting and merging traffic streams. As all distributions within the QBD-based approach are already matched by appropriate phase-type distributions, it would be reasonable to stay within the phase-type domain during all computations and to avoid additional transformation steps. Approaches to realize this will be discussed in Section 7.4.

## 7.3 Embedding iSPNs

In Section 7.2, substantial improvements to the QNA method could be reached by considering network nodes as QBD processes. These QBDs were derived by matching all involved distributions (characterized by their first two moments) by appropriate phase-type distributions. This way, complex

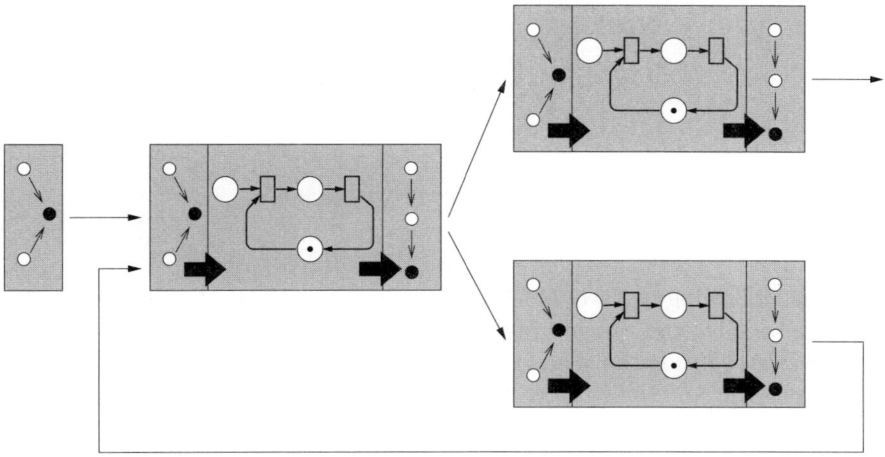

**Fig. 7.9.** The modeling framework that results from employing iSPN node models and using phase-type descriptions for all occurring traffic streams.

service time distributions could already be accounted for; however services must comply to the renewal assumption, which is often not given in the context of complex servers. Furthermore, the question arises how complex phase-type descriptions of a server can be derived.

In this section, we therefore suggest to employ the high-level iSPN approach established in Chapter 5 for the specification of node models. As the underlying Markov chain is no longer restricted to those Markov chains underlying PH|PH|1 queueing stations, the approaches presented in Section 7.2 can no longer be applied. As a result, the following two issues need to be tackled:

- At the iSPN level, interfaces must be identified in order to specify where arriving jobs are to be received (or "queued"), and what transitions correspond to job departures from the node.
- Approaches need to be developed for deriving the departure process of an iSPN node subject to a given arrival process.

For the time being, we assume that arrival and departure processes are described in terms of phase-type interarrival- and interdeparture time distributions. The modeling framework that then results from employing iSPN-based node descriptions is depicted in Figure 7.9: Incoming traffic streams are described in terms of a phase-type distribution, and are subsequently processed by an iSPN node model. The analysis of this system leads to a departure process description, which is again of phase-type. After splitting and merging operations are applied, these traffic streams form the new arrival streams to subsequent nodes.

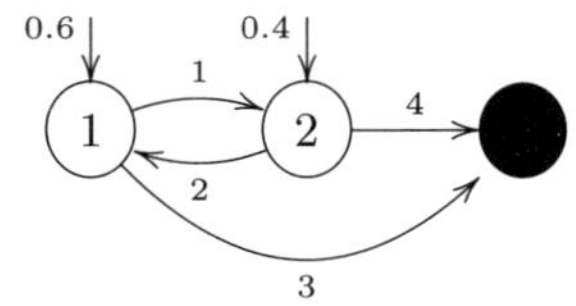


**Fig. 7.10.** An example arrival phase-type distribution.

In the following two sections, we focus on the analysis of individual nodes in order to solve the two above-mentioned issues, corresponding to the two thicker arrows inside the shaded boxes in Figure 7.9. For inter-node relationships, merging and splitting of traffic streams needs to be considered; this issue will be covered later in Section 7.4.

### 7.3.1 Job Arrivals

Assuming that the arrival process to an iSPN node is given in terms of a phase-type distribution, it needs to be specified how and where these jobs are to be queued in the iSPN node model.

To accomplish this, we suggest to build an overall iSPN model that comprises both the description of the arrival process and the node model. Both parts are connected by a designated place in the node model which receives the jobs generated by the arrival process. Note that this is exactly the approach taken in Chapter 6, where several example systems were investigated subject to different arrival processes. For example, the phase-type description of PSSPs (see Figure 6.21(a) on page 177) was transformed to a Petri net description, and then connected to the (exponential) server in Figure 6.23 by putting arriving jobs into the unbounded buffer place.

We now formally generalize this approach for arbitrary arrival phase-type distributions. To this end, we consider a phase-type interarrival time distribution $(\boldsymbol{\alpha}, \mathbf{A})$, with $\boldsymbol{\alpha} = (\alpha_1, \cdots, \alpha_N)$, $\mathbf{A} = [a_{i,j}]$ and $\mathbf{A}^0 = (a_1^0, \cdots, a_N^0)^{\mathrm{T}}$, where $i, j \in \{1, \cdots, N\}$. Assuming that the designated arrival place in the iSPN is $\mathsf{P}_{\mathrm{arr}}$, an overall iSPN can be constructed in three steps as shown below. To illustrate these steps, we apply them to the example phase-type distribution illustrated in Figure 7.10. The resulting iSPN is shown in Figure 7.11.

*Step 1: Transient Part.* Transient transitions of the phase-type distribution are integrated in the iSPN by introducing a set of additional places $\mathsf{P}_i$ for each transient state of the phase-type distribution, and by adding transitions $\mathsf{T}_{i,j}$ which reflect transitions between them.

Thus, for each positive non-zero rate $a_{i,j}$ in the transition matrix $\mathbf{A}$ of the arrival phase-type distribution, a transition $\mathsf{T}_{i,j}$ is added to the overall iSPN, and its input and output arcs and its firing rate are set accordingly:

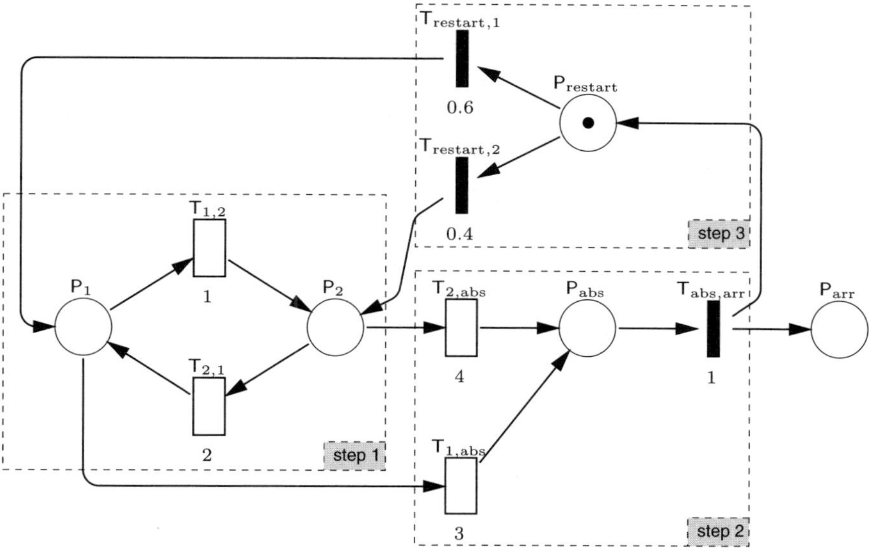


**Fig. 7.11.** The three main steps for connecting the phase-type interarrival time distribution in Figure 7.10 to an iSPN input place.

- $P := P \cup \{\mathsf{P}_1, \cdots, \mathsf{P}_N\}$
- $T := T \cup \{\mathsf{T}_{i,j} | a_{i,j} > 0\}$
- $I := I \cup \{(\mathsf{P}_i, \mathsf{T}_{i,j}) | a_{i,j} > 0\}$,
  $O := O \cup \{(\mathsf{T}_{i,j}, \mathsf{P}_j) | a_{i,j} > 0\}$
- $\forall i, j : a_{i,j} > 0 \Rightarrow (W(\mathsf{T}_{i,j}) := a_{i,j} \wedge Pr(\mathsf{T}_{i,j}) := 0)$.

*Step 2: Absorption.* After accounting for the phase-type distribution's transient behaviour, the absorption case must also be accounted for. This is accomplished by introducing a place $\mathsf{P}_{\text{abs}}$ for representing the phase-type distribution's absorbing state. Positive rates in $\mathbf{A}^0$ are represented by transitions $\mathsf{T}_{i,\text{abs}}$ which are connected to places $\mathsf{P}_i$ and $\mathsf{P}_{\text{abs}}$. Furthermore, in order to generate an arrival upon absorption, an immediate transition $\mathsf{T}_{\text{abs,arr}}$ connects the absorption place to the iSPN's designated arrival place $\mathsf{P}_{\text{arr}}$:

- $P := P \cup \{\mathsf{P}_{\text{abs}}\}$
- $T := T \cup \{\mathsf{T}_{i,\text{abs}} | a_i^0 > 0\} \cup \{\mathsf{T}_{\text{abs,arr}}\}$
- $I := I \cup \{(\mathsf{P}_i, \mathsf{T}_{i,\text{abs}}) | a_i^0 > 0\} \cup \{(\mathsf{P}_{\text{abs}}, \mathsf{T}_{\text{abs,arr}})\}$,
  $O := O \cup \{(\mathsf{T}_{i,\text{abs}}, \mathsf{P}_{\text{abs}}) | a_i^0 > 0\} \cup \{(\mathsf{T}_{\text{abs,arr}}, \mathsf{P}_{\text{arr}})\}$
- $\forall i : a_i^0 > 0 \Rightarrow (W(\mathsf{T}_{i,\text{abs}}) := a_i^0 \wedge Pr(\mathsf{T}_{i,\text{abs}}) := 0)$,
  $W(\mathsf{T}_{\text{abs,arr}}) := 1 \wedge Pr(\mathsf{T}_{\text{abs,arr}}) := 1)$.

*Step 3: Restart.* Finally, after an arrival has been generated, the phase-type distribution needs to be restarted. This is accomplished by adding a place $\mathsf{P}_{\text{restart}}$ as second output place of $\mathsf{T}_{\text{abs,arr}}$. After an arrival, its purpose is to place a new token in one of the places $\mathsf{P}_1, \cdots, \mathsf{P}_N$ according to the initial probability vector $\boldsymbol{\alpha}$ of the arrival phase-type distribution. For this, additional transitions $\mathsf{T}_{\text{restart},i}$ are introduced:

- $P := P \cup \{\mathsf{P}_{\text{restart}}\}$
- $T := T \cup \{\mathsf{T}_{\text{restart},i} | \alpha_i > 0\}$
- $I := I \cup \{(\mathsf{P}_{\text{restart}}, \mathsf{T}_{\text{restart},i}) | a_i^0 > 0\}$,
  $O := O \cup \{(\mathsf{T}_{\text{restart},i}, \mathsf{P}_i) | a_i^0 > 0\} \cup \{(\mathsf{T}_{\text{abs,arr}}, \mathsf{P}_{\text{restart}})\}$
- $\forall i : \alpha_i > 0 \Rightarrow (W(\mathsf{T}_{\text{restart},i}) := \alpha_i \wedge Pr(\mathsf{T}_{\text{restart},i}) := 1)$.

We remark that the initial number of tokens in all newly introduced places equals zero, except for the restart place $\mathsf{P}_{\text{restart}}$. This place initially holds one token to start the generation of job arrivals; subsequently, this token "cycles" through the constructed sub-iSPN to continuously generate arrivals.

Now, the complete iSPN node model including the arrival process description can be analyzed to obtain node-specific performance measures. A key issue which needs to be determined is the interdeparture time distribution; this is the focus of the next section.

### 7.3.2 Departure Process Derivation

Similar to introducing a designated arrival place as an "interface" for external jobs which are flowing into an iSPN node, we propose that the iSPN model also contains a set of designated *departure transitions*. Each firing of such a transition will then be associated to a job departure from the node. Clearly, this leads to a set of designated transitions between states in the underlying QBD as well.

In the following, we assume that the underlying QBD has been constructed and that the "departure" transitions in this QBD have been determined. We now describe how a phase-type description of the interdeparture time distribution can be derived; this is accomplished in two steps. First, we deal with the special case that departures always correspond to transitions to lower QBD levels; then, we discuss how to deal with situations where this assumption does not hold.

**Case 1: Departures Lead to Lower QBD Levels.** For the time being, we assume that the underlying QBD of the considered iSPN contains no boundary states states (i.e., $N_0 = 0$, see our remark on page 54); this leads to a QBD with the following generator:

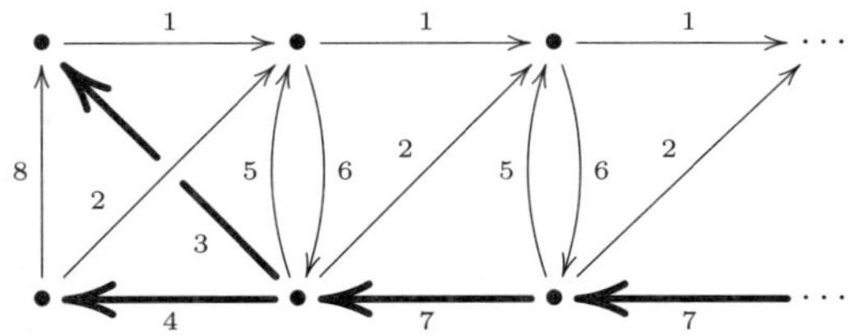


**Fig. 7.12.** An example QBD process from which the interdeparture time distribution is derived; departure transitions are emphasized.

$$\mathbf{Q} = \begin{pmatrix} \mathbf{B}_{1,1} & \mathbf{A}_0 & & & \\ \mathbf{B}_{2,1} & \mathbf{A}_1 & \mathbf{A}_0 & & \\ & \mathbf{A}_2 & \mathbf{A}_1 & \mathbf{A}_0 & \\ & & \ddots & \ddots & \ddots \end{pmatrix}. \tag{7.34}$$

Note that all submatrices are square and of dimension $N$ in this case. In the following, we illustrate our considerations using the example QBD shown in Figure 7.12.

If the QBD is observed at those time instants immediately following a departure, then the time to the next departure can be described in terms of a phase-type distribution. This distribution depends on the level which has been reached by the last departure: the boundary level or a on-boundary level. We consider these two cases below.

- If the last departure event led to repeating level $i$, where $i \geq 1$, then the time to the next level-decreasing transition does not depend on the actual level number. This is due to the fact that for all repeating levels $i$ with $i \geq 2$, downward transitions always take place according to the same matrix $\mathbf{A}_2$ (see also Eq. (7.34)).
  In this case, the time to the next departure (or level-down transition) is described by a phase-type distribution with $N$ states, corresponding to the states in the repeating levels of the QBD. Absorption takes place from those states that correspond to states in the QBD which lead to a lower level. As all these transitions are contained in matrix $\mathbf{A}_2$, this leads to the absorption vector $\mathbf{A}^0 = \mathbf{A}_2\mathbf{1}$.
  The same argument holds in case the last departure led to repeating level $i = 1$. Then, departures take place according to $\mathbf{B}_{2,1}$; however, $\mathbf{B}_{2,1}\mathbf{1}$ must be equal to $\mathbf{A}_2\mathbf{1}$, since the row sums in all blocks-rows in (7.34) that correspond to levels $i \geq 1$ must be identical (due to constant diagonal entries of $\mathbf{A}_1$), so that the same absorption vector is obtained.
  The transient transitions of the phase-type distribution naturally correspond to transitions that take place within the repeating levels, described

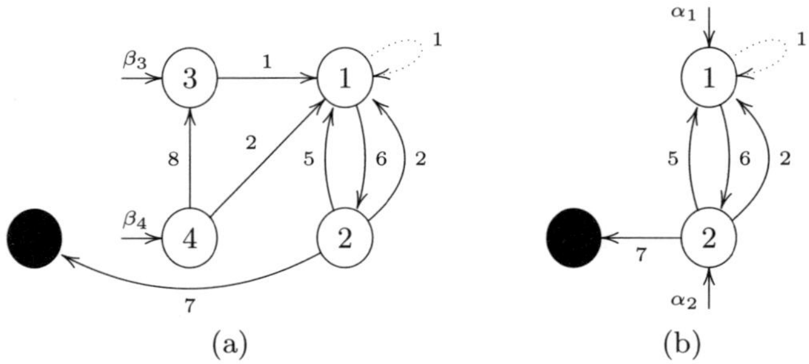


**Fig. 7.13.** Phase-type distribution for the time to the next departure under the condition that the last departure led to the boundary level (a) and under the condition that it did not reach the boundary level (b).

by $\mathbf{A}_1$. Finally, transitions to upper levels (according to $\mathbf{A}_0$) need only to be accounted for in the phase-type distribution if they lead to a different state within a repeating level. Therefore, transient transitions are described by $\mathbf{A}_1 + \mathbf{A}_0$, and the following phase-type distribution is obtained:

$$\mathcal{D}_1 = ((\alpha_1, \cdots, \alpha_N), \mathbf{A}_1 + \mathbf{A}_0). \tag{7.35}$$

The derivation of the initial probability vector $(\alpha_1, \cdots, \alpha_N)$ will be discussed later.
For the example QBD in Figure 7.12, the resulting phase-type distribution is illustrated in Figure 7.13(b). Note that the dotted arc indicates a transition to an upper level that does not change the state within a repeating level; it is only depicted for illustration.

- If the last departure transition led to the boundary level (repeating level 0), then a transition to repeating level 1 must occur before the next departure can take place (corresponding to an arrival event). This leads to a phase-type distribution of order $2N$, where, after a transient phase according to $\mathbf{B}_{1,1}$, a second transient phase is entered according to $\mathbf{A}_0$. This second transient phase, and absorption from it, takes place according to the considerations presented above for departures from non-boundary levels. This leads to the following phase-type distribution:

$$\mathcal{D}_0 = ((\beta_1, \cdots, \beta_N, 0, \cdots, 0), \begin{pmatrix} \mathbf{B}_{1,1} & \mathbf{A}_0 \\ \mathbf{0} & \mathbf{A}_1 + \mathbf{A}_0 \end{pmatrix}). \tag{7.36}$$

Note that the initial probabilities for the states belonging to the second transient phase are zero due to the assumption that the last departure led to the boundary level.
For the example QBD in Figure 7.12, these considerations lead to the phase-type distribution in Figure 7.13(a). Note that in the example, the

states have been renumbered so that the phase-type distribution in Figure 7.13(b) can easily be recognized.

What remains to be determined are the initial probabilities for these two phase-type distributions. They can be derived from considering the steady-state distribution of the embedded Markov chain which is obtained by looking at the original QBD at time instants *immediately following a departure.* We will denote this steady-state distribution by the infinite vector $(\mathbf{y}_0, \mathbf{y}_1, \cdots)$.

In [152, Theorem 3.5.1], the embedded Markov chain of a QBD observed *immediately after left transitions* is considered, and expressions for its steady-state solution are provided. This is the reason why we assumed that job departures from the considered iSPN correspond to transitions to lower levels: we can now use these expressions to derive the vector $(\mathbf{y}_0, \mathbf{y}_1, \cdots)$. According to [152], it is given by

$$\mathbf{y}_0 = c\mathbf{v}_0\mathbf{R}\mathbf{B}_{2,1} \tag{7.37}$$

$$\mathbf{y}_i = c\mathbf{v}_0\mathbf{R}^{i+1}\mathbf{A}_2, \qquad i \geq 1 \tag{7.38}$$

where

$$c = \left[\mathbf{v}_0(\mathbf{I} - \mathbf{R})^{-1}\mathbf{A}_0\mathbf{1}\right]^{-1}. \tag{7.39}$$

Here, $\mathbf{v}_0$ and $\mathbf{R}$ correspond to the usual steady-state solution of the QBD as derived with the methods in Chapter 4.

Clearly, the initial probability distribution for the phase-type distribution $\mathcal{D}_0$ is obtained by renormalizing $\mathbf{y}_0$, since $\mathbf{y}_0$ holds the probability of being in any of the boundary states after a departure took place. Thus, $(\beta_1, \cdots, \beta_N) = \frac{1}{|\mathbf{y}_0|}\mathbf{y}_0$ holds.

For $\mathcal{D}_1$, the probability vectors $\mathbf{y}_i, i \geq 1$ of all non-boundary levels need to be added. We denote this sum by $\mathbf{y}_{1+}$, and obtain

$$\mathbf{y}_{1+} = \sum_{i=1}^{\infty} \mathbf{y}_i \tag{7.40}$$

$$= c\mathbf{v}_0 \left( \sum_{i=1}^{\infty} \mathbf{R}^{i+1} \right) \mathbf{A}_2 \tag{7.41}$$

$$= c\mathbf{v}_0\mathbf{R}^2(\mathbf{I} - \mathbf{R})^{-1}\mathbf{A}_2. \tag{7.42}$$

Renormalizing, the initial probability vector for $\mathcal{D}_1$ is then obtained as $(\alpha_1, \cdots, \alpha_N) = \frac{1}{|\mathbf{y}_{1+}|}\mathbf{y}_{1+}$.

To obtain the overall interdeparture time distribution of the considered QBD, the distributions $\mathcal{D}_0$ and $\mathcal{D}_1$ need to be mixed with weighting factors according to the probability mass in $\mathbf{y}_0$ and $\mathbf{y}_{1+}$, respectively. Since the phase-type representation of $\mathcal{D}_1$ is a subset of the one for $\mathcal{D}_0$, this mixture can conveniently be described by using non-zero initial probabilities for the second transient phase of $\mathcal{D}_0$. Since the weights in the mixing operation cancel the renormalization steps performed on $\mathbf{y}_0$ and $\mathbf{y}_{1^+}$, the overall interdeparture phase-type distribution $\mathcal{D}$ is obtained as

$$\mathcal{D} = ((\mathbf{y}_0, \mathbf{y}_{1^+}), \begin{pmatrix} \mathbf{B}_{1,1} & \mathbf{A}_0 \\ \mathbf{0} & \mathbf{A}_1 + \mathbf{A}_0 \end{pmatrix}) \tag{7.43}$$

$$= (c\mathbf{v}_0\mathbf{R}(\mathbf{B}_{2,1}, \mathbf{R}(\mathbf{I}-\mathbf{R})^{-1}\mathbf{A}_2), \begin{pmatrix} \mathbf{B}_{1,1} & \mathbf{A}_0 \\ \mathbf{0} & \mathbf{A}_1 + \mathbf{A}_0 \end{pmatrix}). \tag{7.44}$$

It should be noted that $\mathcal{D}$ is an *exact* description of the interdeparture time distribution; no approximate steps have been employed in its derivation. In particular, as it is obtained from looking at the embedded QBD at departure instants, this approach does not suffer from neglecting the state of the arrival process at departure instants. This is in contrast to the approximation made in the approach presented in Section 7.2.2, which can imply significant errors, as shown in Figure 7.8.

We note that the assumption of $N_0 = 0$ does not restrict the results presented here. While it simplifies the presentation, the approach in [152, Theorem 3.5.1] can be extended to account for non-empty boundaries as well. However, non-empty boundaries will increase the order of the obtained phase-type distribution, as the (non-regular) transitions from and to the boundary level need to be accounted for in a similar way as we dealt with the non-regular behaviour of the border level.

It is important to note that our phase-type description of the interdeparture time distribution is exact, but still represents an approximation of the departure *process*, as it is approximated by a renewal process. Thus, this description is not capable to account for correlations between successive departures. This becomes clear when considering the fact that the time to the next departure depends on whether the last departure led to the boundary level of the QBD or not. If correlations are important, one approach to improve this situation is to resort to MAP (Markovian Arrival Process, see Appendix A.3) descriptions of the output process instead of employing a phase-type description. A suitable MAP can be derived in a very similar way to the approach shown here, but in addition, it can provide an additional amount of "memory" for remembering whether the last departure led to the boundary level or not. This approach can even be extended to account for more levels, albeit at the cost of requiring more states for representing the distribution. Therefore, such an approach would provide a means to balance the complexity of the output process description against the degree to which correlations need to be accounted for.

**Case 2: Mapping Departures to Lower-Level Transitions.** An important assumption which has been made in the above considerations for obtaining the interdeparture time distribution is that those transitions in the QBD which correspond to the firing of a designated departure transition in

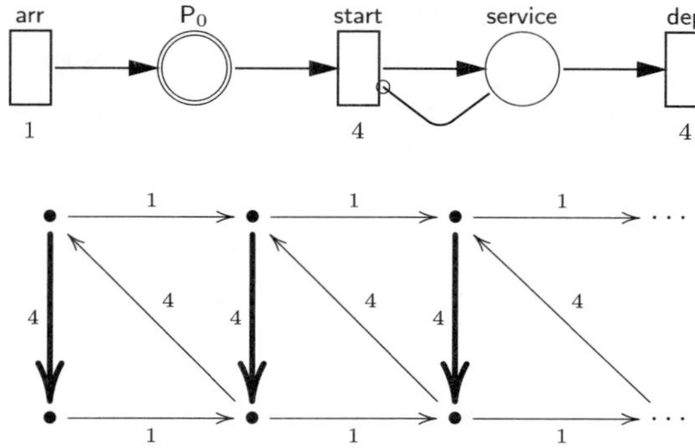


**Fig. 7.14.** The iSPN for a M|E$_2$|1 system and the underlying QBD. Note that departures via transition dep do not map to lower-level transitions in the QBD. The lower left state of the QBD corresponds to the initial marking as indicated in the Petri net above.

the iSPN always lead to lower QBD levels. However, this is a situation which will not be given in most practical cases.

In the algorithms for transforming iSPNs to QBDs which have been developed in Chapter 5, all markings in one QBD level correspond to the same number of tokens in the unbounded iSPN place $P_0$. If departures coincide with transitions to lower QBD levels, this means that a departure is "generated" as soon as token is taken out of the unbounded place $P_0$. Now, in most modeling applications, the place $P_0$ represents some sort of input buffer from which jobs are removed for further processing. Thus, a departure is usually not generated as soon as a token has been taken out of $P_0$, but only *after its processing has been completed.* As an example, consider the iSPN specification of a M|E$_2$|1 queueing station and its underlying QBD in Figure 7.14. As can be observed, transitions to lower QBD levels take place as soon as the first service phase starts; those transitions which correspond to actual departures from the iSPN via transition dep occur *within* a QBD level.

From these considerations, we conclude that

1. departure transitions at the iSPN level usually do not lead to lower levels in the underlying QBD; instead,
2. departure transitions take place *within* a QBD level in most cases.
3. Departure transitions at the iSPN level rarely lead to higher levels in the underlying QBD. While new tokens may be put into the unbounded place by the iSPN (server) model itself, this most often represents some kind of re-scheduling a job for processing (as an example, see the iSPN discussed in Section 5.1.4). In practical applications, such a step will rarely be used to characterize a job *departure.*

In the first case, the presented approach for deriving the interdeparture time distribution can be applied directly. The third case will not be considered

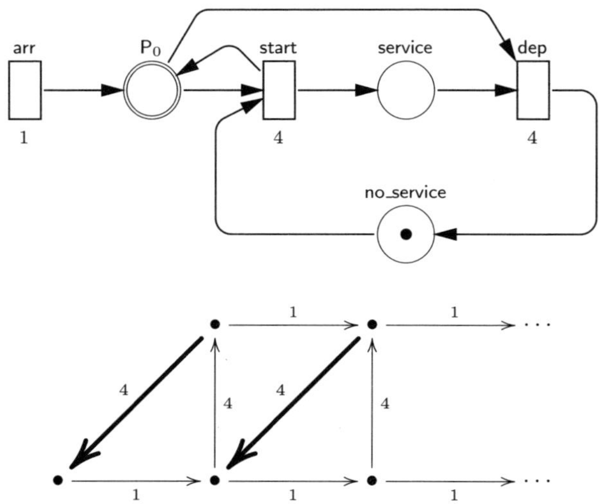


**Fig. 7.15.** Modified iSPN for a M|E$_2$|1 system and its underlying QBD. Departures via transition dep now correspond to lower-level transitions in the QBD.

here, as it is of low practical importance and since it is difficult to tackle as well. For dealing with the second case, we will now discuss three different approaches: adapting the iSPN model, restructuring the underlying QBD process, and allowing for more general departure transitions in the QBD.

*Adapting the iSPN Model.* The reason why departures in iSPNs like the one shown in Figure 7.14 usually do not to correspond to transitions to lower QBD levels is that a token that represents a job to be processed is *immediately* removed from the input buffer as soon as servicing starts. Subsequently, the corresponding token circulates through the service part of the iSPN. The actual departure of the job from the system then takes place at a later time.

It is possible to circumvent this situation at the iSPN level by removing the job from the input buffer only after the service has actually been completed. For the M|E$_2$|1 example, this approach is illustrated by the iSPN in Figure 7.15. Similar to the initial iSPN in Figure 7.14, the input place is checked for the presence of any tokens that need to be serviced. However, if there are tokens to be processed, they are not removed immediately; instead, an additional token for tracking the current servicing status is taken from the new place no_service. The job is removed from the input buffer only after the last service phase has been completed. As can be observed in the underlying QBD in Figure 7.15, this leads to a situation where job departures correspond to left transitions in the underlying QBD, so that the procedure described under Case 1 can be applied.

Note that the approach has the additional benefit that the actual number of jobs in the system can be determined directly by looking at the buffer

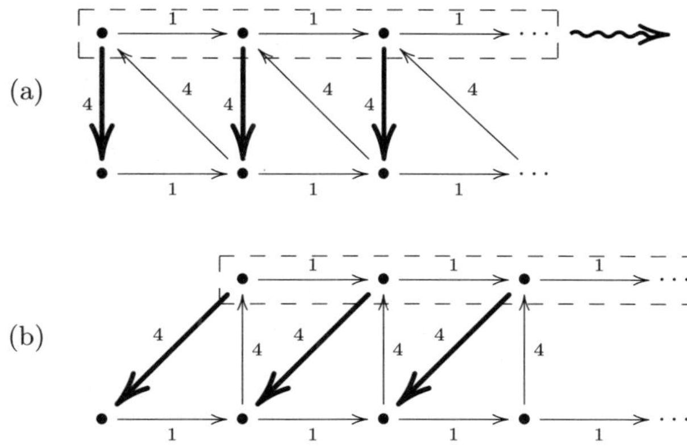


**Fig. 7.16.** An example application of the shifting operation.

place ($\mathsf{P}_0$ in the example). Thus, it is no longer necessary to consider all places which are involved in servicing as well. In practice, this also simplifies the analysis of the iSPN from the modeler's point of view in many cases.

However, even though the proposed method represents a feasible approach to obtain the desired QBD structure, it requires additional understanding and effort on behalf of the system modeler. The modeler must be aware of properties of the underlying QBD in order to ensure that all transitions that remove tokens from the unbounded place do correspond to job departures. Thus, to a certain extent, the suggested approach contradicts the main idea of using iSPNs as an abstraction of QBD processes designed to hide the special properties of this class of stochastic processes from the modeler.

*Restructuring the QBD.* Instead of changing the high-level iSPN model of a system, another approach to obtain a QBD structure where job departures correspond to transitions to lower levels is based on directly operating on the state space of the QBD so that the desired structure is obtained.

We will refer to this approach as *shifting*, as it involves moving "rows" of (an infinite number of) states in the QBD to the right. In situations where node departure transitions in a QBD take place within the same QBD level, the basic idea is that all states from which such a departure transition originates are moved to the next higher level. For the example QBD obtained from the $\mathrm{M}|\mathrm{E}_2|1$ iSPN in Figure 7.14, this procedure is illustrated in Figure 7.16. As can be observed, intra-level transitions corresponding to job departures in Figure 7.16(a) are transformed to left-transitions in the QBD by shifting the upper row of states to the right.

For more complex systems, the application of shifting is more complicated. In addition to transforming intra-level transitions to level-down transitions in the QBD, it must be guaranteed that there are no other transitions to lower QBD levels that do not correspond to job departures. To accomplish this task, further shifting operations may be necessary. In fact, this would already be the case when considering the QBD underlying a M|E$_3$|1 system, where it is necessary to shift two rows to the right (corresponding to the first two service stages).

Also, care must be taken when performing the shifting step. First, shifting itself may introduce additional transitions to lower levels which are not desired, and which need to be removed by further shifting steps. Secondly, shifting moves states in the QBD to higher levels, which can lead to non-skip-free QBDs with jump sizes larger than one iff the following two conditions are fulfilled:

- the shifted states are reached from states in lower levels, and
- the states from which the shifted states are reached are not shifted themselves.

The second condition is the reason why no problem was encountered in the example in Figure 7.16, as all lower level predecessor states that led to the departure states (with rate 1) have been shifted as well. Conversely, a non-skip-free QBD would be obtained if the departure states were reached from states in the lower row of the QBD in Figure 7.16.

Summarizing, shifting can help to transform the QBD to the desired form in many cases; however, it remains to be investigated under what conditions the (repeated) application of shifting operations can be successful.

*Arbitrary Departure Transitions.* The most powerful and challenging approach for dealing with the requirement that node departures coincide with transitions to lower QBD levels is to entirely drop this requirement. To accomplish this, the approach presented in [152, Theorem 3.5.1] needs to be extended to derive the steady-state distribution of the embedded QBD not only at left transitions, but at arbitrarily selected transitions.

Clearly, this represents an extension for which further investigations need to be performed. However, once appropriate approaches are available, they provide a number of substantial benefits:

- No assumptions regarding the structure of the iSPN or its underlying QBD need to be made.
- At the iSPN level, different sets of designated departure transitions can be introduced to specify different interdeparture time distributions according to different job classes or routing directions. Therefore, the development of algorithms for splitting an overall departure traffic stream is no longer necessary, and further approximations can be avoided.

As mentioned above, further research is need in this area. However, the fact that the position of departure transitions in the underlying QBD of an iSPN will exhibit the same regular structure as all other transitions is encouraging.

### 7.3.3 Conclusion

Several approaches for embedding iSPN nodes in a parametric decomposition framework have been developed. While the integration of phase-type interarrival time distributions can be performed in a straightforward manner, the derivation of the departure stream of a node is more difficult.

Under the assumption that job departures correspond to left transitions in the underlying QBD process of an iSPN, an exact result has been obtained to derive a phase-type representation of the interdeparture time distribution. Also, different approaches were discussed to deal with situations where this assumption is not given; however, further work is still needed for a general solution that does not assume special properties of the iSPN model.

It should be mentioned that the derived phase-type representations for the departure process can become very large, as their order equals at least the number of states in two repeating levels of the underlying QBD. As these distributions will be integrated as arrival processes in successor nodes in the network, this can lead to iSPN with very large underlying QBDs. We will further comment on this issue in the next section, where splitting and merging operations are discussed.

## 7.4 Splitting and Merging Traffic Streams

In this section, we focus on the necessary steps to account for inter-node traffic streams to complete the framework depicted in Figure 7.9. As we concentrate on phase-type distributions and MAPs for describing these traffic streams, appropriate splitting and merging operations for this class of renewal processes will be presented in Sections 7.4.1 and 7.4.2, respectively. From the considerations, it will become clear that the size of the involved traffic stream descriptions will increase, which represents a problem if a fixed-point iteration approach is employed. Therefore, approaches for dealing with this issue are discussed in Section 7.4.3.

### 7.4.1 Splitting

As illustrated in Section 7.1, a parametric decomposition framework requires that the departure traffic that leaves a node is split into several partial streams that are routed towards subsequent nodes. Similar to the QNA approach, we assume that splitting is performed according to a set of independent routing probabilities.

In accordance with the results of Section 7.3.2, we assume that the interdeparture times are given as a phase-type distribution $\mathcal{D} = (\boldsymbol{\alpha}, \mathbf{A})$. If routing to some other node is performed according to a probability $p$, then we need to derive a description of the traffic stream that represents only a fraction $p$ of the overall departure stream. The key idea to compute the corresponding

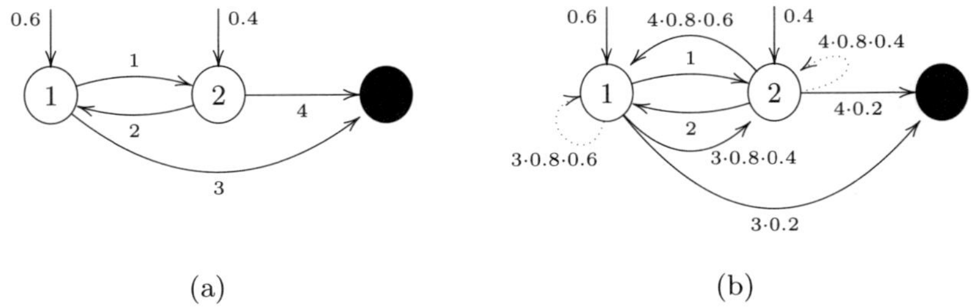


**Fig. 7.17.** Splitting the example phase-type distribution (a) according to the probability 0.2 yields the phase-type distribution (b). Note that the dotted arcs are only added for illustrating "pseudo absorption" from and to the same state.

interdeparture time distribution is to change the original distribution $\mathcal{D}$ so that the final transitions that lead to the absorbing state are only performed with probability $p$. With probability $1 - p$, the distribution is "restarted" by taking appropriate transitions to the initial states, but without actually entering the absorbing state.

As an example, consider splitting the phase-type distribution depicted in Figure 7.17(a) according to the splitting probability 0.2. As can be observed in the resulting phase-type distribution in Figure 7.17(b), the rates of the original transitions to the absorbing state have been multiplied by 0.2, and additional transitions have been added for "restarting" the distribution according to the initial probability vector $(0.6, 0.4)$ at 0.8 times the original absorption rates.

For a general distribution $\mathcal{D} = (\boldsymbol{\alpha}, \mathbf{A})$ and a splitting probability $p$, this approach can conveniently be described by the phase-type distribution

$$\mathcal{D}_p = (\boldsymbol{\alpha}, \mathbf{A} + (1-p)\mathbf{A}^0\boldsymbol{\alpha}). \tag{7.45}$$

We remark that the proposed splitting operation is exact for the considered class of departure processes with phase-type renewal times. Note that the results are in accordance to the (exact) formula (7.6) used by Whitt to determine the second moment of a split departure stream.

The arguments employed for the phase-type case can also be used for splitting MAP traffic streams. When splitting according to probability $p$, the rate of marked transitions is changed to $p$ times the original transition rates, and the remaining factor $(1 - p)$ of the original marked transition rates are replaced by non-marked transitions. For a MAP with description $(\mathbf{Q}_1, \mathbf{Q}_2)$, this leads to a split MAP with representation $(\mathbf{Q}_1 + (1-p)\mathbf{Q}_2, p\mathbf{Q}_2)$.

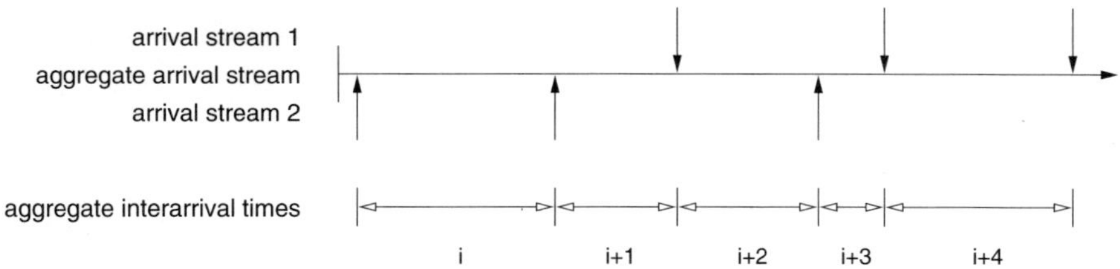


**Fig. 7.18.** Superposition of two arrival streams.

### 7.4.2 Merging

When several traffic streams are routed towards one node in a parametric decomposition framework, the resulting aggregate traffic stream needs to be considered in the node analysis step.

In contrast to the splitting operations, the superposition of several traffic streams with phase-type interarrival time distributions leads to an aggregate stream where interarrivals no longer form a renewal process. This becomes clear when considering the superposition of two arrival streams, as illustrated in Figure 7.18. After an arrival occurred due to one of the two superposed traffic streams (say, stream 2), the time to the next departure depends on the *residual* time to absorption in the other interarrival time distribution (stream 1). If this time is larger than the time to absorption in the stream that generated the first arrival, then the interarrival time is distributed according to the interarrival time of stream 2 (see interval $i$ in Figure 7.18). However, if the residual time of the other stream is smaller, then the time to the next arrival is distributed according to the residual time in the other stream (see e.g. intervals $i+1$ and $i+3$ in Figure 7.18).

It is, however, possible to derive an exact description of the aggregate arrival process in terms of a MAP. Considering two arrival processes with phase-type representation $(\boldsymbol{\alpha}, \mathbf{A})$ and $(\boldsymbol{\beta}, \mathbf{B})$, their respective MAP representations $(\mathbf{A}, \mathbf{A}^0\boldsymbol{\alpha})$ and $(\mathbf{B}, \mathbf{B}^0\boldsymbol{\beta})$ are used to let each distribution pass through its stages independently of the other. Using tensor products, this is conveniently described by the MAP

$$(\underbrace{\mathbf{A} \otimes \mathbf{I} + \mathbf{I} \otimes \mathbf{B}}_{\mathbf{Q}_1}, \underbrace{\mathbf{A}^0\boldsymbol{\alpha} \otimes \mathbf{I} + \mathbf{I} \otimes \mathbf{B}^0\boldsymbol{\beta}}_{\mathbf{Q}_2}). \tag{7.46}$$

We note that this MAP description of the aggregate process can also be used to derive a phase-type description of the aggregate arrival stream, thereby neglecting correlations among successive arrivals. It is obtained by replacing marked transitions by transitions to the phase-type distribution's absorbing state, leading to $\mathbf{Q}_1$ as transition matrix. To obtain the initial probabilities, the state probabilities of the MAP (7.46) at time instants immediately following a marked transition need to be considered. According to [29, Proposition 2], these are given by the steady-state distribution vector $\boldsymbol{\pi}$ of an embedded

Markov chain with transition probability matrix $\mathbf{P} = -\mathbf{Q}_1^{-1}\mathbf{Q}_2$, so that

$$\boldsymbol{\pi}\mathbf{P} = \boldsymbol{\pi} \tag{7.47}$$

$$\boldsymbol{\pi}\mathbf{1} = 1 \tag{7.48}$$

holds. Summarizing, this leads to the phase-type distribution $(\boldsymbol{\pi}, \mathbf{Q}_1)$ for the interarrival times of the superposed traffic stream.[1]

This approximation of the aggregate interarrival process by a renewal process is, however, not even necessary on our context. The MAP description in (7.46) can directly be integrated in the embedding process described in Section 7.3.1 to yield an overall iSPN that includes the aggregate MAP arrival process. This is easily accomplished by performing the three suggested steps for *each* of the individual arrival phase-type distributions. Thus, a construction like the one depicted in Figure 7.11 is added for each arrival stream.

Regardless of whether a MAP or a phase-type description is used to model the aggregate arrival process, the size of the resulting descriptions can become a problem in practice. Since both descriptions are derived from the tensor products in (7.46), their order equals the product of the orders the individual arrival processes. In the following section, we will discuss approaches to deal with this issue.

### 7.4.3 Dealing With the Distributional Explosion

A key problem that needs to be tackled when employing iSPN node descriptions and phase-type or MAP descriptions of the occurring traffic streams is that the order of the interarrival time distributions can grow very large. This is due to the following two reasons:

- As pointed out in Section 7.3, a phase-type description of the departure stream at a node comprises at least $2N$ states, where $N$ is the order of the underlying QBD of the iSPN model (which includes the arrival process description). For more complex boundaries or MAP descriptions that account for correlated interdeparture times, even more states will be necessary.
- The superposition of several traffic streams which are routed towards one iSPN node leads to a MAP or phase-type description of an order that equals the product of the orders of the interarrival time distributions of the superposed streams (see Section 7.4.2).

The problem that arises from these two points is that the integration of an interarrival time distribution of order $N$ into an iSPN model (as described in Section 7.3.1) increases the order of the underlying QBD process by a factor of $N$. Even in the context of a simple feed-forward network, the order of

[1] Note that this phase-type distribution corresponds exactly to what is termed the stationary interval by Whitt [195]. In Eq. (7.4), it is complemented with results from the *asymptotic method* in order to account for long-term correlations between arrival events in QNA's renewal framework.

the QBD of the next iSPN is thus increased by a factor of at least $2N$, as the departure process of the first node represents its arrival process. Clearly, the problem becomes even worse in an overall analysis framework that is based on fixed-point iteration: the distributions become substantially larger in each iteration step, as the distributions in one iteration are used to build the distributions of the next step.

Summarizing, with the exception of pathologically simple systems, the occurring QBD process will very quickly approach a size which is numerically intractable. The purpose of this section is therefore to discuss how the orders of the occurring distributions can be reduced. Once a procedure for reducing the order of the occurring distributions to some constant maximum size has been found, the "propagation" of larger and larger orders can be avoided. Clearly, this maximum order will be chosen so that the largest iSPN node model can still be analyzed when subject to arrival streams of that order.

**When to Reduce?** Two different approaches are conceivable for decreasing the order of a superposed arrival stream that is to be processed by an iSPN node:

- Reduction of the component arrival streams *before* performing the merging operation, or
- reducing the aggregate arrival stream that is derived *after* superpositioning the individual streams.

The advantage of the first approach is that the order of the individual distributions that need to be reduced is smaller than the order of the superposed arrival process. All reduction procedures must at least be of complexity $\mathcal{O}\left(N^2\right)$ (where $N$ denotes the order), since all transitions need to be considered in a reduction process. Therefore, the reduction of $m$ component arrival streams of order $N$ is of order $\mathcal{O}\left(m \cdot N^2\right)$, compared to $\mathcal{O}\left(m^2 \cdot N^2\right)$ if it is applied to the aggregate arrival stream.

The advantage of reducing the aggregate process, however, is that relations between different arrival streams can be taken into account. Since the reduction procedure operates on an "overall view", it can recognize correlations or replace several similar distributions by one distribution.

**Moment-Based Reduction.** Clearly, one objective of a reduction algorithm is that the mean interarrival time of the reduced process is identical to that of the original arrival process. This view can be extended so that complex arrival process descriptions are replaced by smaller ones which also approximate higher moments of the interarrival time distributions as good as possible.

For the case of phase-type interarrival times, moments of the original interarrival time distributions can easily be computed. Afterwards, a new interarrival time distribution of lower order can be obtained by applying appropriate matching algorithms like those presented in [14, 125]. However, further investigations are needed in order to extend this approach towards

MAP descriptions and to study the computational complexity of the employed matching procedures.

**Spectrum-Based Reduction.** Here, the fact is used that the transition matrices of a MAP and a phase-type distribution are closely related to the associated interarrival time distributions. Based on the eigenvalues and eigenvectors of these transition matrices, new arrival processes can be constructed that capture the properties of the original ones as good as possible. Taking phase-type interarrival time distributions as an example, a Taylor series similar to that of the exponential function can obtained (see [152]). As this series involves the powers of the phase-type distribution's transition matrix, the impact of larger eigenvalues on the distribution is much larger than that of smaller eigenvalues. Therefore, approaches are conceivable that reduce the order of the original phase-type representation by approximating it with a smaller one that matches the largest eigenvalues of the original representation.

In this context, interesting relations exist to approaches where stochastic processes are considered in the *frequency-domain*. In these approaches, the eigenvalues of the MAP transition matrices play a central role. As an example, a special class of MAPs has successfully been employed in [130] for modeling video traffic, where considerations in the frequency domain have been made in order to match real traffic measurements with the used MAP class. Further work is still needed on investigating how approaches like this can be used to derive more compact descriptions of given arrival processes.

**Graph-Based Reduction.** Both MAPs and phase-type distributions have an elegant graphical representation. By introducing equivalence classes of the states of these representations, their order can be reduced. To accomplish this, approaches need to be developed for building these equivalence classes. In this area, three approaches are basically conceivable:

- Graph-based algorithms which identify equivalences based on structural properties of the graphical arrival process representation.
- Approaches that account for the probabilistic "relevance" of the states in the distribution. As an example, the quasi-stationary probability distribution (see Section 4.6.3) among the transient states of a phase-type distribution can be used to estimate the importance of different states by providing the probability that a state is reached under the condition that absorption has not taken place yet.
- Heuristic optimization procedures can be employed to build equivalence classes. For example, in [179] genetic algorithms were successfully applied to find equivalence classes of states in stochastic automata. For such optimization-based approaches, suitable cost functions like those proposed in [100] need to be employed.

**Reduction Based on the Stationary Interval/Asymptotic Method.** In [29], so-called *Super-Erlang* (SE) chains are introduced as a special class of MAPs specifically designed towards matching the superposition process resulting from several renewal processes with phase-type renewal time distributions. The approximation procedure accounts both for the stationary interval variance and the asymptotic variance, similar to the approximation formulas derived by Whitt for the QNA method. The authors obtain very promising results, however, the approach is currently not suitable for the superposition of arbitrary arrival streams (for example, it does not apply if the variation coefficients of the interarrival time distributions of all arrival streams are larger than one).

### 7.4.4 Conclusion

With the treatment of the merging and splitting operations in this section, all necessary steps for embedding iSPNs in a parametric decomposition approach have been tackled. It has been shown that the traffic splitting operation can be performed in an exact fashion for phase-type renewal processes and MAPs. Concerning the merging of traffic streams, an exact result has been derived in terms of a MAP description of the aggregate traffic stream.

The suggested merging and splitting operations represent a significant improvement over the approaches employed by the original QNA method and its variants as presented in Section 7.2:

- All traffic streams among nodes are described as MAPs or phase-type renewal processes. This allows a much more accurate description of traffic streams than the original QNA approach, which is based on two moments only.
- Our operations on traffic streams are capable to account for non-renewal processes. Thus, it is no longer necessary to neglect correlations among successive arrival and departure events, a property which has a large impact on system performance.
- The merging operation yields *exact* results, compared to the original QNA formulas for appoximating the first two moments of the superposed streams.

However, even though the proposed merging and splitting operations are exact, it is important to note that the resulting MAPs and phase-type distributions quickly reach a size that can not be dealt with numerically in the node analysis step. Therefore, the exact traffic stream descriptions need to be approximated by smaller representations. Further work is still needed in this area in order to develop suitable approaches.

Together with the results of Section 7.3, the methods developed in this section can be used to derive an iterative fixed-point solution approach for the analysis of networks of iSPN nodes. The main steps of this algorithm are depicted in Figure 7.19. Concerning the stopping criterion of the algorithm, suitable measures need to be defined for calculating the "distance"

Initialize all node departure streams to an initial process (e.g., use simple Poisson streams).

- Perform the following steps for all iSPN node models:
  - Split the departure traffic streams to obtain the arrival processes for the successor nodes (→Section 7.4.1)
  - Merge all incoming traffic streams (→Section 7.4.2) and perform a reduction procedure (→Section 7.4.3).
  - Build an overall iSPN model that contains the arrival process (→Section 7.3.1).
  - Solve the iSPN's underlying QBD, and determine the departure process for the next iteration (→Section 7.3.2).

Repeat until convergence of the traffic streams.

**Fig. 7.19.** The overall solution algorithm for the analysis of networks of iSPN nodes.

between traffic stream descriptions that were determined in successive iterations. To this end, the same measures that were defined in Section 7.4.3 can be employed.

In spite of the fact that a similar fixed-point approach has been applied successfully for the solution of QNA models with finite buffers (see Section 7.2.2), it remains still to be shown that the algorithm in Figure 7.19 converges as well. It is also worth noting that in addition to iSPN nodes, the resulting algorithm is also capable to account for "conventional" QNA nodes (both with finite and infinite buffers) by using the original approximation formulas that have been presented in Section 7.2 for the node analysis and departure stream derivation. In these cases, transformations will be necessary between the improved traffic description in terms of MAPs or phase-type distributions and the moment-based description employed by QNA.

Finally, we remark that the proposed algorithm is perfectly suited for a distributed implementation (e.g., using clusters of workstations), as the computationally most expensive part (the analysis of the iSPN node models) can be performed in isolation without any relation to other computing nodes. The same applies to the merging and splitting operations, which can be performed locally for each iSPN node after the associated traffic stream descriptions have been distributed.

## 7.5 Summary and Concluding Remarks

In this chapter, the efficient high-level iSPN-based node modeling approach of Chapter 5 has been embedded in a parametric decomposition framework in order to allow the concurrent analysis of interconnected nodes modeled by iSPNs. The decision for the parametric decomposition approach and a discussion of existing work in this area has been presented in Section 7.1. Based on this, the QNA method has been chosen as a starting point due to its good performance and its close relations to matrix-geometric methods, as described in Section 7.2.

From these considerations, it has been concluded that the following areas needed to be tackled for embedding iSPNs in network-oriented context:

- Suitable interfaces needed to be defined for characterizing how an arrival traffic stream "enters" an iSPN node, and how the modeler specifies departures at the iSPN level.
- Algorithms must be developed for deriving the departure traffic stream leaving an iSPN node.
- To match the increased node modeling flexibility, an improved traffic stream description and associated merging and splitting operations needed to be defined.

The first two points have been tackled in Section 7.3. The external interfaces of an iSPN node are specified by definining designated arrival places and departure transitions. For deriving the departure stream of an iSPN node, an exact approach has been presented, leading to a phase-type interdeparture time distribution. The approach can also be extended for approximating correlations among successive departures by employing MAP departure process descriptions. The proposed departure process derivation approach currently assumes that job departures from an iSPN node correspond to transitions to lower levels in the QBD underlying the iSPN. Several approaches were suggested for assuring this; however, the development of a general solution that does not require special considerations on behalf of the modeler still needs further work.

Concerning the third point, a MAP or phase-type traffic stream description was adopted, as it allows for a more accurate description of traffic streams compared to the moment-based approach taken in QNA, and because such a description naturally results from the departure process derivation procedure in Section 7.3. Suitable approaches for splitting and merging such traffic streams have been presented in Section 7.4, again leading to exact results. However, it has been pointed out that the size of the derived traffic stream descriptions will quickly lead to numerical problems in the node analysis step. Thus, techniques need to be developed for approximating these traffic stream descriptions by descriptions of a lower order. Several conceivable approaches for achieving this have been proposed, but further work needs still to be done in this area. Apart from the dpearture process derivation at a node (where the approximation of interdeparture-time correlation can be made arbitrarily small), this reduction step represents the *only* approximation in our framework. Therefore, careful investigations need to be conducted here. Once such reduction approaches are available, the proposed methods readily lead to a fixed-point iteration approach for the analysis of networks of iSPN nodes.

Summarizing, the open requirements for a network-oriented modeling and performance evaluation framework mentioned in the introduction of this chapter have been tackled:

- R2 (Non-Local Interactions) are accounted for by employing detailed phase-type or MAP descriptions of the occurring traffic flows. When MAPs are employed, the proposed solution framework can also account for correlated traffic streams; in view of the impact of correlations on performance, this represents a key advantage compared to existing approaches.
- R5 (Model Evaluation) can easily be extended towards network-oriented performance measures by adopting the approach taken in QNA, where network-wide results are derived from properties of the individual node models. For example, delay measures that were derived from the solution of the iSPN nodes can be added to yield network-wide delay measures. Also, if waiting time distributions at a node are available in phase-type form, convolution can easily be applied to obtain the distribution of network-wide delays.

# 8. Conclusions and Outlook

In this work, a model-based performance evaluation environment for large telecommunication networks has been developed. The purpose of such an environment is to guide the network design process in a way that important design issues can be resolved *before* a network has been installed. In the preceding chapters, it has become clear that the development of a suitable modeling framework is a complex task; the aim of this chapter is to review the most important insights gained during this development and to outline areas for further research.

## Summary of Major Results

Prior to actually developing a performance evaluation approach for telecommunication networks, a thorough assessment of the requirements towards such a framework has been performed. This was accomplished by investigating two of the most relevant network architectures and by closely cooperating with an industrial partner in order assure that the practitioner's point of view is taken into account. In view of the requirements that were identified, the development of a simple prototype modeling environment and its comparison to existing approaches has shown that two key areas needed to be tackled in this work: (i) the expressiveness of network node models needs to be increased, and (ii) the relationships among different nodes need to be described in more detail.

The reason why existing approaches often resort to very high-level abstractions of network nodes is that the analysis step quickly becomes computationally infeasible. To resolve this situation, quasi-birth-and-death (QBD) processes were suggested as a special class of stochastic processes that extends classical queueing-theoretic models with a large degree of additional modeling flexibility. Still, QBD processes are of Markovian nature, and a variety of very efficient algorithms exist for their analysis. To guide the selection of an appropriate solution algorithm, a thorough comparison study has been performed not only on a theoretical level, but also in a numerical study based on an example application. Compared to other investigations, care has been taken to not only account for the computational efficiency of the algorithms, but also to consider the obtained solution quality. Also, we accounted for complex boundaries and non-skip-free QBDs due to the relevance of these issues in practical applications.

While QBDs represent a good mathematical foundation which allows for the quick analysis of complex nodes, the creation of node models directly at the QBD level is a tedious task. Therefore, an approach has been developed that allows modelers to create complex QBD models without knowledge of the special properties of this class of stochastic processes. This has been accomplished by developing an approach based on stochastic Petri nets (SPNs), which represents a well-known high-level description technique in the modeling community. The formal definition of our class of *infinite-state* SPNs (iSPNs) enabled us to show their equivalence to QBD processes; furthermore, special care has been taken to discuss issues related to efficiently transforming iSPN models to the underlying QBDs. The applicability of the iSPN approach for complex node models has then been illustrated by investigating case studies in the areas of the World Wide Web and TCP/IP interaction, ATM connection management, and self-similar traffic patterns.

After iSPNs were introduced as a powerful means for the model-based analysis of individual network nodes in Part II of this work, we focused in Part III on how this approach can be extended to account for *networks* of interacting nodes. To this end, a parametric decomposition approach has been suggested, where traffic streams among different nodes are described as phase-type renewal processes and Markovian arrival processes (MAPs). Appropriate procedures for merging and splitting traffic streams have been developed. Also, approaches for embedding arrival streams into iSPN node models and for determining the departure process leaving the nodes have been developed. Apart from the departure process derivation step (where the correlation structure can be approximated as close as desired), all these procedures provide *exact* results. Especially the MAP-based traffic stream description represents a substantial improvement over existing approaches in this area, which often employ a moment-based approach, thereby reducing complex interarrival-time distributions to a few moments only. Furthermore, such approaches are not capable to describe correlated traffic streams, the impact of which can be crucial in practical applications.

For solving the complex system of non-linear and non-explicit relationships that result from these considerations, an algorithm based on fixed-point iteration has been suggested. A similar approach has been employed successfully in a similar context for extending the capabilities of the QNA method, however, further experiments are still needed to show the feasibility of this approach in our context. Furthermore, the number of states occurring in the traffic stream descriptions will grow in each iteration step, which might lead to numerical problems in the node analysis step. A feasible solution to this issue is to reduce the traffic stream descriptions to descriptions with less states, and several conceivable approaches have been suggested for dealing with this.

### Future Research

The presented performance modeling and evaluation methodology fulfills the five main requirements that were identified in the beginning of this work. However, it has also been shown that several improvements are still possible. Three main areas can be identified that need to be tackled in the future, covering the implementation of the presented framework, the further development of the methodology, and further application studies.

Concerning the implementation, a powerful modeling and evaluation tool for individual network nodes has been introduced with SPN2MGM (see Appendix B). In its current form, SPN2MGM does not yet account for modeling non-skip-free QBDs and buffer resets; as the necessary extensions have been described in detail in Chapter 4, their implementation can be performed in a straightforward fashion, thereby further extending the expressiveness of the considered model class. Similarly, SPN2MGM currently does not include a dynamic derivation of a near-optimal $j_{\min}$ as described in Section 5.3.1.

For other improvements, the theoretical background needs to be extended. Many of the presented methods for the analysis of QBDs are also suitable for investigating *finite* QBDs. We thus consider the extension of the iSPN framework for modeling finite QBDs as an important task for further increasing the power of our modeling framework.

For completing the analysis framework for networks of iSPN nodes as described in Chapter 7, further work is needed for reducing the orders of the occurring traffic stream descriptions. Once suitable approaches have been found, a tool environment needs to be developed "on top" of SPN2MGM to allow for the analysis of entire networks. Here, a challenging task is to parallelize the overall algorithm in order to allow for a distributed solution, and to embed the approach in an optimization framework.

It has been shown that the iSPNs represent a convenient means for the specification of very large QBDs, which can easily exceed the size that can be dealt with by current numerical solution algorithms for QBDs. From a modeling point of view, the development of methods for the efficient analysis of even larger QBDs therefore represents a desirable extension for increasing the amount of detail that can be accounted for in the iSPN node modeling step. Such algorithms might be based on exploiting special structures that appear in QBDs (see e.g. the approach suggested in [48]) or on the development of approximate solution techniques (see also our remarks in Section 4.6.1). Another interesting approach in this area that deserves further investigation is to take advantage of the efforts that have been spent on the solution of eigenvalue problems. Suitable algorithms from this area might be used to substantially improve the applicability of the spectral expansion method towards much larger QBDs. A promising approach in this area has been mentioned in Appendix A.1.

In Chapter 7, an iterative solution algorithm has been suggested for the analysis of networks of iSPN node models. For its application, the convergence properties of this algorithm need to be investigated. A formal proof of convergence shall be difficult, as such a proof has also not yet been found for a similar approach that has been taken for introducing finite buffers in the QNA method, despite the fact that the inter-node traffic streams there are described in a much simpler way than in our proposed framework. For providing a formally secured basis to both of these methods, further work is required in this area.

Next to that, extensive application studies need to be conducted for assessing the benefits of a MAP/iSPN-based network modeling approach with respect to the solution accuracy compared to applicable conventional algorithms. Also, comparisons to simulation results and to measurements performed in real networks need to be performed. Especially the last issue suggests to accomplish these application-oriented studies in close cooperation with an industrial partner.

### A Résumé

This work presented an approach towards a model-based performance evaluation environment for the analysis of modern telecommunication networks, thereby providing a means for the detailed description and analysis of individual network nodes *and* entire networks. Special care has been taken to not only take advantage of the benefits provided by the underlying mathematical model class, but also to provide convenient access to these models to modelers which are not experts in the field. In view of the increasing economical importance of the communications sector, such an approach represents a key component towards optimizing network infrastructures for efficiently providing better services.

# A. Linear Algebra and Probability Theory Primer

## A.1 Polynomial Eigenvalue Problems

In Section 4.3.3, the spectral expansion method has been presented as an efficient approach for deriving the steady-state solution of a QBD process. From a computational point of view, the heart of this approach is the solution of a polynomial eigenvalue problem. Since only little literature is available on this type of eigenvalue problems, we comment on its solution in this appendix.

In Section A.1.1, we briefly review the characteristics of polynomial eigenvalue problems. Section A.1.2 then describes an approach for transforming polynomial eigenvalue problems to linear eigenvalue problems, so that common solution algorithms can be applied. In Section A.1.3, we briefly comment on alternative solution approaches.

### A.1.1 Definition

In Section 4.3.3, the consideration of a skip-free QBD process led to the following quadratic eigenvalue problem (see Eq. (4.33)):

$$\boldsymbol{\psi}(\mathbf{A}_0 + \lambda\mathbf{A}_1 + \lambda^2\mathbf{A}_2) = \mathbf{0}. \tag{A.1}$$

For non-skip-free QBDs, polynomial eigenvalue problems of a higher order $n$ are obtained. In this general form, the problem is to find non-zero (left) eigenvectors $\boldsymbol{\psi} \in \mathbb{C}^N$ and associated eigenvalues $\lambda \in \mathbb{C}$ so that

$$\sum_{l=0}^{n} \boldsymbol{\psi}\lambda^l \mathbf{A}_l = \mathbf{0} \tag{A.2}$$

holds, where $\mathbf{A}_0, \cdots, \mathbf{A}_n \in \mathbb{R}^{N\times N}$.

### A.1.2 Linearization

The common approach for solving the polynomial eigenvalue problem (A.2) is to transform it to a linear eigenvalue problem and then to apply well known solution approaches like QR decomposition to the linearized problem. For quadratic eigenvalue problems, a linearization procedure was described in [113]; here, we present a generalization of the approach towards polynomial eigenvalue problems of order $n$ that has originally been developed in [156].

The basic idea of the linearization procedure is to extend the dimension of the eigenvectors so that they not only hold the desired eigenvector $\boldsymbol{\psi}$, but also the products $\boldsymbol{\psi}\lambda, \cdots, \boldsymbol{\psi}\lambda^{n-1}$. Thus, the size of the matrices and vectors increases from $N$ to $N \cdot n$, leading to a linear eigenvalue problem of the form

$$\boldsymbol{\phi}\mathbf{A} = \boldsymbol{\phi}\mu \tag{A.3}$$

with $\boldsymbol{\phi} \in \mathbb{C}^{Nn}$, $\mathbf{A} \in \mathbb{C}^{Nn \times Nn}$ and $\mu \in \mathbb{C}$. For convenience, we partition the eigenvectors $\boldsymbol{\phi}$ into $N$-dimensional subvectors $\boldsymbol{\phi}_i$, so that $\boldsymbol{\phi} = (\boldsymbol{\phi}_0, \cdots, \boldsymbol{\phi}_{n-1})$ holds.

The linearization procedure depends on whether the inverses of the coefficient matrices $\mathbf{A}_n$ and $\mathbf{A}_0$ of the polynomial problem exist. Three cases are distinguished, corresponding to $\mathbf{A}_n$ being nonsingular, $\mathbf{A}_0$ being nonsingular, or both of them being singular.

**Case 1: $\mathbf{A}_n$ is Nonsingular.** In this case, the polynomial problem (A.2) can be multiplied by $\mathbf{A}_n^{-1}$, yielding

$$\sum_{l=0}^{n-1} \boldsymbol{\psi}\lambda^l \mathbf{A}_l \mathbf{A}_n^{-1} + \boldsymbol{\psi}\lambda^n = \mathbf{0}. \tag{A.4}$$

We now define the vectors $\boldsymbol{\phi}_0, \cdots, \boldsymbol{\phi}_{n-1}$ so that the following equations hold:

$$\boldsymbol{\phi}_0 = \boldsymbol{\psi}\lambda^0 = \boldsymbol{\psi} \tag{A.5}$$

$$\boldsymbol{\phi}_1 = \boldsymbol{\psi}\lambda^1 = \boldsymbol{\phi}_0\lambda \tag{A.6}$$

$$\boldsymbol{\phi}_2 = \boldsymbol{\psi}\lambda^2 = \boldsymbol{\phi}_1\lambda \tag{A.7}$$

$$\vdots$$

$$\boldsymbol{\phi}_{n-1} = \boldsymbol{\psi}\lambda^{n-1} = \boldsymbol{\phi}_{n-2}\lambda. \tag{A.8}$$

Using this, Eq. (A.4) can be rewritten as

$$\sum_{l=0}^{n-1} \boldsymbol{\phi}_l \mathbf{A}_l \mathbf{A}_n^{-1} + \boldsymbol{\phi}_{n-1}\lambda = \mathbf{0}. \tag{A.9}$$

All equations (A.5)-(A.9) can be summarized in matrix form, yielding

$$(\boldsymbol{\phi}_0, \cdots, \boldsymbol{\phi}_{n-1}) \begin{pmatrix} \mathbf{0} & & & & -\mathbf{A}_0\mathbf{A}_n^{-1} \\ \mathbf{I} & \mathbf{0} & & & -\mathbf{A}_1\mathbf{A}_n^{-1} \\ & \mathbf{I} & \mathbf{0} & & -\mathbf{A}_2\mathbf{A}_n^{-1} \\ & & \ddots & \ddots & \vdots \\ & & & \mathbf{I} & -\mathbf{A}_{n-1}\mathbf{A}_n^{-1} \end{pmatrix} = (\boldsymbol{\phi}_0, \cdots, \boldsymbol{\phi}_{n-1})\lambda. \tag{A.10}$$

Clearly, this represents a linear eigenvalue problem conforming to Eq. (A.3). The eigenvalues obtained from its solution are identical to those of the polynomial eigenvalue problem, and the associated eigenvectors are obtained by considering the first subvector $\boldsymbol{\phi}_0$ of $\boldsymbol{\phi}$ only (see Eq. (A.5)).

**Case 2: $\mathbf{A}_n$ is Singular and $\mathbf{A}_0$ is Nonsingular.** If $\mathbf{A}_n$ is singular, it can not be inverted in order to transform the polynomial eigenvalue problem to Eq. (A.4). However, a solution similar to the previous case is possible if $\mathbf{A}_0$ is nonsingular. This is accomplished by reversing the sequence of the coefficient matrices in the polynomial problem (A.2).

To this end, note that if $\mathbf{A}_0$ is nonsingular, then the polynomial eigenvalue problem can not have any zero eigenvalues $\lambda = 0$, since Eq. (A.2) would then imply

$$\boldsymbol{\psi}\mathbf{A}_0 = \mathbf{0} \tag{A.11}$$

and thus $\boldsymbol{\psi} = \mathbf{0}$. However, as eigenvectors must be non-zero, this contradicts the assumption that zero is an eigenvalue.

Therefore, we can safely assume that $\lambda \neq 0$ holds, and multiply the original polynomial problem (A.2) by $1/\lambda^n$, yielding

$$\sum_{l=0}^{n} \boldsymbol{\psi}\lambda^{l-n}\mathbf{A}_l = \mathbf{0} \tag{A.12}$$

$$\sum_{l=0}^{n} \boldsymbol{\psi}\lambda^{-l}\mathbf{A}_{n-l} = \mathbf{0}. \tag{A.13}$$

Introducing a variable transformation $\mu = 1/\lambda$, this leads to

$$\sum_{l=0}^{n} \boldsymbol{\psi}\mu^{l}\mathbf{A}_{n-l} = \mathbf{0}, \tag{A.14}$$

which is a polynomial eigenvalue problem where the last coefficient matrix is nonsingular. Thus, the approach for the first case can be applied. Note that the derived eigenvalues $\mu$ need to be backtransformed by $\lambda = 1/\mu$ in order to obtain the eigenvalues $\lambda$ of the original problem.

**Case 3: Both $\mathbf{A}_n$ and $\mathbf{A}_0$ are Singular.** Similar to the approach taken for the second case, the idea here is to construct a new polynomial eigenvalue problem which is related to the initial one. However, instead of simply reversing the order of the coefficient matrices $\mathbf{A}_i$, a different approach is needed that ensures that the last coefficient matrix of the newly constructed polynomial eigenvalue problem is non-singular, so that the approach for nonsingular $\mathbf{A}_n$ can be taken.

The approach taken here is to derive new coefficient matrices as a linear combination of the original ones. This is accomplished by using a variable transformation from $\lambda$ to $\mu$ (similar to the second case), but with the relationship

$$\mu = \frac{\theta + \lambda}{\theta - \lambda} \qquad \text{or} \qquad \lambda = \theta \frac{\mu - 1}{\mu + 1}, \tag{A.15}$$

where $\theta$ is an arbitrary value that is not an eigenvalue of the original polynomial eigenvalue problem (this can easily be verified by inserting the selected $\theta$ as $\lambda$ in Eq. (A.2)). Expressing $\lambda$ in terms of $\mu$ in the original polynomial eigenvalue problem, we obtain

$$\sum_{l=0}^{n} \psi \mathbf{A}_l \theta^l \frac{(\mu - 1)^l}{(\mu + 1)^l} = 0. \tag{A.16}$$

In order to remove the fraction that involves the transform variable $\mu$, this is multiplied by $(\mu + 1)^n$, yielding

$$\sum_{l=0}^{n} \psi \mathbf{A}_l \theta^l (\mu + 1)^{n-l} (\mu - 1)^l = 0. \tag{A.17}$$

Clearly, the coefficient matrix of a term $\mu^l$ now consists of a superposition of all original coefficient matrices $\mathbf{A}_i$. We denote these coefficient matrices by $\mathbf{A}'_l$; once they have been determined, Eq. (A.17) can be rewritten as

$$\sum_{k=0}^{n} \psi A'_k \mu^k = \mathbf{0}. \tag{A.18}$$

In order to derive the new coefficient matrices $\mathbf{A}'_k$, we need to investigate the factor at which each $\mathbf{A}_i$ is multiplied by $\mu^l$ in Eq. (A.17). This is accomplished by considering the factors occurring in Eq. (A.17), using the binomial theorem, and collecting all factors related to the powers of $\mu$:

$$\begin{aligned} \theta^l (\mu + 1)^{n-l} (\mu - 1)^l &= \theta^l (1 + \mu)^{n-l} (1 - \mu)^l (-1)^l \\ &= \theta^l \left[ \binom{n-l}{0} + \binom{n-l}{1} \mu + \cdots + \binom{n-l}{n-l} \mu^{n-l} \right] \\ &\quad \cdot \left[ \binom{l}{0} - \binom{l}{1} \mu + \cdots \pm \binom{l}{l} \mu^l \right] (-1)^l \\ &= \theta^l \sum_{k=0}^{n} \mu^k \underbrace{\sum_{j=\max\{0,k-l\}}^{\min\{k,n-l\}} \binom{n-l}{j} \binom{l}{k-j} (-1)^{k-j+l}}_{:=e(l,k)} . \end{aligned} \tag{A.19}$$

The term $\theta^l e(l,k)$ thus denotes the factor at which the matrix $\mathbf{A}_l$ contributes to the coefficient matrix of $\mu^k$ in Eq. (A.17). For the entire set of coefficient matrices, we thus obtain

$$\mathbf{A}'_k = \sum_{l=0}^{n} \mathbf{A}_l \theta^l e(l, k). \tag{A.20}$$

By selecting $\theta$ so that the new coefficient matrix $\mathbf{A}'_n$ is nonsingular, the new polynomial problem in Eq. (A.18) can be solved according to the procedure presented for the first case. After an eigenvalue $\mu$ has been determined, a backtransformation according to Eq. (A.15) needs to be performed in order to derive the eigenvalues of the original problem.

### A.1.3 Other Solution Approaches

With the linearization approach presented in Section A.1.2, any standard algorithm for the solution of linear eigenvalue problems can be employed for solving polynomial eigenvalue problems. However, it is important to note that the coefficient matrices appearing in the linearized eigenvalue problem can grow considerably large, especially for non-skip-free QBDs.

For very large linear eigenvalue problems, iterative solution algorithms like the Arnoldi method or the Jacobi-Davidson method are usually employed (for an overview, see e.g. [192]). These iterative solution algorithms, however, are specifically designed for very sparse eigenvalue problems. In the spectral expansion context, the coefficient matrix of the linearized problem in Eq. (A.10) is usually not sparse: the rightmost block-column is usually dense due to the way the block-entries are computed. Unless the skip-size of the QBD is very large (i.e., the share of the rightmost block-column in the coefficient matrix of Eq. (A.10) becomes very small), iterative solution algorithms are not suitable.

However, an interesting approach has been suggested in [180], where the iterative Jacobi-Davidson method is directly applied to polynomial eigenvalue problems. As the coefficient matrices appearing in the spectral expansion approach are usually sparse, the direct application of this iterative solution method seems promising.

## A.2 Phase-Type Distributions

Continuous phase-type distributions (see e.g. [152]) represent a powerful approach for constructing complex distributions from simple exponential distributions. This is accomplished by defining them as the time to absorption in a finite continuous-time Markov chain. Due to the fact that phase-type distributions are based on exponential distributions, they can easily be integrated in a Markovian environment and are amenable to efficient solution algorithms. Furthermore, phase-type distributions form a dense subset of all distributions on $[0; \infty)$, and can they can therefore be used to approximate any distribution arbitrarily close [13].

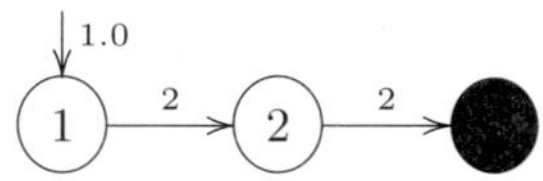


**Fig. A.1.** An example phase-type distribution that represents an Erlang-2 distribution.

In our presentation, we adopt the notation of [152] to denote a phase-type distributions of order $N$ by a pair

$$(\boldsymbol{\alpha}, \mathbf{A}) \qquad \text{with} \qquad \boldsymbol{\alpha} \in \mathbb{R}^N, \mathbf{A} \in \mathbb{R}^{N \times N}, \tag{A.21}$$

where the vector $\boldsymbol{\alpha}$ holds the initial probabilities of the non-absorbing (or transient) states, and $\mathbf{A}$ contains the transition rates between these states.

Phase-type distributions have an intuitive graphical representation; in this work, the transient states are denoted by circles that contain the number of the states, and the absorbing state is denoted by a filled circle. The initial probabilities are illustrated on arcs without originating states; they are labelled by the corresponding initial probabilities. If these arcs are omitted, the corresponding initial probability is zero. As an example, consider the phase-type representation of an Erlang-2 distribution with mean 1 in Figure A.1.

For a transition matrix $\mathbf{A}$, we implicitly define the *column* vector $\mathbf{A}^0$ which contains the rates from the transient states to the absorbing state; it is given by $\mathbf{A}^0 = -\mathbf{A}\mathbf{1}$. In this work, we also use the fact that the $n$-th moment $m_n$ of a phase-type distribution $(\boldsymbol{\alpha}, \mathbf{A})$ is readily derived by

$$m_n = (-1)^n n! (\boldsymbol{\alpha}\mathbf{A}^{-n}\mathbf{1}). \tag{A.22}$$

Note that that to compute $m_n$, the explicit inversion of $\mathbf{A}$ is not necessary (see the "trick" presented in Section 4.2.1).

## A.3 Markovian Arrival Processes

Markovian Arrival Processes (MAPs) can be considered as an extension of phase-type renewal processes where the initial state probabilities of the renewal phase-type distribution depend on the state that initiated the last event (MAPs are thus no longer renewal processes, but belong to the more general class of point processes). Due to this property, MAPs are suitable for modeling correlations among the time to successive observable events (like job arrivals or departures).

While MAPs can be described in several ways, we adopt the notation used in [29] (where MAPs are termed *generalized phase processes*, GPHs) due to its close relation to the description of phase-type distributions. The description

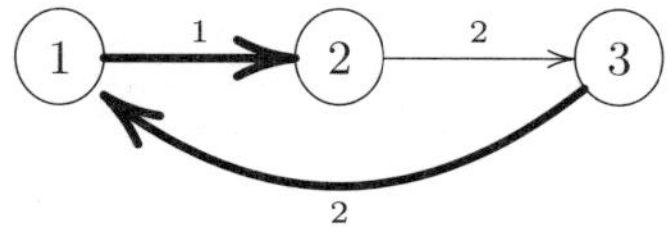


**Fig. A.2.** An example MAP that alternatingly generates exponential and Erlangian interarrival times. Marked transitions are emphasized.

is based on a finite irreducible continuous-time Markov chain with generator $\mathbf{Q}$. This matrix is partitioned into two matrices, so that $\mathbf{Q} = \mathbf{Q}_1 + \mathbf{Q}_2$ holds. The entries of $\mathbf{Q}_2$ contain a subset of the (off-diagonal) elements of $\mathbf{Q}$ in order to specify a set of *marked* transitions. Each time the Markov chain takes one of these marked transitions, an observable event (like a job arrival) is generated.

As an example, consider the MAP description of an arrival process where the interarrival times are alternatingly distributed exponentially and Erlangian in Figure A.2. The corresponding MAP description is then given by the following pair of matrices

$$(\mathbf{Q}_1, \mathbf{Q}_2) = \left( \begin{pmatrix} -1 & 0 & 0 \\ 0 & -2 & 2 \\ 0 & 0 & -2 \end{pmatrix}, \begin{pmatrix} 0 & 1 & 0 \\ 0 & 0 & 0 \\ 2 & 0 & 0 \end{pmatrix} \right). \tag{A.23}$$

As mentioned above, MAPs represent a superclass of phase-type renewal processes. Thus, a renewal process with phase-type renewal time according to $(\boldsymbol{\alpha}, \mathbf{A})$ can also be expressed in terms of a MAP; it has the representation $(\mathbf{A}, \mathbf{A}^0\boldsymbol{\alpha})$.

# B. Tool Description

In this section, we describe the tool environment SPN2MGM that has been developed for the analysis of iSPN node models. All numerical studies presented in this work have been performed using SPN2MGM.

Large parts of the SPN2MGM package are based on work accomplished in [118] and [73]. The extensions performed in the scope of this work dealt with the development of an option-based shell interface for controlling the program execution and a graphical interface for the design of iSPNs. Furthermore, the implementation of the existing QBD solution methods has been optimized, and the spectral expansion method, Naoumov's method and the invariant subspace approach have been integrated. Furthermore, routines for computing the solution quality (as defined in Chapter 4) have been added.

A strength of SPN2MGM is that it strictly follows the iSPN philosophy of hiding all mathematical details related to the underlying QBD processes behind a comfortable iSPN interface. We will therefore start our presentation with a description of the user interfaces of SPN2MGM in Section B.1. In Section B.2 we then comment on the implementation of SPN2MGM.

## B.1 User Interface

SPN2MGM provides a variety of user interfaces for specifying iSPN models and the desired model parameters, for controlling the model solution, and for obtaining the results of the model analysis. For an overview, see Figure B.1.

Each of these interfaces will be described in this section; Section B.1.1 describes the model specification interfaces, Section B.1.2 focuses on the execution control, and Section B.1.3 describes the output formats of SPN2MGM.

### B.1.1 Model Specification

The SPN2MGM program reads all information related to the iSPN model, the desired performance measures, and the definition of variables that are used in the specification part from three input files. These files can either be created automatically by using a convenient graphical iSPN editor, or they can be created manually, which is often suitable for batch operation.

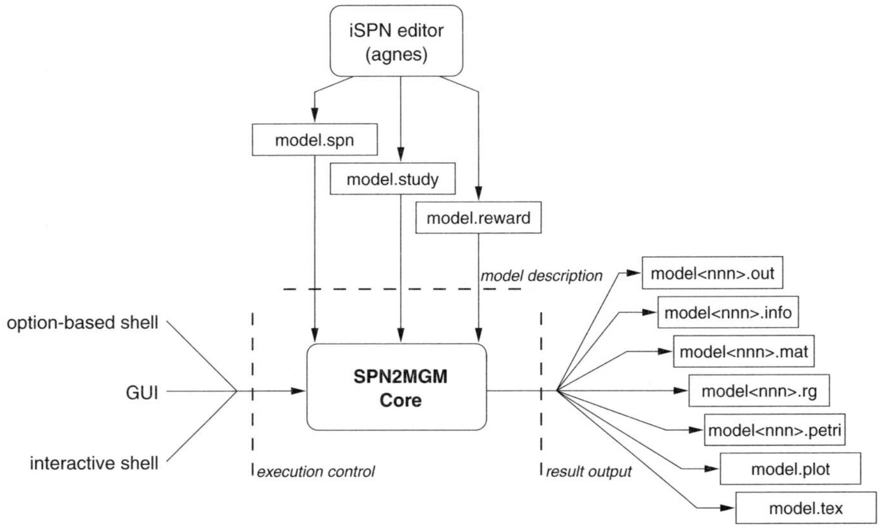


**Fig. B.1.** Interfaces of SPN2MGM.

The graphical iSPN editor is based on an adaptation of the generic net editor agnes (see Section B.2); its usage is very intuitive. After an iSPN model has been created, the textual model description expected by the core SPN2MGM tool is generated by using the provided export function. For a screenshot of the editor, see Figure B.2.

Below, we provide a summary of the format and the purpose of the three text files in terms of examples; for a more formal description, see [73].

**Variable Definition (study-file).** In this part of the specification, constants and variables are declared. These can be used in other parts of the specification, so parametric modifications of a basic model can be controlled centrally in this file. It is possible to assign a range of values to a single variable in order to evaluate an iSPN model for a number of different parameter values. All model specification files may contain comment lines starting with an asterisk ("*"):

```
* single value constants:
erlang_stages   value 4
* a list of values:
lambda          list  0.1 0.2 0.3 0.4
* a range of values (from 10 to 100 step 10)
buffersize      loop  10 100 10
end
```

The keyword "end" indicates the end of the variable definition file.

**Petri Net Definition (model-file).** The syntax employed for the specification of iSPNs is similar to the one used in the tool *Sharpe* [174]. The

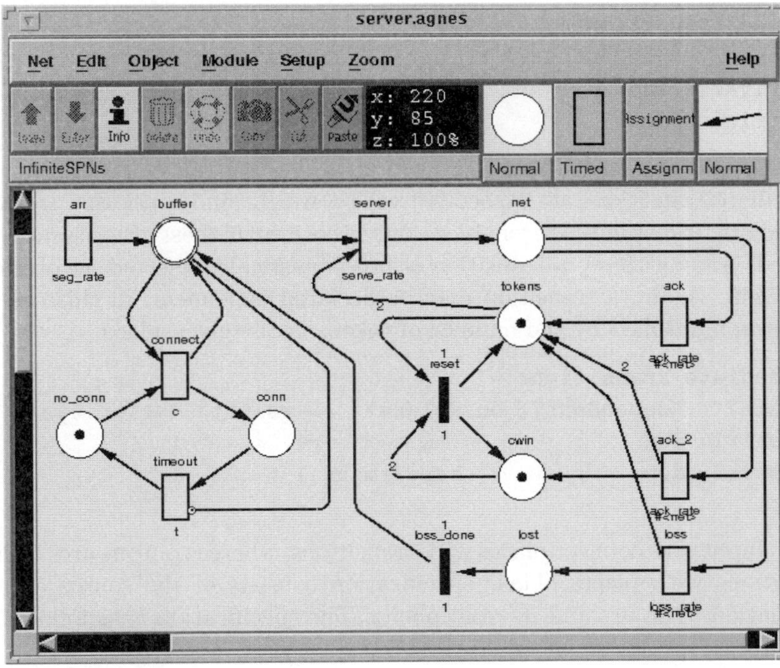


**Fig. B.2.** A screenshot of the graphical iSPN editor.

specification comprises the definition of places, timed transitions, immediate transitions, input arcs, output arcs, inhibitor arcs, enabling functions and reward functions. These parts have to be given in the mentioned order, and each part must be terminated with the keyword `end`.

*Places.* Each place is associated with a symbolic name and the initial number of tokens in this place. The definition may use the variables introduced in Section B.1.1. Places are defined as follows:

```
* places:
* <name>   <tokens>
  buffer   0
  servers  inital_servers
end
```

Note that the *first* place defined here is the place with unlimited capacity, representing e.g. an unbounded buffer.

*Transitions.* Timed transitions are defined by their symbolic name and a rate. Furthermore, they are declared either independent or dependent of a place; in the latter case, a transition's firing rate is multiplied by the number of tokens in that place:

```
* timed transitions:
* <name>    <dependent?> <...on place> <firing rate>
  arrival   ind                        lambda
  service   dep          servers       mu
end
```

Immediate transitions are associated with a weight and a priority. If several immediate transitions are enabled, only those of highest priority are considered. Among these, a transition is probabilistically selected according to its weight. Again, a transition may be declared dependent; in this case, its weight is multiplied by the number of tokens in the given place.

```
* immediate transitions:
* <name>   <dependent?> on <place>    <weight> <priority>
  repair1  ind                        1        1
  repair2  dep             repairmen  1        1
end
```

*Arcs.* Input arcs connect places with transitions, whereas output arcs connect transitions with places. Their specification consists of the source and the destination of an arc and its multiplicity. The specification is as follows:

```
* input arcs:
* <from place> <to trans.> <multiplicity>
  always       arrival     1
end
* output arcs:
* <from trans.> <to place> <multiplicity>
  arrival       always     1
end
```

Furthermore, inhibitor arcs can be defined to deactivate a (timed or immediate) transition if the number of tokens in their originating place is larger than or equal to the inhibitor arc's multiplicity:

```
* inhibitor arcs:
* <from place> <to trans.> <multiplicity>
  block        arrival     1
end
```

*Enabling functions.* Enabling functions relate a (timed or immediate) transition to an explicit condition which is necessary to activate the transition. The condition checks whether a place contains less than, exactly, or more than a given number of tokens:

```
* enabling functions:
* <transition> <place>  <less/eq./more> <#tokens>
  repair       servers        <          1
end
```

*Reward functions.* Reward functions simplify the computation of complex probability measures and mean values. They are used to add or subtract the number of tokens in different places or to add constant values to the number of tokens in a place. This way, the computation of the number of customers in a system where customers are distributed over several places can easily be performed. By default, each place of the Petri net is associated with a reward function `PM<name>` (for *p*lace *m*arking), yielding the number of tokens in that place. Custom reward functions are introduced as follows:

```
* reward functions:
* <name>       (  <place/constant/variable> <+/-> [...]  )
  customers    (  stage1                      +    stage2 )
end
```

In all parts of the Petri net specification it is possible to use a `loop` macro to easily create a large number of similar components:

```
* use loop macro to create a large number of places
  loop 1 6 place%d 1
```

This defines the six places `place1`,···,`place6`, each containing one token initially.

**Measure specification (reward-file).** This section determines which performance measures shall be computed. The definition is based on the reward functions defined above. SPN2MGM can compute the steady-state probability of a reward function being within certain bounds, as well as the expected value of a reward function:

```
* reward based probability measures and mean values:
  P ( PMbuffer > 0 )
  P ( PMbuffer = 0 )
  P ( 5 < customers < 10 )
  P ( stage1 > stage2 )
  E ( customers )
end
```

Furthermore, the user can request that SPN2MGM outputs the computation time needed for the different tasks performed by SPN2MGM. This is obtained by using the following special measures:

```
t ( Parsing )
t ( RG )
t ( Kappa )
t ( R )
t ( Boundary )
t ( Probs )
t ( Means )
```

These measures correspond to the time needed to parse the model specification, to determine the reachability graph, the parameter $j_{\min}$, the matrix $\mathbf{R}$, to solve the set of boundary equations, to compute the requested probabilities, and to compute the desired mean values.

SPN2MGM is also capable to create graphs and tables which result from evaluating a model for a variety of parameters. This is accomplished by adding the keyword `output` to one line in the variable specification section as well as to at least one line in the measure specification file. For example,

```
* studyfile: a range of values
arrival_rate  loop  1 100 3 output


* rewardfile: plot buffer-empty probability
P ( PMbuffer = 0 ) output
```

will create a table (in LaTeX-format) and a graph (raw ASCII-format suitable for e.g. GnuPlot) relating the value of the variable `arrival_rate` to the probability of place `buffer` being empty.

### B.1.2 Execution Control

The execution control interface represents the central control point for conducting an iSPN analysis. Its purpose is to select the desired QBD solution algorithm as well as algorithm-specific parameters (like the stopping criterion for iterative solvers). Furthermore, options can be selected for providing additional information on the iSPNs and their underlying QBDs.

SPN2MGM provides three execution control interfaces. A graphical interface and an interactive command line interface have been implemented in [73]. More details on this intuitive interface can be found in [73]. Here, we only note that SPN2MGM automatically starts the graphical execution control window when invoked with no command line arguments. The interactive shell interface is accessed by providing the single command line argument `shell`.

In addition to these interfaces, an option-based command line interface has been integrated for the batch operation of SPN2MGM. Its synopsis is as follows:

```
$ spn2mgm [options]
```

where `[options]` allows to control SPN2MGM according to the options summarized in Table B.1.

### B.1.3 Output Format

The results of an iSPN analysis are written to a set of output files. Note that some output file are only generated when the appropriate option has

**Input Options**

| | |
|---|---|
| -b *basename* | set basename for study/model/reward/outputfile; the extensions .study/.spn/.reward are appended |
| -s *study*<br>-m *model*<br>-r *reward*<br>-o *output* | these options set the study/model/reward/output file name individually; they can also be used after -b |
| -a *algorithm* | select solution algorithm; available algorithms are<br>– lr - Latouche/Ramaswami (default)<br>– ss - successive substitution<br>– nau - Naoumov's improved LR method<br>– se - spectral expansion<br>– is - invariant subspace<br>– lr2 - Latouche/Ramaswami, Telpack implementation |
| -i *iters* | maximum number of iterations; default: 20 |
| -l *levels* | additional number of levels to investigate; default: 10 |
| -e *epsilon* | precision; default: 1e-08 |

**Output Options**

| | |
|---|---|
| -R | dump the iSPN's reduced reachability graph to *output*<*n*>.rg |
| -V | dump the iSPN's reachability graph with vanishing markings to *output*<*n*>.rg |
| -P | dump information on iSPN transitions and places to *output*<*n*>.petri |
| -M | dump all matrices and **b** to *output*<*n*>.mat |
| -I | output various information to *output*<*n*>.info |
| -S | select synchronous parameter advancement |
| -d | print additional debugging information to stderr |
| -h, -? | print a short summary of the available options and the current default values |

**Table B.1.** Parameter selection in the option-based SPN2MGM shell interface.

been selected in the execution control interface. For each combination of parameters that results from the variable specification in the study-file, a distinct output file is generated; its name is derived by appending a numerical suffix <n> to the base name.

The different output files contain the following information:

- *model*<*n*>.out
  contains the values of all performance measures as defined in the reward specification file.

- *model*<*n*>`.info`
  contains information related to the execution of the solution algorithm, e.g., the number of iterations and the residual error after each iteration, the repeating level size of the QBD and the value of $j_{\min}$ that has been used. Also, the actual parameter values used in this run are kept here.
- *model*<*n*>`.mat`
  contains the QBD's transition matrices and the boundary solution vector $\mathbf{b}$.
- *model*<*n*>`.rg`
  contains the iSPN reachability graph with and without vanishing marking up to level $j_{\min}$.
- *model*<*n*>`.petri`
  contains information on the transitions and places of the iSPN model.
- *model*`.plot`
  contains the values of all performance measures on a line-by-line basis for all parameter combinations so that they can easily plotted.
- *model*`.tex`
  contains the values of all performance measures for all parameter in a tabular format that can be typeset with the LaTeX text processing system.

## B.2 Implementation

The main parts of SPN2MGM have been implemented in the C++ programming language. For all numerical computations, we resort to calling efficient subroutines from the high-performance linear algebra package LAPACK [10]. For implementing the spectral expansion method, the solution classes from the libse library [156] have been employed, which use LAPACK functions as well.

Access to the invariant subspace method has been realized by calling original C functions from the telpack [2] package (version 1.0.1a). telpack also calls LAPACK for all linear algebra operations.

The iSPN editor shown in Figure B.2 is an adaptation of the powerful generic net editing system agnes [120] for iSPNs. agnes has been implemented in C++ and uses Motif for the realization the graphical user interface.

# C. Model Specifications

This appendix contains the specification files for all models presented in this work; they can be used directly as input to the corresponding tools to reproduce the numerical results. Note that in some cases external scripts were used to automatically produce a variety of model configurations, and the specification files given here just represent one particular configuration.

## C.1 An IN Model Based on M|G|1 Node Models

This section contains a sample specification for application example presented in Section 3.3; the file is suitable for direct input to our prototype tool.

```
# VPN study, demonstrate heavy load at SCP to motivate
# migration of SDF to SSP

# first, number N of physical nodes ( SCP, SSP, E, I1, I2 )
# and number K of virtual nodes
# ( SCF, SSF, SDF, SRF, IF1, IF2, IF3, IF4 )

5 8

# physical node speeds:
# SCP  SSP  E       I1       I2
  1000 1500 1000000 10000000 10000000

# workload matrix L: K lines (--> virtual nodes) with N entries each:

30 0  0  0   0
0  40 0  0   0
60 0  0  0   0
0  40 0  0   0
0  0  50 0   0
0  0  0  200 0
0  0  50 0   0
0  0  0  0   50

# number of services, then their arrival rates, and
# then number of building blocks:

1
```

```
2.0

15

# for each service: for each building block the number
#                   of virtual device invocations ( = matrix R^T )

1 1 0 0 1 0 0 0
1 1 0 0 1 0 0 0
1 1 0 0 1 0 0 0
1 0 1 0 0 1 0 0
1 0 1 0 0 1 0 0
1 0 1 0 0 1 0 0
1 0 1 0 0 1 0 0
1 1 0 1 0 0 1 1
1 1 0 1 0 0 1 1
1 1 0 0 1 0 0 0
1 1 1 0 1 1 0 0
1 1 1 0 1 1 0 0
1 1 1 0 1 1 0 0
1 0 1 0 0 1 0 0
1 1 1 0 1 0 0 0

# for each service: probability of entering each building block

 0.1 0.9
 1.0
 0.2 0.25 0.25 0.25 0.05
0.02 0.18 0.25 0.25 0.25
     0.18
     0.18

# number of measures to be computed

1

# for each measure: number of involved service and then 0/1-vector
# indicating involved building blocks, followed by additional
# delay value

0
0 0 0 0 1 0 0 0 0 0 1 0 0 0 0
0.0

# EOF
```

## C.2 A Checkpointing Transaction Processing System

This section contains the SPN2MGM specification files for the example iSPN presented in Section 5.1.4. Note that the same iSPN has been used to gen-

| $K$ | Traffic intensity: 0.1 | 0.2 | 0.3 | 0.4 | 0.5 | 0.6 | 0.7 | 0.8 | 0.9 | 0.95 |
|---|---|---|---|---|---|---|---|---|---|---|
| 40 | 0.798 | 1.597 | 2.395 | 3.194 | 3.992 | 4.791 | 5.590 | 6.388 | 7.187 | 7.586 |
| 80 | 0.694 | 1.388 | 2.082 | 2.776 | 3.470 | 4.164 | 4.858 | 5.552 | 6.246 | 6.593 |

**Table C.1.** Values for $\lambda$ for different traffic intensities at $K = 40$ and $K = 80$.

| $\rho$ | Checkpoint interval size $K$: 10 | 20 | 30 | 40 | 50 | 60 | 70 | 80 | 90 | 100 |
|---|---|---|---|---|---|---|---|---|---|---|
| 0.2 | 1.795 | 1.738 | 1.665 | 1.597 | 1.535 | 1.480 | 1.431 | 1.388 | 1.349 | 1.314 |
| 0.9 | 8.080 | 7.821 | 7.494 | 7.187 | 6.910 | 6.662 | 6.442 | 6.246 | 6.071 | 5.915 |

**Table C.2.** Values for $\lambda$ for different values of $K$ at traffic intensities 0.2 and 0.9.

| $\rho$ | Maximum arrival batch size $S$ / checkpoint interval size $K$: 1/100 | 2/50 | 3/33 | 4/25 | 5/20 | 6/16 | 7/14 | 8/12 | 9/11 | 10/10 |
|---|---|---|---|---|---|---|---|---|---|---|
| 0.2 | 1.314 | 1.023 | 0.822 | 0.680 | 0.579 | 0.504 | 0.444 | 0.397 | 0.358 | 0.326 |
| 0.9 | 5.915 | 4.606 | 3.699 | 3.062 | 2.607 | 2.270 | 2.000 | 1.788 | 1.613 | 1.469 |

**Table C.3.** Values for $\lambda$ for different values of $S$ at traffic intensities 0.2 and 0.9.

erate the QBD processes which were employed in the comparison study in Section 4.5.

### C.2.1 Parameterization

In the comparison study presented in Section 4.5, we employed the same model parameters as used in scenario 1 in [92]. More specifically, $f = 0.13$ and $r = 11$ were used as failure and restart rates, respectively. The checkpointing rate was set to $c = 10$, and services are accomplished at $\mu = 10$.

As we were interested in the model at traffic intensities for various checkpointing intervals $K$ and batch arrival sizes $S$, the arrival rate $\lambda$ has been changed appropriately. The results of Section 4.5 can be reproduced by selecting $\lambda$ according to Tables C.1, C.2 and C.3.

### C.2.2 Variable Definitions

```
**
** SPN2MGM checkpointing/batch arrival variable specification file
**

** single value constants
```

```
lambda  value   0.822173838721934
mu      value   10
f       value   0.13
r       value   11
c       value   21
K       value   33
S       value   3
end
```

### C.2.3 Petri Net Specification

```
**
** SPN2MGM checkpointing WITH batch arrivals Petri Net specification
** file example for max. batch arrival size 3
**

** places
** name / initial marking
batches         0
not_checked     0
up              1
check_buf       0
down            0
remain          0
select          0
end


** timed transitions
** name / dependent? / rate
serve           ind     mu
reserve         ind     mu
fail            ind     f
arrival         ind     lambda
do_check        ind     c
restart         ind     r
end


** immediate transitions
** name / dependent? / weight / priority
startcheck      ind     1       1
loop 1 S T%d    ind     1       1
unfold          ind     1       1
fold            ind     1       1
end


** input arcs
** source trans. / dest. place / multiplicity
not_checked     startcheck      K
up              fail            1
up              startcheck      1
```

```
check_buf       do_check        1
down            reserve         1
not_checked     reserve         1
down            restart         1
remain          serve           1
loop 1 S select T%d             1
remain          fold            S
batches         unfold          1
up              serve           1
end


** output arcs
** source trans. / dest. place / multiplicity
arrival         select          1
serve           not_checked     1
fail            down            1
startcheck      check_buf       1
do_check        up              1
reserve         down            1
restart         up              1
loop 1 S T%d    remain          %d
fold            batches         1
unfold          remain          S
serve           up              1
end


** inhibitor arcs
** source place / dest. trans. / multiplicity
not_checked     restart         1
remain          unfold          1
end


** enabling functions
fold            remain > S
end


** reward functions
end
```

### C.2.4 Definition of Reward-Based Measures

```
**
** SPN2MGM checkpointing/batch arrival reward specification file
** example for max. batch arrival size 3
**

** reward based probability measures and mean values
P ( PMup > 0 )          output
E ( PMbatches )         output
```

```
E ( PMremain )          output
end
```

## C.3 Connection Management for Video Traffic

The following subsections contain the study-, Petri net, and reward-definition files for the study presented in Section 6.1. They are suitable for use in the SPN2MGM tool.

### C.3.1 Variable Definitions

```
**
** SPN2MGM OCDR/Video traffic variable specification file
**

** single value constants
str     value   10
vtr     value   150
mu      value   150

** lists of values
rtr     list    40      80

** ranges of values
lambda  loop    0       300     10      output
end
```

### C.3.2 Petri Net Specification

```
**
** SPN2MGM OCDR/Video traffic Petri Net specification file
**

** places
** name / initial marking
busy            0
connected       0
buffer          0

* places for traffic generation
V1              1
loop 2 5 V%d    0

* places for connection setup
S1              0

* places for connection release
loop 1 7 R%d    0
no_conn         1
```

```
end


** timed transitions
** name / dependent? / rate
serve           ind     mu

* transitions for traffic generation
loop 1 5 VT%d   ind     vtr
arr             ind     lambda

* transitions for connection setup
ST1             ind     str

* transitions for connection release
loop 1 8 RT%d   ind     rtr
end


** immediate transitions
** name / dependent? / weight / priority
setup                   ind     1       1
loop 0 7 start%d        ind     1       1
end


** input arcs
** source trans. / dest. place / multiplicity
busy                    serve   1
* input arcs for traffic generation
loop     1 5 V%d        VT%d    1

* input arcs for connection setup
buffer                  setup   1
S1                      ST1     1

* input arcs for connection release
loop     0 7 buffer     start%d 1
loop     1 7 R%d        start%d 1
nextloop 1 7 R%d        RT%d    1
no_conn                 setup   1
connected               RT1     1
connected               start0  1
end


** output arcs
** source trans. / dest. place / multiplicity
serve                   connected       1

* output arcs for traffic generation
nextloop 1 4 VT%d       V%d             1
VT5                     V1              1
```

```
arr                        buffer          1

* output arcs for connection setup
setup                      S1              1
ST1                        connected       1
ST1                        buffer          1

* output arcs for connection release
loop 0 7 start%d           busy            1
loop 1 7 RT%d              R%d             1
RT8                        no_conn         1
end


** inhibitor arcs
** source place / dest. trans. / multiplicity
loop 3 5 V%d    arr     1
end


** enabling functions
end


** reward functions
Customers       ( buffer + busy + S1 )
Connected       ( busy + connected + R1 + R2 + R3 + R4 + R5 + R6 + R7 )
end
```

### C.3.3 Definition of Reward-Based Measures

```
**
** SPN2MGM OCDR/Video traffic reward measure specification file
**

** reward based probability measures and mean values
E ( Customers )          output
P ( Connected > 0 )      output
end
```

## C.4 WWW Traffic and TCP/IP Congestion Control

This section holds the SPN2MGM specification files for the study presented in Section 6.2. Note that the variable definition part corresponds to one particular model configuration; in our experiments, this file was generated automatically by a Perl [193] script.

### C.4.1 Variable Definitions

```
**
```

```
** SPN2MGM OCDR/Video traffic variable specification file
**

** single value constants only;
** these files are generated automatically for
** each model configuration by an external perl script
c               value   50
t               value   0.4
ack_rate        value   39.4904458598726
idle_rate       value   0.1
weight_s        value   60
weight_m        value   40
fs_rate         value   530
fm_rate         value   350
is_rate         value   3280
im_rate         value   7300
rtt_rate        value   50
serve_rate      value   919.43735868746
loss_rate       value   0.318471337579618
max_win         value   8
end
```

### C.4.2 Petri Net Specification

```
**
** SPN2MGM WWW/Slow Start Petri Net specification file
**

** places
** name / initial marking
* connection management and server places first:
buffer          0
no_conn         1
conn            0
tokens          1
cwin            1
net             0

* additional server place for deriving connection setup rate
conn_enable     0

* places for workload subnet:
idle            1
start           0
frame_s         0
img_s           0
frame_m         0
img_m           0
lost            0
end


** timed transitions
```

```
** name / dependent? / rate
* connection management and server transitions first:
server          ind             serve_rate
connect         ind             c
timeout         ind             t
ack_1           dep     net     ack_rate
ack_2           dep     net     ack_rate
loss            dep     net     loss_rate

* timed transitions for workload subnet:
f1              ind     rtt_rate
f2              ind     rtt_rate
i1              ind     rtt_rate
i2              ind     rtt_rate
T1              ind     fs_rate
T2              ind     is_rate
T3              ind     fm_rate
T4              ind     im_rate
end


** immediate transitions
** name / dependent? / weight / priority
select_s        ind     weight_s        1
select_m        ind     weight_m        1
loss_done       ind     1               1
* immediate trans. for deriving connection setup rate
do_conn         ind     1               1
* reset and reset2 have the some functionality; we have two
* for realizing the 'ored' enabling functions
reset           ind     1               1
reset2          ind     1               2
end


** input arcs
** source trans. / dest. place / multiplicity
* server-side arcs
buffer          server          1
buffer          do_conn         1
no_conn         do_conn         1
conn_enable     connect         1
conn            timeout         1
cwin            reset           2
cwin            reset2          2
tokens          reset           2
tokens          reset2          2
tokens          server          1
net             ack_1           1
net             ack_2           1
net             loss            1
lost            loss_done       1
```

```
* workload-side arcs
start           select_s        1
idle            leave_idle      1
frame_s         f1              1
img_s           i1              1
frame_m         f2              1
img_m           i2              1
start           select_m        1
frame_s         T1              1
img_s           T2              1
frame_m         T3              1
img_m           T4              1
end


** output arcs
** source trans. / dest. place / multiplicity
* server-side arcs
connect         conn            1
do_conn         buffer          1
do_conn         conn_enable     1
timeout         no_conn         1
reset           tokens          1
reset           cwin            1
reset2          tokens          1
reset2          cwin            1
server          net             1
ack_1           tokens          1
ack_2           tokens          2
ack_2           cwin            1
loss            lost            1
loss            tokens          1
loss_done       buffer          1

* client-side arcs
select_s        frame_s         1
select_m        frame_m         1
leave_idle      start           1
f1              img_s           1
f2              img_m           1
i1              idle            1
i2              idle            1
T1              frame_s         1
T2              img_s           1
T3              frame_m         1
T4              img_m           1
T1              buffer          1
T2              buffer          1
T3              buffer          1
T4              buffer          1
end
```

```
** inhibitor arcs
** source place / dest. trans. / multiplicity
buffer          timeout         1
net             timeout         1
end


** enabling functions
** transitions / value / condition / value
server          conn =  1
ack_1           cwin =  max_win
ack_2           cwin <  max_win
reset           conn =  0
reset2          lost =  1
end


** reward functions
customers ( buffer + net )
end
```

### C.4.3 Definition of Reward-Based Measures

```
**
** SPN2MGM OCDR/Video traffic reward measure specification file
**

** reward based probability measures and mean values
P ( PMconn = 1 )        output
P ( PMtokens > 0 )      output
P ( PMconn_enable = 1 ) output
P ( PMframe_s = 1 )     output
P ( PMframe_m = 1 )     output
P ( PMimg_s = 1 )       output
P ( PMimg_m = 1 )       output
E ( PMcwin )            output
E ( PMbuffer )          output
E ( PMlost )            output
E ( PMnet )             output
E ( PMtokens )          output
E ( customers )         output
end
```

## C.5 Pseudo-Self-Similar Arrival Processes

Here, the specification files for the example PSSP|M|1 system depicted in Figure 6.23 are given. The arrival process is a 4-state PSSP where the rates have been chosen to comply with the parameters $a = 10$ and $q = 2$, thus it is the continuous-time analogue of the discrete phase type distribution shown in Figure 6.21(b). It has a mean interarrival time of 1.875 and a Hurst parameter of approximately 0.7.

### C.5.1 Variable Definitions

```
**
** PSSP|M|1 variable specification file
**

** single value constants
q12     value   0.1
q13     value   0.01
q14     value   0.001
q1a     value   0.89
q2a     value   0.2
q3a     value   0.04
q4a     value   0.008
mu      value   1
end
```

### C.5.2 Petri Net Specification

```
**
** PSSP|M|1 Petri Net specification file
**

** places
** name / initial marking
buffer          0
S1              1
S2              0
S3              0
S4              0
absorbed        0
end


** timed transitions
** name / dependent? / rate
T12     ind     q12
T13     ind     q13
T14     ind     q14
T1a     ind     q1a
T2a     ind     q2a
T3a     ind     q3a
T4a     ind     q4a
serve   ind     mu
end


** immediate transitions
** name / dependent? / weight / priority
arr     ind     1       1
end
```

```
** input arcs
** source trans. / dest. place / multiplicity
S1              T1a             1
S1              T12             1
S1              T13             1
S1              T14             1
S2              T2a             1
S3              T3a             1
S4              T4a             1
absorbed        arr             1
buffer          serve           1
end


** output arcs
** source trans. / dest. place / multiplicity
T1a             absorbed        1
T2a             absorbed        1
T3a             absorbed        1
T4a             absorbed        1
T12             S2              1
T13             S3              1
T14             S4              1
arr             buffer          1
arr             S1              1
end


** inhibitor arcs
end


** enabling functions
end


** reward functions
end
```

### C.5.3 Definition of Reward-Based Measures

```
**
** PSSP|M|1 reward measure specification file
**

** reward based probability measures and mean values
E ( PMbuffer )          output
end
```

# Notation and Abbreviations

## Notation

Throughout this work, vectors are denoted by bold lowercase letters (like $\mathbf{a}$) and matrices are represented in bold uppercase ($\mathbf{A}$); by default, all vectors are row vectors. Petri net places and transitions are denoted in a sans serif font ($\mathsf{P}$, $\mathsf{T}$). Furthermore, the following notation is used:

| Symbol | Meaning |
|---|---|
| $\#\mathsf{P}$ | Number of tokens in Petri net place $\mathsf{P}$ |
| $\lvert\lvert\cdot\rvert\rvert_1$ | Maximum column sum norm of a matrix |
| $\lvert\lvert\cdot\rvert\rvert_\infty$ | Maximum row sum norm of a matrix |
| $\otimes$ | Kronecker product |
| $\mathbf{1}, \mathbf{1}_m$ | A *column* vector of all 1s, of dimension $m$ |
| $\mathbf{A}^0$ | The *column* vector derived from matrix $\mathbf{A}$ as $\mathbf{A}^0 = -\mathbf{A}\mathbf{1}$ |
| $\mathbf{A}^*, \mathbf{a}^*, a^*$ | Complex conjugate of a matrix/vector/scalar |
| $\mathbf{A}^\mathrm{T}$ | Transpose of matrix $\mathbf{A}$ |
| $\mathbf{I}, \mathbf{I}_m$ | Unity matrix, unity matrix of given dimension $m$ |
| $\mathrm{Img}_S\, \mathbf{A}$ | Image of the linear transformation induced by $\mathbf{A}$ under set $S$, $\{\mathbf{x}\mathbf{A} \mid x \in S\}$ |
| $\bar{\mathbf{m}}$ | Submarking of an iSPN (excluding the number of tokens in the unbounded place) |
| $Rs(\mathcal{P})$ | Reachability set of Petri net $\mathcal{P}$ |
| $\bar{\mathbf{S}}$ | The set of all submarkings associated to the markings in $S$ |
| $\mathrm{sp}(\mathbf{A})$ | Spectral radius (absolute value of maximum eigenvalue) of matrix $\mathbf{A}$ |

## Abbreviations

| | |
|---|---|
| ARIMA | Autoregressive Integrated Moving Average |
| ASE | Application Specific Element |
| ATM | Asynchronous Transfer Mode |
| BCSM | Basic Call State Model |
| CCSN | Common Channel Signaling Network |
| CR | Cyclic Reduction |

| | |
|---|---|
| CS | Capability Set |
| CTMC | Continuous-Time Markov Chain |
| DPE | Distributed Processing Environment |
| FBM | Fractional Brownian Motion |
| FCFS | First Come, First Served |
| FE | Functional Entity |
| FGN | Fractional Gaussian Noise |
| GPH | Generalized Phase Process |
| GSM | Global System for Mobile Communications |
| GSPN | Generalized Stochastic Petri Net |
| HTML | Hypertext Markup Language |
| HTTP | Hypertext Transfer Protocol |
| IN | Intelligent Network |
| INAP | Intelligent Network Application Protocol |
| INCM | Intelligent Network Conceptual Model |
| IP | Internet Protocol |
| IPP | Interrupted Poisson Process |
| IS | Invariant Subspace |
| ISDN | Integrated Services Digital Network |
| ISUP | ISDN User Part |
| ITU | International Telecommunication Union |
| KLB | Krämer-Langenbach-Belz |
| kTN | Kernel Transport Network |
| LAPACK | Linear Algebra Package |
| LCFS | Last Come, First Served |
| LR | Logarithmic Reduction |
| MAP | Markovian Arrival Process |
| MMBP | Markov-Modulated Bernoulli Process |
| MMPP | Markov-Modulated Poisson Process |
| MRE | Maximum Relative Entropy |
| MSS | Maximum Segment Size |
| MTP | Message Transfer Part |
| NA | Naoumov |
| NCCE | Native Computing and Communications Environment |
| OCDR | On-demand Connection with Delayed Release |
| ODP | Open Distributed Processing |
| OMAP | Operations and Management Part |
| OSI | Open Systems Interconnection |
| PASTA | Poisson Arrivals see Time Averages |
| PCS | Personal Communication System |
| PE | Physical Entity |
| PH | Phase-Type |
| PS | Processor Sharing |
| PSSP | Pseudo Self-Similar Phase-Type |

| | |
|---|---|
| QBD | Quasi-Birth-and-Death |
| QNA | Queueing Network Analyzer |
| RBM | Reflected Brownian Motion |
| SBD | Sequential Bottleneck Decomposition |
| SCCP | Signaling Connection Control Part |
| SCF | Service Control Function |
| SCP | Service Control Point |
| SDF | Service Data Function |
| SDH | Synchronous Digital Hierarchy |
| SDL | Specification and Description Language |
| SDP | Service Data Point |
| SE | Spectral Expansion |
| SIB | Service-Independent Building block |
| SLP | Service Logic Program |
| SN | Short Number |
| SPA | Stochastic Process Algebra |
| SPN | Stochastic Petri Net |
| SRF | Specialized Resource Function |
| SRN | Stochastic Reward Net |
| SS | Successive Substitution |
| SS7 | Signaling System No. 7 |
| SSF | Service Switching Function |
| SSP | Service Switching Point |
| TCAP | Transaction Capabilities Application Part |
| TCP | Transmission Control Protocol |
| TINA | Telecommunications Infrastructure Networking Architecture |
| TUP | Telephone User Part |
| UDP | User Datagram Protocol |
| UPT | Universal Personal Telecommunications |
| URL | Uniform Resource Locator |
| VPN | Virtual Private Network |
| WWW | World Wide Web |

# Bibliography


1. I. J. B. F. Adan, J. Wessels, and W. H. M. Zijm. A compensation approach for two-dimensional Markov processes. *Advances in Applied Probability*, 25:783–817, 1993.
2. A. Agrawal, N. Akar, N. C. Oğuz, and K. Sohraby. *Telpack v1.0a Manual.* Department of Computer Science Telecommunications, University of Missouri-Kansas City, Kansas City, February 1997.
3. M. Ajmone Marsan, G. Balbo, G. Chiola, and G. Conte. Generalized stochastic Petri nets revisited: Random switches and priorities. In *Proc. 2nd International Workshop on Petri Nets and Performance Models*, pages 44–53, Madison, Wisconsin, 1987. IEEE Computer Society Press.
4. M. Ajmone Marsan, G. Balbo, and G. Conte. *Performance Models of Multiprocessor Systems.* MIT Press, Cambridge, 1986.
5. M. Ajmone Marsan, G. Conte, and G. Balbo. A class of generalized stochastic Petri nets for the performance evaluation of multiprocessor systems. *ACM Transactions on Computer Systems*, 2(2):93–122, 1984.
6. N. Akar, N. C. Oğuz, and K. Sohraby. Matrix-geometric solutions of M|G|1-type Markov chains: a unifying generalized state-space approach. *IEEE Journal on Selected Areas in Communications*, 16(5):626–639, June 1998.
7. N. Akar and K. Sohraby. Finite and infinite QBD chains: A simple and unifying algorithmic approach. In *Proc. IEEE Infocom 97.* IEEE Computer Society Press, 1997.
8. N. Akar and K. Sohraby. An invariant subspace approach in M|G|1 and G|M|1 type Markov chains. *Communications in Statistics/Stochastic Models*, 13(3):251–257, 1997.
9. A. T. Andersen and B. F. Nielsen. A Markovian approach for modeling packet traffic with long-range dependence. *IEEE Journal on Selected Areas in Communications*, 16(5):719–732, June 1998.
10. E. Anderson, Z. Bai, C. Bischof, J. Demmel, J. Dongarra, J. Du Croz, A. Greenbaum, S. Hammarling, A. McKenney, S. Ostrouchov, and D. Sorensen. *LAPACK Users' Guide.* Society for Industrial and Applied Mathematics, 1994.
11. N. Anerousis, R. Gopalakrishnan, R. Kalmanek, A. E. Kaplan, W. T. Marshall, P. P. Mishra, P. Z. Onofryk, K. K. Ramakrishnan, and C. J. Sreenan. TOPS: An architecture for telephony over packet networks. *IEEE Journal on Selected Areas in Communications*, 17(1):91–108, January 1999.
12. M. F. Arlitt and C. L. Williamson. Internet Web servers: Workload characterization and performance implications. *IEEE/ACM Transactions on Networking*, 5(5):631–645, 1997.
13. S. Asmussen. *Applied Probability and Queues.* John Wiley, New York, 1987.
14. S. Asmussen, O. Nerman, and M. Olsson. Fitting phase-type distributions via the EM-algorithm. *Scandinavian Journal of Statistics*, 23:419–441, 1996.

15. M. Bafutto. *Modellierung, Verkehrsanalyse und Planung von Zentralkanal-Signalisiernetzen im ISDN mit besonderer Berücksichtigung neuer Dienste der Mobilkommunikation und des Intelligenten Netzes.* PhD thesis, Institute of Communications Switching and Data Technics, University of Stuttgart, Germany, 1995.
16. M. Bafutto, P. J. Kühn, and G. Willmann. Capacity and performance analysis of signaling networks in multivendor environments. *IEEE Journal on Selected Areas in Communications*, 12(3):490–500, April 1994.
17. M. Bafutto and M. Schopp. Network performance and capacity figures of intelligent networks based on the ITU-TS IN capability set 1. In T. Margaria, editor, *Proc. International Workshop on Advanced Intelligent Networks (AIN'96)*, pages 15–29, Passau, Germany, March 1996.
18. W. J. Barr, T. Boyd, and Y. Inoue. The TINA initiative. *IEEE Communications Magazine*, 33(3):70–76, March 1993.
19. F. Baskett, K. M. Chandy, R. R. Muntz, and F. Palacios. Open, closed, and mixed networks of queues with different classes of customers. *Journal of the ACM*, 22(2):248–260, April 1975.
20. F. Bause and P. S. Kritzinger. *Stochastic Petri Nets: An Introduction to the Theory.* Vieweg Verlag, 1996.
21. N. Bean, P. K. Pollet, and P. G. Taylor. The quasistationary distributions of level-independent quasi-birth-and-death processes. *Communications in Statistics/Stochastic Models: Special Issue in Honour of Marcel Neuts*, 14(1):389–406, 1997.
22. J. Beran. Statistical methods for data with long-range dependence. *Statistical Science*, 7:404–427, 1992.
23. J. Beran, R. Sherman, M. S. Taqqu, and W. Willinger. Long-range dependence in variable-bit-rate video traffic. *IEEE Transactions on Communications*, 43(2–4):1566–1579, February/March/April 1995.
24. A. Bertozzi and J. McKenna. Multidimensional residues, generating functions, and their application to queueing networks. *SIAM Review*, 35(2):239–268, June 1993.
25. D. Bertsimas. An analytic approach to a general class of G|G|s queueing systems. *Operations Research*, 38(1):139–155, January/February 1990.
26. D. A. Bini and B. Meini. On cyclic reduction applied to a class of Toeplitz-like matrices arising in queueing problems. In W. J. Stewart, editor, *Computations with Markov Chains*, pages 21–38. Kluwer Academic Publishers, 1995.
27. D. A. Bini and B. Meini. On the solution of a nonlinear matrix equation arising in queueing problems. *Siam J. Matrix Anal. Appl.*, 17:906–926, 1996.
28. D. A. Bini and B. Meini. Using displacement structure for solving non-skip-free M|G|1 type Markov chains. In A. Alfa and S. Chakravarthy, editors, *Advances in Matrix Analytic Methods for Stochastic Models*, pages 17–37. Notable Publications, 1998.
29. G. R. Bitran and S. Dasu. Approximating nonrenewal processes by Markov chains: Use of Super-Erlang (SE) chains. *Operations Research*, 41(5):903–923, September/October 1993.
30. J. P. C. Blanc. Performance analysis and optimization with the power-series algorithm. In L. Donatiello and R. Nelson, editors, *Performance Evaluation of Computer & Communication Systems—Joint Tutorial Papers of Performance '93 & Sigmetrics '93*, LNCS 729, pages 53–80. Springer, 1993.
31. P. P. Bocharov. Analysis of the queue length and the output flow in single server with finite waiting room and phase type distributions. *Problems of Control and Information Theory*, 16(3):211–222, 1987.

32. P. P. Bocharov and V. Naumov. Matrix-geometric stationary distribution for the PH|PH|1|r queue. *Elektronische Informationsverarbeitung und Kybernetik (EIK)*, 22(4):179–186, 1986.
33. T. Bolognesi and E. Brinksma. Introduction to the ISO specifiation language LOTOS. *Computer Networks and ISDN Systems*, 14, 1987.
34. J.-Y. Le Boudec. The asynchronous transfer mode: a tutorial. *Computer Networks and ISDN Systems*, 24:279–309, 1992.
35. G. E. P. Box and G. M. Jenkins. *Time Series Analysis: Forecasting and Control.* Holden Day, San Francisco, 2nd edition, 1976.
36. C. A. Brackett. Dense wavelength division multiplexing networks: Principles and application. *IEEE Journal on Selected Areas in Communications*, 8(6):948–963, August 1990.
37. R. Braden. *Extending TCP for Transactions – Concepts*, November 1992. Internet Request for Comments, RFC 1379.
38. R. Braden. *T/TCP – TCP Extensions for Transactions, Functional Specification*, July 1994. Internet Request for Comments, RFC 1644.
39. D. Brocker. Messung und Modellierung komplexer Verkehrsstrukturen in Hochgeschwindigkeitsnetzen. Master's thesis, Laboratory for Performance Evaluation and Distributed Systems, CSD, RWTH Aachen, Germany, September 1998.
40. M. Cazzaniga, A. Garavelli, and A. Robrock. Implementation of SS7: Italtel's experience. *IEEE Communications Magazine*, 28(7):84–88, July 1990.
41. R. Chakka. *Performance and Reliability Modelling of Computing Systems Using Spectral Expansion.* PhD thesis, University of Newcastle upon Tyne, 1995.
42. R. Chakka. Spectral expansion solution for some finite capacity queues. *Annals of Operations Research*, 79:27–44, 1998.
43. M. Chapman and S. Montesi. Overall concepts and principles of TINA. TINA-C deliverable, Telecommunications Information Networking Architecture Consortium (TINA-C), February 1995.
44. T. Chiotis, F. Stanatelopoulos, and B. Maglaris. Traffic source models for realistic ATM performance modelling. In *Proc. 5th IFIP Workshop on Performance Modelling and Evaluation of ATM Networks*, Ilkley, UK, July 1997.
45. T. Chiotis, C. Stathis, and B. Maglaris. The impact of self-similarity on the statistical multiplexing of MPEG video data. In *Proc. 6th IFIP Workshop on Performance Modelling and Evaluation of ATM Networks*, Ilkley, UK, July 1998.
46. G. Ciardo, J. Muppala, and K. Trivedi. On the solution of GSPN reward models. *Performance Evaluation*, 12(4):237–253, 1991.
47. G. Ciardo, J. K. Muppala, and K. S. Trivedi. Analyzing concurrent and fault-tolerant software using stochastic reward nets. *Journal of Parallel and Distributed Computing*, 15:225–269, 1992.
48. G. Ciardo, A. Riska, and E. Smirni. An aggregation-based solution method for M|G|1-type processes. In B. Plateau, W. J. Stewart, and M. Silva, editors, *Proc. 3rd International Workshop on the Numerical Solution of Markov chains (NSMC'99)*, pages 21–40, Zaragoza, Spain, September 1999. Prensas Universitarias de Zaragoza.
49. J. W. Cohen and O. J. Boxma. *Boundary Value Problems in Queueing Systems Analysis.* North-Holland, Amsterdam, 1983.
50. P. Coolen-Schrijner, A. G. Hart, and P. K. Pollett. Quasistationarity of continuous-time markov chains with positive drift. *Journal of the Australian Mathematical Society, Series B*, 1999. To appear.

51. D. R. Cox. Long-range dependence: A review. In H. A. David and H. T. David, editors, *Statistics: An Appraisal*, pages 55–74. Iowa State University Press, 1984.
52. M. E. Crovella and A. Bestavros. Self-similarity in World Wide Web traffic: Evidence and possible causes. *IEEE/ACM Transactions on Networking*, 5(6):835–846, December 1997.
53. J. G. Dai, V. Nguyen, and M. I. Reiman. Sequential bottleneck decomposition: An approximation method for generalized Jackson networks. *Operations Research*, 42(1):119–136, January/February 1994.
54. J. N. Daigle and D. M. Lucantoni. Queueing systems having phase-dependent arrival and service rates. In W. J. Stewart, editor, *Numerical Solution of Markov Chains*, pages 161–202. Marcel Dekker, 1991.
55. P. Davids. *ATLAS Version 6.0*. Communication and Distributed Systems Department, Communication and Distributed Systems Department, CSD, RWTH Aachen, Germany, 1993. Available at `http://www-i4.informatik.rwth-aachen.de/~atlas`.
56. Defense Advanced Research Projects Agency. *Internet Protocol – Protocol Specification*, September 1981. Internet Request for Comments, RFC 791.
57. Defense Advanced Research Projects Agency. *Transmission Control Protocol – Protocol Specification*, September 1981. Internet Request for Comments, RFC 793.
58. Deutsche Telekom Group: Report January 1 to September 30, 1999.
59. N. M. van Dijk. *Queueing Networks and Product Forms: A Systems Approach*. Wiley, October 1993.
60. N. G. Duffield and N. O'Connell. Large deviations and overflow probabilities for the general single server queue, with applications. *Math. Proc. Cam. Phil. Soc.*, 118:363–374, 1995. Available at `ftp://www.stp.dias.ie/DAPG/dapg9330.ps`.
61. D. E. Duffy, A. A. McIntosh, M. Rosenstein, and W. Willinger. Statistical analysis of CCSN/SS7 traffic data from working subnetworks. *IEEE Journal on Selected Areas in Communications*, 12(3):544–551, April 1994.
62. M. Elixmann and J. Fischer. Service creation environment – a development environment for services in the intelligent network. *Philips Telecommunication Review*, 50(1), 1992.
63. J. Enssle. The impact of VBR MPEG video traffic on ATM multiplexer performance and its evaluation. In D. Kouvatsos, editor, *Performance modelling and evaluation of ATM networks*, volume 3. Chapman & Hall, 1997.
64. A. Erramilli, O. Narayan, and W. Willinger. Experimental queueing analysis with long-range dependent packet traffic. *IEEE/ACM Transactions on Networking*, 4(2):209–223, April 1996.
65. A. Erramilli and R. P. Singh. An application of deterministic chaotic maps to model packet traffic. *Queueing Systems*, 20(I–II):171–206, 1995.
66. R. V. Evans. Geometric distribution in some two-dimensional queueing systems. *Operations Research*, 15:830–846, 1967.
67. Y. Fan and N. D. Georganas. Performance analysis of ATM switches with self-similar input traffic. *Computer Systems Science and Engineering Journal*, 12(2):95–102, March 1997.
68. A. Fasbender. *Messung und Modellierung der Dienstgüte paketvermittelnder Netze*. PhD thesis, Communication and Distributed Systems Department, RWTH Aachen, Germany, 1998.
69. R. Fielding, J. Gettys, J. Mogul, H. Frystyk, and T. Bernes-Lee. *Hypertext Transfer Protocol – HTTP/1.1*, January 1997. Internet Request for Comments, RFC 2068.

70. G. Florin and S. Natkin. *Les Réseaux de Petri Theorie Techniques de Calcul et Applications.* PhD thesis, Université de Paris VI, Paris, June 1985.
71. G. Florin and S. Natkin. One place unbounded stochastic Petri nets: Ergodicity criteria and steady-state solution. *Journal of Systems and Software*, 1(2):103–115, 1986.
72. H. J. Fowler and W. E. Leland. Local area network traffic characteristics, with implications for broadband network congestion management. *IEEE Journal on Selected Areas in Communications*, 9(7):1139–1149, September 1991.
73. C. Frank. Bewertung von stochastischen Petrinetzen mit Hilfe der Matrix-geometrischen Methode. Master's thesis, Laboratory for Performance Evaluation and Distributed Systems, CSD, RWTH Aachen, Germany, January 1997.
74. L. J. Franken and B. R. Haverkort. The performability manager. *IEEE Network*, 8(1):24–32, January/February 1994.
75. K. A. Frenkel. Allan L. Scherr—Big Blue's time-sharing pioneer. *Communications of the ACM*, 30(10):824–828, 1987.
76. M. Fujioka and Y. Wakahara. Consideration on common channel signaling evolution for global intelligent networking. *IEEE Journal on Selected Areas in Communications*, 12(3):510–516, April 1994.
77. H. R. Gail, S. L. Hantler, and B. A. Taylor. Non-skip-free M|G|1 and G|M|1 type Markov chains. *Advances in Applied Probability*, 29:733–758, 1997.
78. J. J. Garrahan, P. A. Russo, K. Kitami, and R. Kung. Intelligent network overview: CCITT's IN recommendation and related activities in other organizations. *IEEE Communications Magazine*, 31(3):30–36, March 1993.
79. M. W. Garrett and W. Willinger. Analysis, modeling and generation of self-similar VBR video traffic. In *Proc. ACM SIGCOMM '94*, volume 24 of *Computer Communications Review*, pages 269–280, London, October 1994.
80. C. Gbaguidi, J.-P. Hubaux, G. Pacifici, and A. N. Tantawi. Integration of Internet and telecommunications: An architecture for hybrid services. *IEEE Journal on Selected Areas in Communications*, 17(9):1563–1579, September 1999.
81. W. J. Gordon and G. J. Newell. Closed queueing systems with exponential servers. *Operations Research*, 15:254–265, March 1967.
82. C. Görg, M. Guntermann, and S. Kleier. Future systems for personal mobility services: Design, performance evaluation, and implementation. *IEEE Journal on Selected Areas in Communications*, 15(8):1672–1683, October 1997.
83. M. Grossglauser and J.-C. Bolot. On the relevance of long-range dependence in network traffic. In *Proc. ACM SIGCOMM '96*, volume 26 of *Computer Communications Review*, pages 15–24, October 1996.
84. L. Gün. Experimental results on matrix-analytical solution techniques – extensions and comparisons. *Communications in Statistics/Stochastic Models*, 5:669–682, 1989.
85. R. Händel, M. N. Huber, and S. Schröder. *ATM Networks: Concepts, Protocols, Applications.* Addison-Wesley, 1994.
86. J. M. Harrison and V. Nguyen. The QNET method for two-moment analysis of open queueing networks. *Queueing Systems*, 6:1–32, 1990.
87. B. R. Haverkort. Approximate analysis of networks of PH|PH|1|K queues: Theory and tool support. In H. Beilner and F. Bause, editors, *Quantitative Evaluation of Computing and Communication Systems*, LNCS 977, pages 239–253. Springer, 1995.
88. B. R. Haverkort. Matrix-geometric solution of infinite stochastic Petri nets. In *Proceedings of the 1st International Computer Performance and Dependability Symposium*, pages 72–81. IEEE Computer Society Press, 1995.

89. B. R. Haverkort. SPN2MGM: Tool support for matrix-geometric stochastic Petri nets. In *Proceedings of the 2nd International Computer Performance and Dependability Symposium*, pages 219–228. IEEE Computer Society Press, 1996.
90. B. R. Haverkort. *Performance Evaluation of Computer and Communication Systems.* Wiley, 1998.
91. B. R. Haverkort and A. Ost. Leistungsbewertung funktionaler Architekturen – Offene Fragen. Technical Memorandum, Laboratory for Performance Evaluation and Distributed Systems, CSD, RWTH Aachen, February 1997.
92. B. R. Haverkort and A. Ost. Steady-state analysis of infinite stochastic Petri nets: A comparison between the spectral expansion and the matrix-geometric method. In *Proc. 7th International Workshop on Petri Nets and Performance Models*, pages 36–45, Saint Malo, France, 1997. IEEE Computer Society Press.
93. J. Heidemann, K. Obraczka, and J. Touch. Modeling the performance of HTTP over several transport protocols. *IEEE/ACM Transactions on Networking*, 5(5):616–630, October 1997.
94. G. Heijenk. *Connectionless Communications Using the Asynchronous Transfer Mode.* PhD thesis, Centre for Telematics and Information Technology, University of Twente, the Netherlands, 1995.
95. G. J. Heijenk and B. R. Haverkort. Design and evaluation of a connection management mechanism for an ATM-based connectionless service. *Distributed System Engineering Journal*, 3(1):53–67, 1996.
96. A. Heindl. Approximate analysis of queueing networks with finite buffers and losses by decomposition. Technical Report 1998-8, Institut für technische Informatik, Technical University of Berlin, Germany, May 1998.
97. H. Hermanns, U. Herzog, and V. Mertsiotakis. Stochastic process algebras – between LOTOS and Markov chains. *Computer Networks and ISDN Systems*, 30(9–10):901–924, 1998.
98. D. P. Heyman and T. V. Lakshman. What are the implications of long-range dependence for VBR-video traffic engineering? *IEEE/ACM Transactions on Networking*, 4(3):301–317, June 1996.
99. C. A. R. Hoare. *Communicating Sequential Processes.* Prentice Hall, Englewood Cliffs, New Jersey, 1985.
100. A. Horváth and M. Telek. Approximating heavy tailed behaviour with phase type distributions. In G. Latouche and P. Taylor, editors, *Advances in Algorithmic Methods for Stochastic Models*, pages 191–213. Notable Publications, July 2000.
101. J. R. M. Hosking. Fractional differencing. *Biometrika*, 68:165–176, 1981.
102. J. R. M. Hosking. Modeling persistence in hydrological time series using fractional differencing. *Water Resources Research*, 20(12):1898–1908, 1984.
103. G.-J. van Houtum. *New Approaches for Multi-Dimensional Queueing Systems.* PhD thesis, Eindhoven University of Technology, The Netherlands, 1994.
104. Y. Inoue, D. Guha, and H. Berndt. The TINA consortium. *IEEE Communications Magazine*, pages 130–136, September 1998.
105. ITU-T. Basic reference model of open distributed processing – part 1: Overview and guide to use. Draft recommendation X.901, ITU, Geneva, Switzerland, November 1992.
106. ITU-T Study Group II. GOS parameters and target GOS objectives for IN services. Recommendations E.724, ITU, Geneva, Switzerland, February 1996.
107. ITU-T Study Group XI. Intelligent network. Recommendations Q.120x, ITU, Geneva, Switzerland, 1993.
108. ITU-T Study Group XI. Specifications of signaling system no. 7. Recommendations Q.700–Q.795, ITU, Geneva, Switzerland, 1994.

109. B. Jabbari. Intelligent Network concepts in mobile communications. *IEEE Communications Magazine*, pages 64–69, February 1992.
110. J. R. Jackson. Networks of waiting lines. *Operations Research*, 5:518–521, 1957.
111. V. Jacobson. Congestion avoidance and control. In *Proc. ACM SIGCOMM '88*, volume 18 of *Computer Communications Review*, pages 314–329, 1988.
112. R. Jain and S. A. Routhier. Packet trains–measurements and a new model for computer network traffic. *IEEE Journal on Selected Areas in Communications*, SAC 4(6):986–995, September 1986.
113. A. Jennings, J. J. MacKeown, and S. MacKeown. *Matrix Computation.* Wiley, 2nd edition, 1992.
114. J. Jormakka. Performance analyzer for intelligent network. In J. Harju, T. Karttunen, and O. Martikainen, editors, *Proc. IFIP Workshop on Intelligent Networks*, pages 97–110, Lappeenranta, Finland, August 1994. Chapman & Hall.
115. B. Käs and M. Körkel. Intelligente Netze – spezifische verkehrstheoretische und wirtschaftliche Gesichtspunkte bei der Planung. *Der Fernmelde-Ingenieur*, 44(5):1–27, May 1990.
116. M. Kihl, C. Nyberg, H. Warne, and P. Wollinger. Performance simulation of a TINA network. In *Proc. IEEE GLOBECOM '97*, pages 552–558, Phoenix, AZ, 1997.
117. J. F. C. Kingman. The exponential decay of Markov transition probabilities. *Proc. London Math. Soc.*, 13:337–358, 1963.
118. E. F. J. Klein. SPN2MGM: A tool for solving a class of infinite GSPN models. Master's thesis, University of Twente, The Netherlands, 1995.
119. L. Kleinrock. *Queueing Systems, Vol. 1: Theory.* John Wiley & Sons, 1975.
120. K. Koischwitz. Entwurf und Implementierung einer parametrisierbaren Benutzeroberfläche für hierarchische Netzmodelle. Master's thesis, Technical University of Berlin, Institut für Technische Informatik, 1996.
121. U. Konrads and K. W. Neunast. Internal Communication, March 1997.
122. W. Krämer and M. Langenbach-Belz. Approximate formulae for the delay in the queueing system GI|G|1. In *Proceedings of the 8th International Teletraffic Congress*, pages 235–1/8, 1976.
123. P. Kühn. Approximate analysis of general queuing networks by decomposition. *IEEE Transactions on Communications*, COM-27(1):113–126, January 1979.
124. P. J. Kühn, C. D. Pack, and R. A. Skoog. Common channel signaling networks: Past, present, future. *IEEE Journal on Selected Areas in Communications*, 12(3):383–394, April 1994.
125. A. Lang and J. L. Arthur. Parameter approximation for phase-type distributions. In S. R. Chakravarthy and A. S. Alfa, editors, *Matrix-Analytic Methods in Stochastic Models*, volume 183 of *Lecture Notes in Pure and Applied Mathematics.* Marcel Dekker, September 1996.
126. G. Latouche. A note on two matrices occuring in the solution of quasi-birth-and-death processes. *Communications in Statistics/Stochastic Models*, 3:251–257, 1987.
127. G. Latouche. A simple proof for the matrix-geometric theorem. *Applied Stochastic Models and Data Analysis*, 8:25–29, 1992.
128. G. Latouche. Algorithms for infinite Markov chains with repeating columns. In C. D. Meyer and R. J. Plemmons, editors, *Linear Algebra, Markov Chains, and Queueing Models*, volume 48 of *IMA Volumes in Mathematics and its Applications*, pages 231–265. Springer, August 1993.
129. G. Latouche and V. Ramaswami. A logarithmic reduction algorithm for quasi birth and death processes. *Journal of Applied Probability*, 30:650–674, 1993.

130. W. Lau and S. Li. Statistical multiplexing and buffer sharing in multimedia high-speed networks: A frequency-domain perspective. *IEEE/ACM Transactions on Networking*, 5(3):382–396, June 1997.
131. G. S. Lauer. IN architectures for implementing universal personal telecommunications. *IEEE Network*, 8(2):6–16, March/April 1994.
132. W. E. Leland, M. S. Taqqu, W. Willinger, and D. V. Wilson. On the self-similar nature of Ethernet traffic. In *Proc. ACM SIGCOMM '93*, volume 23 of *Computer Communications Review*, pages 183–193, October 1993.
133. W. E. Leland, M. S. Taqqu, W. Willinger, and D. V. Wilson. On the self-similar nature of Ethernet traffic (extended version). *IEEE/ACM Transactions on Networking*, 2(1):1–15, February 1994.
134. V. O. K. Li and X. Qiu. Personal communication systems. *Proc. IEEE*, 83(9):1210–1243, September 1995.
135. T. Magedanz and R. Popescu-Zeletin. *Intelligent Networks - Basic Technology, Standards and Evolution.* International Thomson Computer Press, July 1996.
136. B. B. Mandelbrot. *The Fractal Geometry of Nature.* Freeman, New York, 1983.
137. B. B. Mandelbrot and J. W. van Ness. Fractional Brownian motions, fractional noises and applications. *SIAM Review*, 10:422–437, 1968.
138. K. T. Marshall. Some inequalities in queuing. *Operations Research*, 16(3):651–665, May/June 1968.
139. B. Meini. Solving QBD problems: the cyclic reduction algorithm versus the invariant subspace method. *Advances in Performance Analysis*, 1:215–225, 1998.
140. D. A. Menascé and V. A. F. Almeida. *Capacity Planning for Web Performance: Metrics, Models, and Methods.* Prentice Hall, 1998.
141. R. Milner. *Communication and Concurrency.* Prentice Hall, London, 1989.
142. I. Mitrani and R. Chakka. Spectral expansion solution for class of Markov models: Application and comparison with the matrix-geometric method. *Performance Evaluation*, 23(3):241–260, September 1995.
143. I. Mitrani, A. Ost, and M. Rettelbach. TIPP and the spectral expansion method. In F. Baccelli, A. Jean-Marie, and I. Mitrani, editors, *Quantitative Methods in Parallel Systems*, ESPRIT Basic research series, pages 99–113. Springer, 1995.
144. A. R. Modarressi and R. A. Skoog. Signaling system no. 7: A tutorial. *IEEE Communications Magazine*, 28(7):19–35, July 1990.
145. M. K. Molloy. Performance analysis using stochastic Petri nets. *IEEE Transactions on Computers*, 31(9):913–917, 1982.
146. A. P. A. van Moorsel and R. J. Boucherie. Metrics for representing performance as rewards in performability models. In *Proc. 4th International Workshop on Performability Modeling of Computer and Communication Systems (PMCCS)*, pages 25–29, Williamsburg, VA, September 1998.
147. A. P. A. van Moorsel and A. Ost. Quasi-stationary distributions of absorbing QBD Markov chains. Technical Memorandum, September 1998.
148. V. Naoumov, U. R. Krieger, and D. Wagner. Analysis of a finite capacity multi-server delay-loss system with a general Markovian arrival process. In S. R. Chakravarthy and A. S. Alfa, editors, *Matrix-Analytic Methods in Stochastic Models*, volume 183 of *Lecture Notes in Pure and Applied Mathematics.* Marcel Dekker, September 1996.
149. B. C. Neuman. *The virtual system model: A scalable approach to organizing large systems.* PhD thesis, University of Washington, Seattle, 1992.
150. K. W. Neunast. Leistungsbewertung funktionaler Architekturen. Technical Memorandum, Philips Research Laboratories Aachen, December 1996.

151. M. F. Neuts. A versatile Markovian point process. *Journal of Applied Probability*, 16:746–779, 1979.
152. M. F. Neuts. *Matrix Geometric Solutions in Stochastic Models: An Algorithmic Approach.* Johns Hopkins University Press, 1981.
153. M. F. Neuts. *Structured Matrices of* M|G|1 *Type and Their Applications.* Marcel Dekker, 1989.
154. I. Norros. A storage model with self-similar input. *Queueing Systems*, 16:387–396, 1994.
155. C. G. Omidyar and A. Aldridge. Introduction to SDH/SONET. *IEEE Communications Magazine*, 31(9):30–33, September 1993.
156. A. Ost. Spezifikation und Analyse unendlicher Systeme mit stochastischen Prozeßalgebren und spektraler Expansion. Master's thesis, University of Erlangen-Nürnberg, Germany, 1995.
157. A. Ost, C. Frank, and B. R. Haverkort. Untersuchungen zum Verbindungsmanagement bei Videoverkehr mit Matrix-geometrischen stochastischen Petrinetzen. In *Proc. 9th ITG/GI Workshop on Measurement, Modeling and Evaluation*, number 1 in Aktuelle Probleme der Informatik, pages 71–85. VDE Verlag, 1997.
158. A. Ost and B. R. Haverkort. Analysis of windowing mechanisms with infinite-state stochastic Petri nets. *ACM Performance Evaluation Review*, 26(2):38–46, August 1998.
159. A. Ost and B. R. Haverkort. Analysis of windowing mechanisms with infinite-state stochastic Petri nets. In W. S. Lai and R. B. Cooper, editors, *Performance and Control of Network Systems II*, Proceedings of SPIE Vol. 3530, pages 178–189, 1998.
160. A. Ost and B. R. Haverkort. Evaluating computer-communication systems using infinite-state stochastic Petri nets. In G. Latouche and P. Taylor, editors, *Advances in Algorithmic Methods for Stochastic Models*, pages 295–314. Notable Publications, July 2000.
161. M. Paterok and O. Fischer. Feedback queues with preemption-distance priorities. *ACM Performance Evaluation Review*, 17(1):136–145, May 1989.
162. V. Paxson and S. Floyd. Wide-area traffic: The failure of Poisson modeling. *IEEE/ACM Transactions on Networking*, 3(3):226–244, June 1995.
163. C. A. Petri. *Kommunikation mit Automaten.* PhD thesis, University of Bonn, Germany, 1962.
164. W. H. Press, S. A. Teukolsky, W. T. Vetterling, and B. P. Flannery. *Numerical Recipes in C: The Art of Scientific Computing.* Cambridge University Press, 2nd edition, January 1993.
165. K. E. E. Raatikainen. A framework for evaluating the performance of IN services. In *Proc. Int. Workshop on Intelligent Networks*, pages 121–129, Lappeenranta, Finland, 1993.
166. S. Robert. *Modélisation markovienne du trafic dans les réseaux de communication.* PhD thesis, EPFL, Lausanne, Switzerland, 1996.
167. S. Robert and J.-Y. Le Boudec. New models for pseudo self-similar traffic. *Performance Evaluation*, 30:57–68, 1997.
168. J. D. Roberts. Linear model reduction and solution of the algebraic Riccati equation by the use of the sign function. *International Journal of Control*, 32:677–687, 1980.
169. W. C. Roehr. Inside SS no. 7: A detailed look at ISDN's signaling system plan. *Data Communications*, pages 120–128, October 1985.
170. O. Rose. Simple and efficient models for variable bit rate MPEG video traffic. *Performance Evaluation*, 30(1–2):69–85, 1997.

171. B. K. Ryu and A. Elwalid. The importance of long-range dependence of VBR video traffic in ATM traffic engineering: Myths and realities. In *Proc. ACM SIGCOMM '96*, volume 26 of *Computer Communications Review*, pages 3–14, October 1996.
172. R. Sadre. Personal communication, January 2000.
173. R. Sadre, B. R. Haverkort, and A. Ost. An efficient and accurate decomposition method for open finite- and infinite-buffer queueing networks. In B. Plateau, W. J. Stewart, and M. Silva, editors, *Proc. 3rd International Workshop on the Numerical Solution of Markov chains (NSMC'99)*, pages 1–20, Zaragoza, Spain, September 1999. Prensas Universitarias de Zaragoza.
174. R. A. Sahner and K. S. Trivedi. A software tool for learning about stochastic models. *IEEE Transactions on Education*, 36(1):56–61, 1993.
175. H. Saito, M. Kawarasaki, and H. Yamada. An analysis of statistical multiplexing in an ATM transport network. *IEEE Journal on Selected Areas in Communications*, 9(3):359–367, April 1991.
176. L. Samson, J. Kerneis, and P.-D. Lansard. Optimal SSP locations in an intelligent network. In *Proc. 15th Int. Switching Symposium*, pages 335–339, Berlin, Germany, April 1995. VDE Verlag.
177. K. Schulz, H. Kaufer, and B. Heilig. The Deutsche Bundespost Telekom implements IN. *IEEE Communications Magazine*, pages 90–96, February 1992.
178. H. Schulzrinne and J. Rosenberg. Internet telephony: Architecture and protocols – an IETF perspective. *Computer Networks and ISDN Systems*, 31(3):237–255, February 1999.
179. M. Schwehm and A. Ost. Inference of stochastic regular grammars by massively parallel genetic algorithms. In L. J. Eshelman, editor, *Proc. 6th Int. Conf. on Genetic Algorithms*, pages 520–527, San Francisco, 1995. Morgan Kaufman.
180. G. Sleijpen, H. A. van der Vorst, and M. van Gijzen. Quadratic eigenproblems are no problem. *SIAM News*, 29(7):8–19, September 1996.
181. W. R. Stevens. *TCP/IP Illustrated, Volume I: The Protocols.* Addison-Wesley, 1994.
182. W. R. Stevens. *TCP Slow Start, Congestion Avoidance, Fast Retransmit, and Fast Recovery Algorithms*, January 1997. Internet Request for Comments, RFC 2001.
183. W. J. Stewart. *Introduction to the Numerical Solution of Markov Chains.* Princeton University Press, 1994.
184. M. S. Taqqu and J. B. Levy. Using renewal processes to generate long-range dependence and high variability. In E. Eberlein and M. S. Taqqu, editors, *Dependence in Probability and Statistics: A Survey of Recent Results*, volume 11 of *Progress in Probability and Statistics*, pages 73–89. Birkhäuser, Boston, 1986.
185. M. S. Taqqu and V. Teverovsky. On estimating the intensity of long-range dependence in finite and infinite variance time series. In R. Adler, R. Feldman, and M. S. Taqqu, editors, *A Practical Guide To Heavy Tails: Statistical Techniques and Applications*, pages 27–53. Birkhäuser, Boston, 1998.
186. TINA-C home page. World Wide Web document. `http://www.tinac.com/`.
187. P. J. Tomaras and D. D. Kouvatsos. MRE hierarchical decomposition of general queueing network models. *Acta Informatica*, 28:265–295, 1991.
188. B. Tsybakov and N. D. Georganas. On self-similar traffic in ATM queues: Definitions overflow probability bound, and cell delay distribution. *IEEE/ACM Transactions on Networking*, 5(3):397–409, June 1997.
189. T/TCP home page. World Wide Web document. `http://www.kohala.com/start/ttcp.html`.

190. B. U. Unger, D. J. Goetz, and S. W. Maryka. Simulation of SS7 common channel signaling. *IEEE Communications Magazine*, 32(3):52–62, March 1994.
191. D. Vere-Jones. Geometric ergodicity in denumerable Markov chains. *Quarterly J. Math. Oxford*, 13:7–28, 1962.
192. H. A. van der Vorst and G. H. Golub. 150 years and still alive: Eigenproblems. In I. S. Duff and G. A. Watson, editors, *The State of the Art in Numerical Analysis*, pages 93–119. Clarendon Press, Oxford, 1997.
193. L. Wall, T. Christiansen, and R. L. Schwartz. *Programming Perl.* O'Reilly, 1996.
194. V. Wallace. *The solution of quasi birth and death processes arising from multiple access computer systems.* PhD thesis, System Engineering Laboratory, University of Michigan, 1969.
195. W. Whitt. Approximating a point process by a renewal process, I: Two basic methods. *Operations Research*, 30(1):125–147, January/February 1982.
196. W. Whitt. The Queueing Network Analyzer. *The Bell System Technical Journal*, 62(9):2779–2815, 1983.
197. N. Widell. STINA – performance simulation of TINA networks. Master's thesis, Department of Communication Systems, Lund Institute of Technology, Lund, Sweden, July 1998.
198. W. Willinger, V. Paxson, and M. S. Taqqu. Self-similarity and heavy tails: Structural modeling of network traffic. In R. Adler, R. Feldman, and M. S. Taqqu, editors, *A Practical Guide To Heavy Tails: Statistical Techniques and Applications*, pages 27–53. Birkhäuser, Boston, 1998.
199. G. Willmann and P. J. Kühn. Performance modeling of signaling system no. 7. *IEEE Communications Magazine*, 28(7):44–56, July 1990.
200. R. W. Wolff. Poisson arrivals see time averages. *Operations Research*, 30(2):223–231, 1982.
201. G. Wolfner and M. Telek. A numerical analysis method of queues with batch arrivals. In *Proc. 5th Int. Conf. on Advanced Computing (ADCOMP)*, Channai, India, December 1997.
202. P. Wollinger. TINA – description and performance simulation. Master's thesis, Department of Communication Systems, Lund Institute of Technology, Lund, Sweden, March 1997.
203. T.-H. Wu, N. Yoshikai, and H. Fujii. ATM signaling transport network architectures and analysis. *IEEE Communications Magazine*, 33(12):90–99, December 1995.
204. W. M. Zuberek. Performance evaluation using unbounded Petri nets. In *Proc. 3rd International Workshop on Petri Nets and Performance Models*, pages 180–186, Kyoto, Japan, 1989. IEEE Computer Society Press.
205. W. M. Zuberek. Timed Petri nets – definitions, properties, and applications. *Microelectronics and Reliability*, 31(4):627–644, 1991.

# Index

Printing and Binding: Stürtz AG, Würzburg